T0192320

Plant-Microbial Interactions and Smart Agricultural Biotechnology

Microbial Biotechnology for Food, Health, and the Environment Series

Series Editor
Ashok Kumar Nadda

Plant-Microbial Interactions and Smart Agricultural Biotechnology
Edited by Swati Tyagi, Robin Kumar, Baljeet Singh Saharan, Ashok Kumar Nadda

For more information about this series, please visit: www.routledge.com

Plant-Microbial Interactions and Smart Agricultural Biotechnology

Edited by
Swati Tyagi, Robin Kumar, Baljeet Singh Saharan,
and Ashok Kumar Nadda

CRC Press
Taylor & Francis Group
Boca Raton London New York

CRC Press is an imprint of the
Taylor & Francis Group, an **informa** business

First edition published 2022
by CRC Press
6000 Broken Sound Parkway NW, Suite 300, Boca Raton, FL 33487-2742

and by CRC Press
2 Park Square, Milton Park, Abingdon, Oxon, OX14 4RN

© 2022 selection and editorial matter, Swati Tyagi, Robin Kumar, Baljeet Saharan, Ashok Kumar Nadda; individual chapters, the contributors

CRC Press is an imprint of Taylor & Francis Group, LLC

Reasonable efforts have been made to publish reliable data and information, but the author and publisher cannot assume responsibility for the validity of all materials or the consequences of their use. The authors and publishers have attempted to trace the copyright holders of all material reproduced in this publication and apologize to copyright holders if permission to publish in this form has not been obtained. If any copyright material has not been acknowledged please write and let us know so we may rectify in any future reprint.

Except as permitted under U.S. Copyright Law, no part of this book may be reprinted, reproduced, transmitted, or utilized in any form by any electronic, mechanical, or other means, now known or hereafter invented, including photocopying, microfilming, and recording, or in any information storage or retrieval system, without written permission from the publishers.

For permission to photocopy or use material electronically from this work, access www.copyright.com or contact the Copyright Clearance Center, Inc. (CCC), 222 Rosewood Drive, Danvers, MA 01923, 978-750-8400. For works that are not available on CCC please contact mpkbookspermissions@tandf.co.uk

Trademark notice: Product or corporate names may be trademarks or registered trademarks and are used only for identification and explanation without intent to infringe.

Library of Congress Cataloging-in-Publication Data

Names: Tyagi, Swati, editor.
Title: Plant-microbial interactions and smart agricultural biotechnology /
edited by Swati Tyagi, Robin Kumar, Baljeet Saharan, Ashok Kumar Nadda.
Description: First edition. I Boca Raton : CRC Press, 2022. I Series:
Microbial biotechnology I Includes index.
Identifiers: LCCN 2021015972 (print) I LCCN 2021015973 (ebook) I ISBN
9781032100418 (hardback) I ISBN 9781032101484 (paperback) I ISBN
9781003213864 (ebook)
Subjects: LCSH: Plant-microbe relationships--Molecular aspects. I
Sustainable agriculture. I Agricultural microbiology.
Classification: LCC QR351 .P583245 2021 (print) I LCC QR351 (ebook) I DDC
579/.178--dc23
LC record available at https://lccn.loc.gov/2021015972
LC ebook record available at https://lccn.loc.gov/2021015973

ISBN: 978-1-032-10041-8 (hbk)
ISBN: 978-1-032-10148-4 (pbk)
ISBN: 978-1-003-21386-4 (ebk)

DOI: 10.1201/9781003213864

Typeset in Times
by Deanta Global Publishing Services, Chennai, India

Contents

Preface

Farmers have great potential to adapt to and accept climate change. The global economy has changed and has been able to develop sustainable agriculture with high rates of growth. This helped us to solve the problem of food shortage. In this age, climate change has brought new and complicated challenges. To combat these challenges, it is necessary to understand the root causes, after which proper implementation of the solutions is required. On the other hand, the overuse of chemical fertilizers and pesticides has caused many environmental problems, including water acidification, greenhouse effect, and depletion of the ozone layer. These problems can be addressed using biopesticides and biofertilizers, which are beneficial, environmentally friendly, natural, and easy to use. Microbial biofertilizers control soilborne diseases, maintain soil structure, and provide nutrients to plants. They play a vital role in sustainable agriculture. Arbuscular mycorrhizal fungi are important soil microbes that form symbiotic associations with most terrestrial plants and are mainly responsible for the uptake of phosphorus. Another group of microbes are biological nitrogen-fixing bacteria, which are universally potent bioinoculants used to promote plant growth. The use of *Rhizobium* in legume crops to support agricultural productivity is increasing day by day. Such types of bioinoculants are easily available to farmers. Moreover, the mass production of these biofertilizers is done in laboratories, and they can easily be stored as liquid or carrier-based bioformulations at room temperature. Phosphate-solubilizing bacteria are also extremely important, as they have been reported to increase P uptake by converting insoluble to soluble forms. *Azotobacter* and *Azospirillum* are two other very important bacteria. The response of these organisms is commonly seen in increasing yields. Apart from these microbes, cyanobacteria also contribute significantly to the nitrogen economy of sustainable agriculture. The tripartite relationship between legumes, rhizobia, and mycorrhiza is the most efficient combination to promote growth and achieve higher productivity in plants, trees, and vegetables. Cyanobacteria are excellent suppliers of nitrogen. Their importance for abundant crop production cannot be ignored and has been felt by farmers. They are easy to multiply and can now be obtained in bottles or packets throughout the year. In fact, the application of microbial bioinoculants is a very effective and natural method to increase and maintain the mineral economy of nature. Their use reduces the need for chemical fertilizers, making them the only effective alternative for self-sustaining farming.

Microbe–host or microbe–microbe interactions are important strategies for colonization and establishment in different agro-climatic conditions/regions. This communication/crosstalk covers all aspects of the microbial communities', e.g., signaling, chemotaxis, genetic exchange, metabolite conversion, intermediary metabolism, and physicochemical changes that lead to genotype selection. The survival rate of microbes in the environment depends on biodiversity because the high genetic variability increases the competitiveness and reduces the chances of intruders settling in the respective climatic conditions. Hence, these interactions are the result of a coordination that results in acclimatization and adroitness and enables the filling of various niches by limiting biological and abiotic stresses or exchanging growth factors and signals. Several mechanisms might be involved in this exchange, such as secondary metabolites, biofilm formation, cellular transduction signaling, quorum sensing system, and siderophores. The final unit of interaction is the expression of each organism's genes in response to environmental stimuli responsible for the production of the biomolecules involved in such associations. Hence, in this book, we focus on the molecular mechanisms involved in such interactions in the molecular strategy used by various microbes, which influence the constitution and structure of microbes.

The challenges are varied and unpredictable, but with the experience of farmers and expertise of our scientists, it is possible to make agriculture smarter. There has been great demand for a book on *Plant-Microbial Interactions and Smart Agricultural Biotechnology*, wherein different issues concerning microbial interactions and smart agriculture have been described.

This book will certainly provide useful information dealing with a diverse group of microbes, beneficial effects, and the bottlenecks in their implementation. Students, researchers, biotechnologists, microbiologists, botanists, soil biologists, industrialists involved in the mass production of biofertilizers, producers, environmentalists, and all other users should find this book extremely useful. The goal of this book is to provide essential information on the use of different microbes and their secondary metabolites for the treatment of various diseases affecting crops. This book also describes the potential of microbes (tiny factories) in improving crop yield, plant health, and the biomolecules' production efficiency for smart agriculture.

Editors

Swati Tyagi is currently working as a Project Scientist at the International Rice Research Institute – South Asia Regional Centre (ISARC), Varanasi, India. She holds more than 5 years' research experience in the area of plant-microbe interaction, genomics, transcriptomics, and next generation sequencing and analysis. Dr. Swati has published more than 10 research papers in journals of national and international repute. She graduated with a degree in microbiology from Kurukshetra University (Haryana) in 2012 and earned her PhD from the Division of Biotechnology, Jeonbuk National University, Republic of Korea. She was awarded Brain Korea 21 plus (BK21Plus) fellowship for her doctoral studies. She also worked as an International Researcher at Jeonbuk National University, Republic of Korea, and as a Post-Doctoral Research Fellow at the genomics division of the National Institute of Agriculture Science, Rural Development Administration, Republic of Korea. During her early career stages, she worked as a Research Assistant and Senior Research Fellow at thte Department of Plant Pathology, Sardar Vallabhbhai Patel University of Agriculture and Technology, Meerut, Uttar Pradesh, India. Her keen research interest is in uncovering the secrets of plant genomes by next generation sequencing and computational analysis, understanding complex plant microbes' interactions, and characterizing and validating the effect of microbial volatile compounds on plants and pathogens. Her work has been published in various internationally reputed journals, namely *Scientific Reports*, *Journal of Biotechnology*, *Plants*, *Food and Crops*, *PeerJ*, *Mitochondria B*, *Critical Reviews in Biotechnologies*, and so on. Dr. Tyagi has also published more than 10 book chapters, and 3 books. She is also a member of the editorial board and reviewer committee of various journals of international repute such as MDPI, Agronomy, IJM, Vaccine, and so on. She has presented her research findings at more than 10 national and international conferences. She has attended more than 50 conferences, workshops, colloquia, seminars, and so on, both in India and abroad. Dr. Tyagi has delivered several invited lectures and has been involved in various social services such as teaching and providing free career counselling to rural area students.

Robin Kumar is presently working as Assistant Professor at Acharya Narendra Dev University of Agriculture and Technology Ayodhya, Uttar Pradesh, India. He did his undergraduate degree, M.Sc., and Ph.D. in Soil Science at Sardar Vallabh Bhai Patel University of Agriculture and Technology, Meerut, Uttar Pradesh, India. He was IRRI research fellow during his M.Sc. studies and CIMMYT young fellowship for his Ph.D. studies. Dr. Kumar worked as a Research Associate at IIFSR, Meerut, India. Dr. Robin is handling projects on integrated farming management system and guiding several undergraduate and master students. His area of specialization is conservation agriculture, integrated farming, and organic farming. Dr. Kumar has published more than 70 research papers, book chapters, and review articles in journals of national and international repute. Dr. Kumar has attended several national and international seminars, conferences, and workshops at which he has presented his research findings.

Baljeet Singh Saharan's area of research includes PGPR, bioremediation, biofertilizers, biosurfactants, bacteriocins, plant-microbial interactions, and molecular microbiology. He completed his M.Sc. and Ph.D. in Microbiology at CCS Haryana Agricultural University Hisar, India. He has worked as Assistant Professor in Microbiology in the Department of Microbiology, SBS, PGI, Dehradun, JCDV, Sirsa, and Kurukshetra University, India. He worked as Associate Professor at Kurukshetra University, India. He joined as Senior Scientist in the Department of Microbiology, CCS HAU, Hisar, in 2019. At present, he is working as Senior Scientist and is in charge of the Biofertilizer Production Centre, Department of Microbiology, CCS HAU, Hisar. He has guided 30 M.Sc. and 17 Ph.D. students as a supervisor. At present, one M.Sc. student and five Ph.D. students are working under his supervision. Dr. Saharan was given the C.V. Raman award (INDO-US) for research in the USA. He has been Visiting Research Scholar at Washington State University, Pullman, WA USA. Dr. Saharan has successfully completed three major research projects financed by the University Grant Commission (UGC), Department of Science & Technology (DST), and Haryana State Council for Science and Technology (HSCST). Dr. Saharan is a recipient of the DAAD (Indo-German) fellowship 2003–2004 for doing post-doctoral research in the Department of Bioremediation (now Department of Environmental Biotechnology), Helmholtz Centre for Environmental Research – UFZ, Leipzig, Germany. He has been given the Raman (Indo-US) fellowship for post-doctoral studies in the Department of Plant Pathology, USDA, Washington State University, Pullman, Washington, USA. He worked on plant growth-promoting rhizobacteria. Dr. Saharan has published more than 70 research papers in journals of national and international repute. He has published two books through international publishers including Springer and CRC Press. He has presented his research findings in more than 40 national/international conferences through funding provided by DST, DBT, HSCST, UGC, KUK, and so on. He was recently nominated to be a member of a national committee by the Director General (Indian Council of Agricultural Research) for the revalidation of results of ZBNF (SPNF – Model) for natural farming. He has attended more than 50 conferences, workshops, refresher courses, colloquia, seminars, and so on, in India and abroad.

Ashok Kumar Nadda is working as an Assistant Professor in the Department of Biotechnology and Bioinformatics, Jaypee University of Information Technology, Waknaghat, Solan, Himachal Pradesh, India. He holds more than 8 years of research and teaching experience in the field of microbial biotechnology, with research expertise focusing on various issues pertaining to nanobiocatalysis, microbial enzymes, biomass, bioenergy, and climate change. Dr. Ashok teaches enzymology and enzyme technology, microbiology, environmental biotechnology, bioresources and industrial products to bachelor's-, master's-, and Ph.D.-level students. He also trains the students in enzyme purification expression, gene cloning, and immobilization onto nanomaterials experiments in his lab. He holds international work experiences in

South Korea, India, Malaysia, and China. He worked as a post-doctoral fellow in the State Key Laboratory of Agricultural Microbiology, Huazhong Agricultural University, Wuhan, China. He also worked as a Brain Pool Researcher/Assistant Professor at Konkuk University, Seoul, South Korea. Dr. Ashok has a keen interest in microbial enzymes, biocatalysis, CO_2 conversion, biomass degradation, biofuel synthesis, and bioremediation. His work has been published in various internationally reputed journals, namely *Chemical Engineering Journal, Bioresource Technology, Scientific Reports, Energy, International Journal of Biological Macromolecules, Science of Total Environment*, and *Journal of Cleaner Production*. Dr. Ashok has published more than 100 scientific contributions in the form of research, reviews, books, book chapters, and others on several platforms in various journals of international repute. His research output includes 70 research articles, 25 book chapters, and 10 books. He is the main series editor of *Microbial Biotechnology for Environment, Energy and Health*, which publishes under Taylor & Francis, CRC Press, USA. He is also a member of the editorial board and reviewer committee of various journals of international repute. He has presented his research findings at more than 40 national and international conferences. He has attended more than 50 conferences, workshops, colloquia, seminars, and so on,in India and abroad. Dr. Ashok is also an active reviewer for many high-impact journals published by Elsevier, Springer Nature, ASC, RSC, and Nature publishers. His research works have gained broad interest through his highly cited research publications, book chapters, conference presentations, and invited lectures.

Contributors

Paul Olusegun Bankole
Department of Pure and Applied
  Botany
College of Biosciences
Federal University of Agriculture
Abeokuta, Nigeria

Ram Naresh Bharagava
Department of Microbiology (DM)
School for Environmental
  Sciences (SES)
Babasaheb Bhimrao Ambedkar
  University (A Central University)
Lucknow, India

Narendra K. Bharat
Department of Seed Science and
  Technology
Dr. Y. S. Parmar University of
  Horticulture and Forestry
Nauni, India

Pradeep Bhatnagar
IIS (deemed to be University)
Mansarovar
Jaipur, India

Kartar Chand
University Institute of Biotechnology
Chandigarh University
Mohali, India

Sushil Sudhakar Changan
ICAR – Central Potato Research
  Institute
Shimla, India

Kumar Nishant Chourasia
ICAR – Central Potato Research
  Institute
Shimla, India

V. K. Dhiman
Dr. Y. S. Parmar University of
  Horticulture and Forestry
Nauni, India

Som Dutt
ICAR – Central Potato Research
  Institute
Shimla, India

Ayanava Goswami
University Institute of Biotechnology
Chandigarh University
Mohali, India

Sanjay Kumar Gupta
Environmental Engineering
Department of Civil Engineering
Indian Institute of Technology Delhi
New Delhi, India

Anyi Hu
CAS Key Laboratory of Urban Pollutant
  Conversion
Institute of Urban Environment Chinese
  Academy of Sciences
Xiamen, China

Young B. Ibiang
Environmental Biotech Unit
Department of Genetics and
  Biotechnology
University of Calabar
Calabar, Nigeria

and

Graduate School of Horticulture
Chiba University
Matsudo, Japan

Syeda Ulfath Tazeen Kadri
Department of Biochemistry
School of Applied Sciences
REVA University
Bangalore, India

and

Department of Biochemistry
School of Sciences
Maharani Cluster University
Bangalore, India

Kanika
Department of Bio and Nanotechnology
Guru Jambheshwar University of
    Science and Technology
Hisar, India

Jaspreet Kaur
University Institute of Biotechnology
Chandigarh University
Mohali, India

Ekta Khare
Department of Microbiology
Chhatrapati Shahu Ji Maharaj
    University
Kanpur, India

Dharmendra Kumar
ICAR – Central Potato Research
    Institute
Shimla, India

Dinesh Kumar
Department of Horticulture
Maharana Pratap Horticultural
    University
Karnal, India

Robin Kumar
Department of Soil Science
Acharya Narendra Dev University of
    Agriculture and Technology
Kumarganj, Aydohya, India

Milan Kumar Lal
ICAR – Central Potato Research Institute
Shimla, India

Mallappa M.
Department of Chemistry
School of Applied Sciences
REVA University
Bangalore, India

Parikshana Mathur
Department of Biotechnology
IIS (deemed to be University)
Mansarovar
Jaipur, India

Payal Mehtani
Department of Biotechnology
IIS (deemed to be University)
Mansarovar
Jaipur, India

Gaurav Mudgal
University Institute of Biotechnology
Chandigarh University
Mohali, India

Sikandar I. Mulla
Department of Biochemistry
School of Applied Sciences
REVA University
Bangalore, India

Ashok Kumar Nadda
Department of Biotechnology and
    Bioinformatics
Jaypee University of Information
    Technology
Waknaghat, India

Shivangi Negi
Department of Seed Science and
    Technology
Dr. Y. S. Parmar University of
    Horticulture and Forestry
Nauni, India

D. Pandey
Central Institute for Subtropical
  Horticulture (CISH)
Lucknow, India

H. Pandey
Dr. Y. S. Parmar University of
  Horticulture and Forestry
Nauni, India

Manisha Parashar
University Institute of Biotechnology
Chandigarh University
Mohali, India

Jagdish Parshad
Department of Microbiology
CCS Haryana Agricultural University
Hisar, India

Nagesh Babu R.
Department of Biochemistry
School of Sciences
Maharani Cluster University
Bangalore, India

Pinky Raigond
ICAR – Central Potato Research Institute
Shimla, India

Baljeet Singh Saharan
Department of Microbiology
Chaudhary Charan Singh Haryana
  Agricultural University
Hisar, India

and

Department of Microbiology
Kurukshetra University
Kurukshetra, India

Mohammed Azharuddin Savanur
MIGAL – Galilee Research Institute
Kiryat Shmona, Israel

Charu Sharma
Department of Biotechnology
IIS (deemed to be University)
Mansarovar
Jaipur, India

I. Sharma
Dr. Y. S. Parmar University of
  Horticulture and Forestry
Nauni, India

Neha Sharma
Division of Crop Improvement
ICAR – Central Potato Research
  Institute
Shimla, India

Nidhi Sharma
Department of Microbiology
CCS Haryana Agricultural
  University
Hisar, India

Parul Sharma
Department of Biotechnology
Dr. Y. S. Parmar University of
  Horticulture and Forestry
Nauni, India

Rajnish Sharma
Department of Biotechnology
Dr. Y. S. Parmar University of
  Horticulture and Forestry
Nauni, India

Swati Sharma
University Institute of Biotechnology
Chandigarh University
Mohali, India

Brajesh Singh
ICAR – Central Potato Research
  Institute
Shimla, India

Gajendra B. Singh
University Institute of Biotechnology
Chandigarh University
Mohali, India

Hemant Sood
Department of Biotechnology and
    Bioinformatics
Jaypee University of Information
    Technology
Waknaghat, India

Adinath N. Tavanappanavar
Department of Biochemistry
School of Applied Sciences
REVA University
Bangalore, India

K. Thakur
Dr. Y. S. Parmar University of
    Horticulture and Forestry
Nauni, India

Rahul Kumar Tiwari
ICAR – Central Potato Research
    Institute
Shimla, India

Swati Tyagi
Division of Genomics
National Institute of Agriculture
    Science
Rural Development Administration (RDA)
Jeonju, Republic of Korea

and

Rice Breeding Platform
International Rice Research Institute
South Asia Regional Centre (ISARC)
Varanasi, India

Amit Kumar Verma
University Institute of Biotechnology
Chandigarh University
Mohali, India

Ankita Vinayak
University Institute of Biotechnology
Chandigarh University
Mohali, India

1 Plant-Microbe Interaction for Sustainable Agriculture

Swati Tyagi, Robin Kumar, Baljeet Singh Saharan, and Ashok Kumar Nadda

CONTENTS

1.1 INTRODUCTION

The world's population is growing exponentially, and, with this increasing population, the rate of urbanization/industrialization has also increased. This has increased the demand for food, which has accelerated the depletion of natural resources (Vejan et al. 2016). To meet this demand, an excessive and irrational use of agrochemicals, such as fertilizers, herbicides, and fungicides, has been adopted in the past, influencing agriculture practices and polluting soil, water, and ecosystems and causing an ecological imbalance. However, these practices cause more serious challenges rather than solving the demand; therefore, it is necessary to improve crop productivity organically or naturally in a sustainable manner in harmony with the ecosystem. These issues pushed agricultural scientists to look for alternative options and understand the underlying molecular mechanism of plant-microbe interactions that can provide fresh perspectives (Tyagi et al. 2021). In the natural environment, plants and microbes constantly interact with each other, and these interactions are highly diverse and can be broadly categorized as favourable, neutral, or harmful, depending on their effect on the plant health and development process (Tyagi, Lee et al. 2020).

DOI: 10.1201/9781003213864-1

1

In past decades, many studies have been conducted to understand the important molecular players involved in plant-microbe interactions, and a lot of information has been discovered; however, this information is not enough, and many queries still need to be addressed. Microorganisms interacting with plants exhibit enormous potential to improve plant health and development as a natural catalyst or to trigger a negative response in the form of disease or stress (van de Mortel et al. 2012). Queries such as how do these microbes interact and exhibit different responses, what are the signalling pathways involved, what makes the interaction harmful or beneficial, what are the immune factors in plant and microbes that affect the overall response in both the interacting partners must be answered to understand the overall interaction process in plants and microbes and will provide the landmark information that can be used to identify and mark the pathogens or beneficial microbes and their interaction effect on the crop plants.

In this context, beneficial microorganisms such as plant growth-promoting rhizobacteria (PGPR) or plant growth-promoting fungi (PGPF) are considered potential elements to improve plant growth and development and, hence, can serve as natural and sustainable alternatives to fertilizers and pesticides (van Loon et al. 1998). In past decades, several PGPFs and PGPRs were identified, characterized, and found to improve plant growth and nutritional quality, soil health, and fertility in a sustained manner. For example, some members of the bacterial genera *Bacillus*, *Trichoderma*, and *Fusarium* prevent plant diseases by subduing plant pathogens, thereby serving as biocontrol agents (Tyagi et al. 2020). Application of PGPR and mycorrhizal fungi enhance plant growth under various stress conditions. Both fungal and bacterial endophytes are found to be active stress relievers of the host plant (Voisard et al. 1989). Studies with many microbial inoculants have demonstrated their beneficial role in plant growth through effective root colonization and induction of plant growth support mechanisms. Overall, the exploitation of beneficial microorganisms and their useful interactions with plants offer promising and eco-friendly strategies in the development of organic agriculture globally (Eckardt 2002). However, microbes can cause a serious threat to agricultural production by causing severe diseases and are considered to be phytopathogens (El-Tarabily et al. 2000). Plant-pathogen studies must be undertaken to identify new pathogens, how they affect plant health, and how the already reported pathogens are developing resistance and evolving. All this information about plant-microbe interaction, whether it is beneficial or harmful, lies in the genetic material, and, thus, the genes involved in these interactions must be identified, characterized, and annotated. There is an urgent need to link the plant studies with "OMICS" to address these questions and provide some useful gene quantitative trait loci (QTLs) controlling specific traits to the breeders that can be used in breeding programs and thus improve the crop varieties in terms of productivity or quality.

1.2   BENEFICIAL PLANT-MICROBE INTERACTIONS

With the start of the green revolution, the use of chemical fertilizers, insecticides, and pesticides increased dramatically to improve agricultural yield and productivity.

However, despite several studies reporting their negative impact on the ecosystems, soil health, and humans, they are still in use (Tyagi et al. 2021). To overcome these potential hazards, the use of microorganisms for crop improvement has been proposed and, over the years, this proposal has been widely accepted by innovative farmers and agriculturists. Several plant growth-promoting microorganisms (PGPM) – including PGPRs, PGPFs, arbuscular mycorrhizae (AMF), and endophytes (bacteria/fungi) – have been reported to influence plant growth positively (Tyagi et al. 2021). These PGPMs either directly secrete plant growth-promoting compounds in the form of volatile organic compounds (VOCs), plant hormones, siderophores, etc. to boost the plant health or use indirect mechanisms such as the release of lytic enzymes, antibiotics, and other defence molecules which antagonise the pathogen growth and promote the plant growth as well as develop resistance against the plant pathogens (Asari et al. 2016). VOCs are compounds with a high vapour pressure that makes them highly diffusible in soil and the plant canopy, thus, making them an ideal molecule for crosstalk between plants and microbes (Asari et al. 2016). VOCs generated by plants and microorganisms can act as signalling molecules activating a series of molecular events that ultimately regulate a wide range of physiological processes of plants and microorganisms (Tyagi et al. 2019). VOCs released by plants determine the type of microbiota that can live in the phytosphere and prime the plant defensive system to upcoming stresses (Tyagi et al. 2018; Tyagi et al. 2019; Tyagi, Lee et al. 2020). Microbes also emit VOCs in response to environmental conditions and can stimulate plant growth and induce resistance/ tolerance to biotic/abiotic anxious factors (Tahir et al. 2017; Tahir et al. 2017). Thus, biogenic VOCs represent a rich and complex chemical vocabulary that can help to uncover the hidden secrets that can be used for modern sustainable agriculture. On the other hand, metal-resistant siderophore-producing bacteria help a plant to thrive under heavy metal stress by alleviating the metal toxicity. Heavy metals, being noxious in nature, enter the food chain and result in the toxicity of plants and animals (Chen et al. 2016). With the expansion of industries, the pollution of these toxic metals is increasing at a very fast pace. The removal of these heavy metals by natural means is the need of the hour. The toxication of soil by these metals can be removed through phytoremediation, mycoremediation, and microbial remediation (Chen et al. 2016). One such way of microbial remediation includes the application of siderophores that are synthesized naturally by microbes and are helpful in forming a complex with heavy metal. Siderophores have been used mostly for clinical studies but have the potential to play a critical role in cleaning the environment (Rajkumar et al. 2010). Siderophores provide heavy-metal detoxification and cleaning of the environment by natural means. Some microbes improve the growth of the plants and serve as biofertilizers (Rajkumar et al. 2010). These microbes form spores and provide a competitive environment for other microbes. Many PGPR/F(s) secrete lytic enzymes such as chitinase, cellulase that can degrade the insoluble organic polymers into soluble compounds that can be utilized by the plants (Tyagi, Lee et al. 2020). Additionally, microbes induce systemic resistance in plants and provide protection against biotic and abiotic stress. Several reports have been published that show the potential and mechanism of action of how microbes secrete different molecules that induce resistance against fungal/bacterial pathogens as well as improve the plant growth and development process (Tyagi et al. 2021). Also, the secondary metabolites released by microbes are taken by plants to resist the abiotic stresses such as drought, salt, etc. These interactions among plants and different classes of microbes are much more important to understand so that they can be utilized as alternatives for chemical based agri-products and to ensure the plant health under unfavourable conditions.

1.3   HARMFUL PLANT-MICROBE INTERACTIONS

In plant-microbe interaction studies, it is reported that fungi are more likely to cause more yield losses than other phytopathogens because of their highly evolving nature. Usually, the fungal phytopathogens are host specific, but survival on alternative hosts has also been reported for many of them, which helps the invading pathogen to breach the plant's immune system and develop resistance (Tyagi, Lee et al. 2020). Other phytopathogens, such as bacteria, viruses, and nematodes, also negatively impact plant health and decrease productivity. These pathogens interact with the plant host, breach the plant's immune system, and develop disease in the host, which, later in the course of infection, kills the plant or reduces productivity. Pathogens secrete effector proteins that interact with plant proteins and initiate the infection process. Also, pathogens secret toxins that shut down the expression of defence-related genes, making them vulnerable to infection. Though previous studies provide informative insights regarding these interactions, much still remains unknown and needs to be addressed (Tyagi et al. 2021).

1.4   MICROBIAL IMMUNITY AGAINST HOST PLANTS

To initiate infection in the host plant, microbes must fight against the plant's defence system. Microbes use different virulence-related biomolecules that interact with the host. In the case of bacteria, type II, III, and IV secretion systems release virulence-related biomolecules that interact with the host to initiate and progress the infection (Tyagi et al. 2021). Similarly, fungi secrete enzymes that degrade the host cell wall/membrane-related molecules and allow the pathogens to enter into the host. On the other hand, viruses' particles enter the host through mechanical and chemical injury caused by external factors or by biological vectors (Tyagi et al. 2021). Once the viral particle enters the host, it can replicate itself and induce the infection response. It is very important to understand the mechanism of how pathogens enter the host and what molecules are involved in infection progression as well as how they initiate the infection. Understanding this molecular crosstalk between plant and microbes will help us to develop resistant or less susceptible plant/crop varieties (Tyagi et al. 2021).

1.5   PLANT IMMUNITY AGAINST INVADING PATHOGENS

In nature, plants, as a defending host, and microbes, as infecting pathogen, are constantly racing to initiate/mitigate infection. Several microbes, including bacteria, fungi, and viruses, cause a number of diseases in plants, and plants have to deal with these invading pathogens to survive (Tyagi et al. 2020). The success rate of any infection is dependent on the susceptibility of the host and the environmental conditions favouring the establishment of infection. To fight invading pathogens, plants have two layers of defence: 1) constitutive and 2) induced. The constitutive defence system includes the physical and chemical barriers that are uniformly present in all plant species and act as the first line of defence (Tyagi et al. 2021). On the other hand, the induced defence system is activated when the pathogen attacks the host and either tries to breach or has breached the first line of defence (Figure 1.1).

Over time, different research groups have introduced different models of plant-microbe interaction to explain plant immunity and response toward the invading pathogens. The most accepted model among them is the zig-zag model (Tyagi et al. 2021). The plant defence system is activated by cell-to-cell communication followed by a complex series of events between the host plant and pathogen or its component. Depending upon the type (bacteria, fungi, virus, etc.) and nature (biotrophic, necrotrophic, etc.) of invading pathogen, the immune system employs different biomolecules (Tyagi et al. 2018). Generally, plants' cell walls or cell membranes have

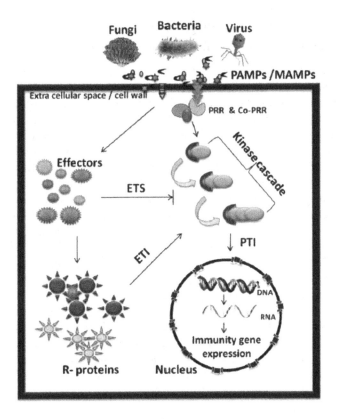

FIGURE 1.1 Plant immune system: The figure illustrates the interaction between plant microbes and the underlying mechanism involved in plant defence.

a different set of receptors often called protein recognition receptors (PRR) or wall-associated kinases (WAKs). On pathogen encounter, this set of receptors identify the pathogen or microbe associated molecular patterns (PAMP/MAMP) and recruit the PAMP/MAPMP-triggered immune (PTI) system to fight the invading pathogen (Tyagi et al. 2018). To breach this defence, pathogens induce effector-triggered susceptibility (ETS), which causes some pathogens to secrete effector proteins that quash the PTI by employing susceptibility (S) proteins that enable the disease to advance. In this scenario, the plant's immune system initiates another line of defence by employing resistance (R) genes. R genes recognize the effectors or avirulence genes from pathogens and activate effector-triggered immunity (ETI) (Tyagi et al. 2021). These two induced defence systems (PTI and ETI) then alter the expression of different sets of genes, such as mitogen-activated protein kinase (MAPKs), plant hormones, transcription factors, and other pathogenesis-related genes that further induce hypersensitive response, reactive oxygen species (ROS) generation, cell-wall modification, stomata closure, or secretion of anti-microbial proteins and compounds, e.g., chitinases, protease inhibitors, defensins, and phytoalexins, etc. (Tyagi et al. 2018).

1.6   USE OF OMICS TECHNIQUES FOR PLANT-MICROBE INTERACTION STUDIES

The OMICS technologies that collect the information in the form of deoxyribonucleotide (DNA), ribonucleotide (RNA), proteins, or metabolites can shed light on the cellular structure and function of an organism. Out of these, the information that is stored in the form of DNA sequences is not much affected by the environmental condition; however, RNA, proteins, and metabolites are strongly influenced by external factors (Crandall et al. 2020; Sharma et al. 2020). The study of DNA sequences can be performed by genomics or next generation sequencing (NGS) techniques. While the study of RNA can be achieved by transcriptomics, proteins by proteomics, and metabolites by metabolomics. These technologies, either alone or in combination with others, can provide the desired information. The advancement of NGS made it possible to uncover the secrets lying within the genome of any organism (Crandall et al. 2020; Sharma et al. 2020). It became easy to sequence the genome of an organism, understand the mechanism of infection, detect the genetic variations, modify the gene/genome of an organism, and provide different prospects for developing resistance against plant pathogens or improving plant growth and development by introducing a specific gene or trait. It is now possible to study the genotype–phenotype relationship with the highest resolution through linkage mapping. Different approaches such as traits associated mapping, QTL mapping, genome-wide associate studies (GWAS), and haplotype analysis have been used to study these associations to improve plant traits and productivity (Crandall et al. 2020; Sharma et al. 2020). On the other hand, transcriptomic studies characterize and quantify the RNAs present in each sample (plant, organ, tissue) in a particular condition and provide the connection between the genotype and phenotype (Crandall et al. 2020; Sharma et al. 2020). RNA sequencing, isoform sequencing, microarray analysis, etc. are different approaches that have been used to capture the transcriptomic information (Crandall et al. 2020; Sharma et al. 2020). In addition to this, proteomic studies are based on the full proteins present in a particular sample in a specific condition. The proteomic studies use two-dimension gel electrophoresis (2DE), liquid chromatography (LC), mass spectroscopy (MS), and coupled approaches such as LC/MS etc. Similarly, metabolomics is the study of by-products/metabolites as distinct molecules involved in a plant in a specific condition (Crandall et al. 2020; Sharma et al. 2020). The techniques employed for metabolomics also include LC/MS, nuclear magnetic resonance spectroscopy (NMR), capillary electrophoresis (CE), and time-of-flight (TOF)-MS devices. On the other hand, advances in molecular biology techniques made it easier to modify the genome of a host either by introducing some genes that can breakdown the toxins, abolish the activity of cell-wall degrading enzymes, excrete antimicrobial compounds, or delete the genes that are susceptible to pathogen attack (Crandall et al. 2020; Sharma et al. 2020). A number of gene editing approaches, such as RNA interface, zinc finger nuclease (ZFN), transcription activator-like effector nuclease (TALEN), CRISPR : Cas9, etc., have been discovered that exhibit potential in the field of genome editing in agriculture and other fields of science (Tyagi et al. 2020). Though these molecular and OMICs techniques seem sophisticated and have proven to be very informative, and a wealth of receptors, genes, genomes, proteins, and metabolic products have been accumulated that can be used to improve the plant-microbe relationship for sustainable agriculture.

1.7 FUTURE PERSPECTIVES

Plant-microbe interaction studies can provide promising solutions for sustainable agriculture. These studies are very important to develop biofertilizers, biopesticides, and bioremediation processes. In the past, several studies were performed that have been very useful, but they are not enough. It is very important to find new genes and targets and understand their mechanisms of action during beneficial or harmful interaction with plants. Therefore, it is necessary to couple the OMICs and biotechnological approaches to understand the genetics of and in-depth knowledge of plant growth, development, disease, traits, and stress (abiotic/biotic) management.

ACKNOWLEDGEMENT

I thank the National Institute of Agriculture Science, Rural Development Administration (RDA) in South Korea for financial assistance and the International Rice Research Institute – South Asia Regional Centre (ISARC) in Varanasi, India, for institutional support.

DISCLOSURE OF POTENTIAL CONFLICTS OF INTEREST

No conflict of interest.

REFERENCES

Asari, S., S. Matzen, M. A. Petersen, S. Bejai, and J. Meijer. "Multiple Effects of Bacillus Amyloliquefaciens Volatile Compounds: Plant Growth Promotion and Growth Inhibition of Phytopathogens." *FEMS Microbiology Ecology* 92, no. 6 (Jun 2016): fiw070.

Chen, Y. M., Y. Q. Chao, Y. Y. Li, Q. Q. Lin, J. Bai, L. Tang, S. Z. Wang, R. R. Ying, and R. L. Qiu. "Survival Strategies of the Plant-Associated Bacterium Enterobacter Sp Strain Eg16 under Cadmium Stress." [In English]. *Applied and Environmental Microbiology* 82, no. 6 (Mar 2016): 1734–44.

Crandall, S. G., K. M. Gold, M. D. Jimenez-Gasco, C. C. Filgueiras, and D. S. Willett. "A Multi-Omics Approach to Solving Problems in Plant Disease Ecology." [In English]. *Plos One* 15, no. 9 (Sep 22 2020).

Eckardt, N. A. "Specificity and Cross-Talk in Plant Signal Transduction: January 2002 Keystone Symposium." *Plant Cell* 14 Suppl (2002): S9–14.

El-Tarabily, K., M. H. Soliman, A. H. Nassar, D. Al-Hassani, K. Sivasithamparam, F. McKenna, and G. Hardy. "Biological Control of Sclerotinia Minor Using a Chitinolytic Bacterium and Actinomycetes." *Plant Pathology* 49 (2000): 573–83. doi:10.1046/j.1365-3059.2000.00494.x.

Rajkumar, M., N. Ae, M. N. V. Prasad, and H. Freitas. "Potential of Siderophore-Producing Bacteria for Improving Heavy Metal Phytoextraction." [In English]. *Trends in Biotechnology* 28, no. 3 (Mar 2010): 142–49.

Sharma, M., S. Sudheer, Z. Usmani, R. Rani, and P. Gupta. "Deciphering the Omics of Plant-Microbe Interaction: Perspectives and New Insights." [In English]. *Current Genomics* 21, no. 5 (2020): 343–62.

Tahir, H. A., Q. Gu, H. Wu, Y. Niu, R. Huo, and X. Gao. "Bacillus Volatiles Adversely Affect the Physiology and Ultra-Structure of Ralstonia Solanacearum and Induce Systemic Resistance in Tobacco against Bacterial Wilt." [In English]. *Scientific Reports* 7 (Jan 16 2017): 40481.

Tahir, H. A., Q. Gu, H. Wu, W. Raza, A. Hanif, L. Wu, M. V. Colman, and X. Gao. "Plant Growth Promotion by Volatile Organic Compounds Produced by Bacillus Subtilis Syst2." *Frontiers in Microbiology* 8 (2017): 171.

Tyagi, S., K. Kim, M. Cho, and K. J. Lee. "Volatile Dimethyl Disulfide Affects Root System Architecture of Arabidopsis Via Modulation of Canonical Auxin Signaling Pathways." *Environmental Sustainability* 2, no. 2 (2019/06/01 2019): 211–16.

Tyagi, S., R. Kumar, A. Das, S. Y. Won, and P. Shukla. "Crispr-Cas9 System: A Genome-Editing Tool with Endless Possibilities." [In English]. *Journal of Biotechnology* 319 (Aug 10 2020): 36–53.

Tyagi, S., R. Kumar, V. Kumar, S. Y. Won, and P. Shukla. "Engineering Disease Resistant Plants through Crispr-Cas9 Technology." [In English]. *GM Crops & Food-Biotechnology in Agriculture and the Food Chain* 12, no. 1 (Jan 2 2021): 125–44.

Tyagi, S., K. J. Lee, P. Shukla, and J. C. Chae. "Dimethyl Disulfide Exerts Antifungal Activity against Sclerotinia Minor by Damaging Its Membrane and Induces Systemic Resistance in Host Plants (Vol 10, 6547, 2020)." [In English]. *Scientific Reports* 10, no. 1 (Oct 20 2020): 1–12.

Tyagi, S., S. I. Mulla, K. J. Lee, J. C. Chae, and P. Shukla. "Vocs-Mediated Hormonal Signaling and Crosstalk with Plant Growth Promoting Microbes." [In English]. *Critical Reviews in Biotechnology* 38, no. 8 (2018): 1277–96.

van de Mortel, J. E., R. C. de Vos, E. Dekkers, A. Pineda, L. Guillod, K. Bouwmeester, J. J. van Loon, M. Dicke, and J. M. Raaijmakers. "Metabolic and Transcriptomic Changes Induced in Arabidopsis by the Rhizobacterium Pseudomonas Fluorescens Ss101." [In English]. *Plant Physiology* 160, no. 4 (Dec 2012): 2173–88.

van Loon, L. C., P. A. Bakker, and C. M. Pieterse. "Systemic Resistance Induced by Rhizosphere Bacteria." *Annual Review of Phytopathology* 36 (1998): 453–83.

Vejan, P., R. Abdullah, T. Khadiran, S. Ismail, and A. Nasrulhaq Boyce. "Role of Plant Growth Promoting Rhizobacteria in Agricultural Sustainability-A Review." *Molecules* 21, no. 5 (Apr 29 2016): 573.

Voisard, C., C. Keel, D. Haas, and G. Dèfago. "Cyanide Production by Pseudomonas Fluorescens Helps Suppress Black Root Rot of Tobacco under Gnotobiotic Conditions." *The EMBO Journal* 8, no. 2 (1989): 351–58.

2 Beneficial Microorganisms in Crop Growth, Soil Health, and Sustainable Environmental Management
Current Status and Future Perspectives

Dharmendra Kumar, Som Dutt, Pinky Raigond, Sushil Sudhakar Changan, Milan Kumar Lal, Rahul Kumar Tiwari, Kumar Nishant Chourasia, and Brajesh Singh

CONTENTS

DOI: 10.1201/9781003213864-2

2.1 INTRODUCTION

The persistent utilization of agrochemicals for improved soil fertility and plant productivity often results in adverse environmental effects, including contamination
of soil, groundwater, air, food, and aquifers (Mushtaq, Faizan, and Hussain 2021).
Currently, agricultural practices depend mainly on chemical inputs (such as fertilizers, pesticides, herbicides, etc.), which, all things being equal, cause a deleterious
effect on the nutritional value of farm products and health of farmworkers and con-
sumers. The excessive and indiscriminate use of these chemicals has resulted in food
contamination, weed and disease resistance, and negative environmental outcomes,
which, together, have a significant impact on human health. The application of these
chemical inputs promotes the accumulation of toxic compounds in the whole ecosystem. The chemical compounds are absorbed by most crops from the soil (Goswami,
Thakker, and Dhandhukia 2016a; Suyal et al. 2016; Riaz et al. 2021; Valette et al.
2020). Synthetic fertilizers contain acid radicals, such as hydrochloride and sulfuric
radicals, and, hence, increase soil acidity and adversely affect soil and plant health.
Highly recalcitrant compounds can also be absorbed by some plants. The continuous
consumption of such crops can lead to systematic disorders in humans. Several pesticides and herbicides have carcinogenicity potential. Hence, eco-friendly methods
of soil and nutrient management are required to maintain sustained crop productivity and ecological stability (Sharma et al. 2020). Fifty years ago, the "green revolution" was launched, combining high-yielding cultivars, inorganic fertilizers, and
pesticides to foster food production. Although the green revolution, urbanization,
and industrialization generally made human life more comfortable and easy and
improved the world's economy, effluents from the industry also affected the health of
the soil, air, water, and atmosphere, leading to overall ecosystem degradation. Heavy
metal contamination from industries poses an adverse impact not only on soil fertility and plant growth but also a serious threat to human health (Vessey 2003; Alori,
Glick, and Babalola 2017b).

   The increasing awareness of health challenges resulting from the consumption of
poor-quality crops has led to a quest for new and enhanced technologies to improve

both the quantity and quality of crops without jeopardizing human health. The impacts were tremendous, however, nowadays, to maintain healthy ecosystems, different approaches have to be employed, such as microbial inoculations (Varah et al. 2020; Lichtfouse 2017). Microbial inoculants can either replace or reduce pesticides, weedicides, or other agrochemicals and clean areas profoundly affected by pollution. The over-use of ecosystem services is also alarming. While the use of regulating services, such as air quality, erosion control, and water purification as well as provisioning services, such as pollination and genetic resources, sharply increase, their conditions steadily decrease. It is estimated that 22 million hectares of soil are adversely affected by chemical contamination worldwide, mostly in Europe but also in Asia (Bai et al. 2008). There are many methods that can be used on a sustainable basis to meet food requirements without compromising ecosystem health. Among these, the use of microbial products is pivotal to ensuring food security within a changing climate (Varah et al. 2020).

Microbial approaches can be used successfully for sustainable agricultural development. These microbes can enhance plant growth and development by improving nutrient availability to plants through various mechanisms, thus, decreasing the dependence on agrochemicals. Beneficial microbial inoculants are potential elements of such management approaches. Studies with many microbial inoculants have demonstrated their beneficial role in plant growth through effective root colonization in the rhizosphere and induction of plant growth support mechanisms. Direct plant growth support by root-associated microbes is mediated through enhanced nutrient acquisition and hormonal activation. The numerous antimicrobial activities of microbes associated with pathogen inhibition are usually linked to superior plant growth and development. Some inoculants have been found to be useful in abiotic stress alleviation, nutritional fortification of edible crops, and sequestration of terrestrial carbon (Varah et al. 2020; Pretty 2008; Timmusk 2012).

Therefore, more attention is being given to sustainable, reliable, stable, suitable, and environmentally friendly crop-production approaches and protection of natural resources from the detrimental effects of agrochemicals, such as fertilizers, weedicides, and fungicides. Sustainable agriculture is vital in today's world not only because it can fulfil humans' need for food and fibres for a longer amount of time but also because it is economically viable and does not damage the ecosystem. The scientifically sound and improved technological interventions – such as sustainable management methods, nitrogen-fixing genetically modified crops, and atmospheric nitrogen fixation exclusive of bacterial symbionts, biopesticides, biofertilizers, bioherbicide, biocontrol agents, improved plant varieties, use of genetically engineered microbes, and use of microbial inoculants as biofertilizers – are pre-requisites for promoting viable and eco-safe agriculture. Furthermore, some other components' abiotic and biotic stresses, such as crop disease resistance, salinity, heat, drought tolerance, tolerance toward metals stress, and crop biofortification, are also helpful in achieving agricultural sustainability (Suyal et al. 2016; Bhattacharyya and Jha 2012a; Maksimov, Abizgil'dina, and Pusenkova 2011; Beneduzi, Ambrosini, and Passaglia 2012b; Bharti et al. 2016).

Plants are entirely dependent upon soil microorganisms that are used as a growth medium, and the synergy between both is vital for their survival. The rhizosphere, the region of soil surrounding the roots, has the highest concentration of microorganisms. The root exudates dictate the microbial communities. The manipulation of the rhizosphere changes microbial diversity and could improve plant performance by influencing water dynamics and enzyme activities. A wide range of microscopic organisms inhabits the rhizosphere: bacteria, algae, fungi, protozoa, and actinomycetes. Of these, bacteria are the most abundant and important group of microorganisms regarding plant growth and productivity. They either live freely in the rhizosphere or in intercellular and intracellular spaces of root tissues, forming symbiotic associations with plants. Fungi play an important role in organic matter decomposition and, therefore, nutrient cycling. Among soil fungi, arbuscular mycorrhizal fungi (AMF) are the most important and widely studied group as potential biofertilizers and biopesticides (Ye et al. 2020; Pretty and Bharucha 2014; Ou et al. 2019; Fasim et al. 2002; Alori and Babalola 2018).

Plant growth-promoting rhizobacteria (PGPR) are potential components for resolving agroecological problems, as they are capable of boosting plant health through its growth promotion, enhanced nutrient acquisition, and disease suppression (Grobelak, Napora, and Kacprzak 2015).

PGPR comprise the following main groups:

(1) Non nodulating PGPR,
(2) AMF,
(3) Nitrogen-fixing rhizobia, and
(4) Disease suppressing rhizobacteria (DSR).

Plant-growth promotion using these microbial inoculants is successful because of their:

(1) Effective root colonization and competency,
(2) Rapid multiplication,
(3) Adaptation to diverse and extreme soil ecosystem, and
(4) Ability to exploit diverse compounds as a source of nutrition.

PGPR are more efficient in rhizosphere colonization and plant growth support, following inoculation onto the surface of seeds. Additionally, they can colonize the root surface (rhizoplane) and within radicular tissues (the root itself). The use of PGPR for the revival of soil and plant health has, thereby, increased agriculture productivity and has been explored in several parts of the globe (Grobelak, Napora, and Kacprzak 2015; Goswami, Thakker, and Dhandhukia 2016b; Bhat et al. 2020; Bhattacharyya and Jha 2012b).

An abundance of useful microbes can be found in plant–soil-associated microecosystems, particularly the rhizosphere and rhizoplane. This microbiome, often involved in antagonistic and mutualistic interactions, ensures augmented plant growth and yield. The probable plant-microbe interactions and interconnected signalling among them occur in the crop rhizosphere and are involved in the stimulation of plant growth, biological suppression of harmful pathogens, and mitigation of crops' abiotic stresses through multifold mechanisms of beneficial rhizosphere microbiota. The positive effects of rhizobacteria on plant growth and soil health are exerted via the mechanism of nitrogen fixation, solubilization of essential nutrients, and supply of plant nutrients, hormonal stimulation, improving root growth, and enhanced uptake of nutrients and water (Armstrong McKay et al. 2019). Additionally, they indirectly enhance crop growth by inhibiting activities of plant-associated harmful fungi through antibiosis, myco-parasitism, detoxification of pathogen virulence, competitive exclusion of nutrients, and stimulation of plant defence. Currently, studies have also documented the potential application of these microbes in the nutritional fortification of edible crops, carbon sequestration through soil improvement, abiotic stress reduction, and reclamation of salt-affected soil. At present, the focus is on biological strategies for sustainable crop production with the aim of reduced chemical usage. To achieve this, different kinds of microbial inoculants possessing novel and unique traits, such as pesticide degradation/tolerance, detoxification of heavy metal, tolerance to salinity, and biocontrol of plant pathogens and insects, are being explored (Olanrewaju, Glick, and Babalola 2017). In comparison with synthetic chemicals, microbial inoculants offer many advantages such as they (1) are environmentally friendly, (2) have a negligible harmful impact on environmental and human health, (3) are specific in their actions, (4) are compelling at a lower concentration, (5) multiply themselves but are controlled by the plant as well as by the indigenous microbial populations, (6) have a biodegradable nature, (7) There is no reason to be concerned about the pathogen preparations, and (8) are compatible with other methods of integrated disease management. Microbial inoculants are natural-based products being widely used to control pests and improve the quality of the soil and crops and, hence, human health (Alori, Glick, and Babalola 2017a; Bargaz et al. 2018).

Although favourable effects from the application of microorganisms in agriculture are extensively documented, there has been variability and inconsistency in their performance under broad experimental conditions. To overcome this, recent advances in the contemporary science of nanomaterials, nanofertilizers, nanobiosensors, nanotechnology, and biotechnology, etc., are exploited (Olanrewaju et al. 2019).

These rhizobacteria can be used in different ways when plant growth promotion is required. The two primary ways through which PGPR can facilitate plant growth and development include direct and indirect mechanisms. Indirect growth promotion occurs when PGPR prevent or reduce some of the harmful effects of plant pathogens by one or more of the several different mechanisms. These include inhibition of pathogens by the production of substances or by increasing the resistance of the host plant against pathogenic organisms. For example, PGPR produce metabolites that reduce the pathogen population and produce siderophores that reduce the iron availability for specific pathogens, thereby causing reduced plant growth. Similarly, PGPR can also increase plant resistance against diseases by changing host-plant vulnerability through a mechanism called induced systemic resistance and, therefore, provide protection against pathogen attack (Olanrewaju et al. 2019; Beneduzi, Ambrosini, and Passaglia 2012b; Mhlongo et al. 2020).

FIGURE 2.1 The different mode of actions of microbes present in soils.

The present book chapter is an attempt to explicate the functions of beneficial and other useful plant-associated microorganisms in the current scenario and their underlying mechanisms in supporting plant growth and nutritional aspects as well as enhancing soil quality and agricultural productivity. In brief, the present scenario highlights research and application of beneficial microbes to develop safe agriculture (Figure 2.1).

2.2 FUNDAMENTALS OF PLANT GROWTH-PROMOTING MICROORGANISMS

Among the most productive soil microbiota, which can support crop growth, are PGPR. They are the subset of plant growth-promoting microorgnisms (PGPMs) and are predominantly involved in plant-microbe interactions. About 2%–5% of the rhizobacteria, which exert a positive effect on plant growth, individually or in co-operation with mycorrhizal fungi, are termed as PGPR. They are characterized by the following inherent characters: (1) effectiveness in root colonization (2); ability to survive, proliferate, and compete with other native microbiota; and (3) accelerate plant growth. The successful root is the colonizing ability of PGPR dependent upon organic compound use, their chemotaxis behaviour, and the production of proteins and lipopolysaccharides (Bhat et al. 2020). The rhizosphere – the surrounding soil of plant roots – is rich in plant-released root exudates and, hence, microbial inhabitants. The population of soil microbes, which are mostly found in an area 50-μm away from the root surface, can go up to 10^9–10^{12}/g soil. Though, interestingly, even if a vast microbial load is there in the rhizosphere, merely 7%–15% of the rhizoplane is occupied by soil microbiota (Calvo, Nelson, and Kloepper 2014).

Upon association, PGPR may colonize a different portion of the host plant. Based on differential colonization, they can be sub-grouped as

(1) Extracellular PGPR (ePGPR), which denote those surviving on the rhizosphere, rhizoplane, or in the spaces between cells of the root cortex. Prominent bacterial genera in this group include *Agrobacterium*, *Arthrobacter*, *Azotobacter*, *Azospirillum*, *Bacillus*, *Burkholderia*, *Caulobacter*, *Chromobacterium*, *Erwinia*, *Flavobacterium*, *Micrococcus*, *Pseudomonas*, and *Serratia*, etc.

(2) Intracellular PGPR (iPGPR), which exist inside root cells, especially in nodules. iPGPR are comprised of *Allorhizobium*, *Azorhizobium*, *Bradyrhizobium*, *Mesorhizobium*, and *Rhizobium* (Glick 2012; Goswami, Thakker, and Dhandhukia 2016a).

The mutualistic association exists between the host plant and associated rhizobacteria. The benefits of the host plant to PGPR include the release of diverse organic products in the form of rhizodeposits, which are utilized by rhizobacteria in the rhizosphere, rhizoplane, or intercellular places. Rhizobacteria metabolize root exudates, compete for the nutrients and space, and alter nutrient composition in the soil and, thus, significantly influence the nutrient supply of plants. Therefore, rhizobacteria function as a master regulator in the turnover of soil nutrients. PGPR can also be termed as yield increasing bacteria (YIB), as they augment plant growth by improving seed germination, seedling growth and vigour, plant stand, and plant biomass, and early flowering and increased yield of grains, fodder, and fruit have been reported in different agro-horticultural crops (Goswami, Thakker, and Dhandhukia 2016b).

2.3 PLANT-MICROBE INTERACTIONS AND THEIR IMPACT ON PLANT GROWTH AND NUTRIENT UPTAKE

2.3.1 ENHANCED NITROGEN ACQUISITION THROUGH NITROGEN FIXATION

Nitrogen (N) is the most important nutrient required for the growth and productivity of living entities. This is because N is the basic building block of plants, animals, and microorganisms. N fixation is the conversion of molecular or atmospheric N into a form that plants can use by N-fixing microorganisms, using an enzyme system called nitrogenase (Olanrewaju et al. 2019).

Many microbial inoculants fixing atmospheric N by symbiotic, non-symbiotic, and associative means have proven to have important roles in N recycling and plant uptake of nitrogenous fertilizer in the crop–soil ecosystem. Several literature review articles have been published on aspects of plant N uptake through symbiotic N_2 fixation in legumes and non-associative N_2 fixation in non-legumes (Bhattacharyya and Jha 2012a). Atmospheric N fixation by rhizobacteria is of agronomical significance and is responsible for the increased N_2 content in crop plants. N-fixing diazotrophs that form well-defined nodules on the root surface convert atmospheric N into a plant-usable organic form, i.e., ammonia. The nitrogenase enzyme complex present in some diazotrophic PGPR is credited for the biological conversion of triple-bonded N into ammonia. The nitrogenase enzyme complex has two parts: (1) component-I, a tetramer encoded by *nif*D and *nif*K genes, and (2) component-II, a homodimer encoded by the *nif*H gene. In diazotrophs, these two components of nitrogenase are conserved in structure, function, and amino acid sequence. At the genetic level, a *nif* gene, which is responsible for N fixation, is found in all symbiotic and non-symbiotic diazotrophs. The *nif* gene usually has seven operons that span over the 20–24 kbp region and is made up of structural genes, regulatory genes, and genes required for activation of the iron (Fe) protein as well as biosynthesis of the metal cofactor. The fixed genes, *nif, fix*, and *nod* genes, regulate the activity of the nitrogenase enzyme in the presence of oxygen, as the symbiotic activation of *nif* genes is dependent upon the low level of oxygen (Alori and Babalola 2018; Bhat et al. 2020). It must be stressed that, along with N_2 fixation, other mechanisms of plant growth promotion also exist in diazotrophic organisms. For example, the proposed modes of plant growth elevation by *Azospirillum spp.* are linked to their phytohormone activation, improved shoot/root growth, nutrient uptake and water adsorption, and extrusion of proton and organic acid. The presence of free-living diazotrophs in the rhizosphere of many crops has contributed to less dependence on the use of N_2 fertilizers (Bhattacharyya and Jha 2012a; Gaby and Buckley 2012).

2.4 PHOSPHORUS (P) SOLUBILIZATION AND MINERALIZATION THROUGH MICROBES

Phosphorus (P) is an essential element that is necessary for plant growth and development, and it is second only to N. P occurs in the soil in both organic and inorganic forms that are not available to plants; however, several PGPR have been reported to mobilize poorly available P via solubilization and mineralization (Akinola and Babalola 2021; Wang et al. 2020; Riaz et al. 2021). The insoluble form of soil P, such as apatite, inositol phosphate, phosphomonoesters, and phosphotriesters, are unavailable for plant nutrition. The plant can only uptake two forms of insoluble phosphorus i.e., monobasic (H_2PO_4) and dibasic (HPO_4). Most of the applied phosphatic fertilizers are re-precipitated into insoluble mineral complexes and, thus, are not available for plant uptake. Therefore, an ecologically safer option is essential for increased bioavailability of P in low P soils for plant growth. The P bioavailability can be increased through the use of phosphate-solubilizing microorganisms. Microbial inoculants such as PGPR and AMF significantly contribute to the solubilization and mineralization of inorganic and organic phosphates in soil (Zhu et al. 2011; Istina et al. 2015; Sharma et al. 2013). The chief mechanism associated with their solubilization potential is the secretion of organic acids. Microbial synthesis of various organic acids – e.g., gluconic acid, citric acid, etc. – and enzymes – such as phosphatases, phytase, phosphonoacetate hydrolase, D-α-glycerophosphate, and C-P lyase – is mainly accountable for the enzyme's capacity of phosphate solubilization and mineralization. Many phosphate solubilizing bacteria (PSB), such as *Achromobacter*, *Azotobacter*, *Beijerinckia*, *Bacillus*, *Burkholderia*, *Erwinia*, *Flavobacterium*, *Microbacterium*, *Rhizobium*, *Pseudomonas*, and *Serratia*, are suitable for converting unavailable phosphates into available phosphates. Besides providing P nutrition, PSB also improve the availability of other trace elements through the synthesis of siderophores, which directly or indirectly enhance plant growth (Bhattacharyya and Jha 2012a; Gaby and Buckley 2012; Ahmadi et al. 2018; Illmer and Schinner 1995; Yadav et al. 2020).

2.5   PHYTOHORMONAL MODULATION AND PLANT GROWTH REGULATION

Phytohormones are the chemical messengers that affect gene expression and transcription levels, cellular division, and plant growth. Phytohormones affect seed germination, the emergence of flowers, sex of flowers, and senescence of leaves and fruits. Plant growth hormones play an important role in its growth regulation and development in the natural environment, especially under stressed conditions. Microbial mediated hormonal modulation is another direct mode through which they boost plant growth. They function as signalling molecules and can influence the biochemical, physiological, and morphological processes in plant growth development. Among them, indole-3-acetic acid (IAA) is the most common, naturally occurring plant growth-regulating hormone of the auxin class. It is mainly responsible for the apical cell division and elongation, tissue differentiation, and stimulation of seed and tuber germination. It facilitates the excess amount of root exudation, thus, providing additional nutrients to the rhizosphere microbiota. Auxin, as a plant promoting hormone, is synthesized by 80% of the rhizosphere bacteria as secondary metabolites. Most well-studied IAA-producing rhizobacteria synthesize it from its precursor amino acid (tryptophan), mainly via tryptophan-independent pathways. The main pathway to synthesize IAA is the acid indole pyruvic pathway, which is found in most PGPR. In *Dioscorea rotundata*, IAA-producing *Bacillus subtilis* strains have been shown to enhance shoot/root length, fresh weight biomass, root to stem ratio, and the number of sprouts. Other frequently studied IAA producers among PGPR genera are *Azospirrilum*, *Azotobacter*, *Aeromonas*, *Burkholderia*, *Enterobacter*, *Pseudomonas*, and *Rhizobium* (Kundan and Pant 2015; Hamzah, Hapsari, and Wisnubroto 2016; Walker et al. 2003).

2.5.1   PRODUCTION OF 1-AMINO CYCLOPROPANE-1-CARBOXYLATE (ACC) DEAMINASE

1-aminocyclopropane-1-carboxylic acid (ACC) is the intermediate precursor of ethylene biosynthesis in higher plants. Plants exposed to several biotic and abiotic stresses experience a significant increase in the production of ethylene levels.

The ethylene hormone, if produced in higher concentration, causes premature leaf abscission, senescence, shoot and root growth stunting, chlorosis, and wilting of flowers. Also, its excessive buildup reduces plants' ability to acquire water and nutrients from the soil, which, in turn, ultimately increases further stress. ACC deaminase-producing rhizobacteria play an imperative role in the regulation of the ethylene level in plants exposed to abiotic stresses. Under stress conditions, bacterial ACC deaminase irreversibly metabolizes plant-produced ACC into α-ketobutyrate and ammonia and, thereby, prevents excessive ethylene buildup in plants. Among PGPR, *Achromobacter*, *Azospirillum*, *Bacillus*, *Pseudomonas*, *Enterobacter*, and *Rhizobium* are the prominent genera involved in the mitigation of plants' abiotic stress through ACC deaminase activity (Glick 2014; Etesami, Alikhani, and Mirseyed Hosseini 2015).

2.5.2 CYTOKININS (CK)

Cytokinins (CK) are another class of plant growth hormones responsible for cell division in apical tissue. CK also influences apical dominance, auxiliary bud growth, and leaf senescence in a plant. Many soil bacteria produce either CK, gibberellins, or both. Many bacteria, such as *Azotobacter spp.*, *Rhizobium spp.*, *Pantoea agglomerans*, *Rhodospirillum rubrum*, *Pseudomonas fluorescens*, *Bacillus subtilis*, and *Paenibacillus polymyxa*, can produce various cytokines (Hayat et al. 2010).

2.5.3 GIBBERELLIN (GA)

Gibberellin (GA) is another phytohormone that has been observed in rhizobacteria. GAs are tetracyclic diterpenoid carboxylic acids with either C20 or C19 carbon skeletons. Even though 136 globberelline structures have been identified, only four have been identified in bacteria. GAs activate important growth processes such as seed germination, stem elongation, flowering, and fruit set. They improve the photosynthesis rate and chlorophyll content. GAs stimulate shoot growth and inhibit root growth via the actions of the GA signalling system and the DELLA repressor that triggers GA-inducing genes (Kang et al. 2014; Martínez, Espinosa-Ruiz, and Prat 2016).

2.6 BIOLOGICAL SUPPRESSION OF HARMFUL CROP PATHOGENS

Many microorganisms demonstrate antifungal and antibacterial activity and are, therefore, used as biopesticides. Microbial inoculants play a critical role in biocontrol technology employed in agricultural ecosystems. The mechanisms of biocontrol exercised by most microbial inoculants could be attributed to the release of extracellular hydrolytic enzymes and competition for nutrients and secondary metabolites toxic to plant pathogens at very low concentrations, while some induce defence responses such as systemic acquired resistance (SAR) in host plants (Beneduzi, Ambrosini, and Passaglia 2012b, 2012a).

These bioagents display a wide variety of mechanisms in suppressing the growth of phytopathogens. Pathogens produce growth-inhibitory metabolites. The production of one or more antifungal metabolites by plant-associated beneficial microorganisms is connected to pathogen growth inhibition. The antibiotics produced by PGPR help to control the negative effects of harmful pathogens on crop growth and, thus, are crucial in crop protection. *Pseudomonas spp.*, a commonly occurring rhizosphere bacteria, is well recognized for producing a wide variety of antifungal antibiotics, such as 2,4 diacetyl phloroglucinol, phenazines, phenazine-1-carboxylic acid, pyoluteorin, pyrrolnitrin, cepaciamide A, rhamnolipids, butyrolactones, amphisin, and pyocyanin (Dey et al. 2004; Gouda et al. 2018).

Bacillus species also produce diverse antibiotics, such as subtilisin A, bacillomycin, iturins, sublancin, mycobacillin, surfactin, and polymyxin, etc. Other antibiotics-producing strains of *Pantoea spp.*, *Lysobacter spp.*, and *Enterobacter spp* are also deployed to control fungal diseases of plants.

Many plants' fungal pathogens and gram-negative and gram-positive bacteria are susceptible to the inhibitory action of these antibiotics. These antifungal metabolites curb pathogen proliferation by disrupting cell wall synthesis, cellular membrane structure, and the process of protein synthesis (Maksimov, Abizgil'dina, and Pusenkova 2011).

Volatile antifungal metabolites include hydrogen cyanide (HCN), alcohols, sulfides, ketones, aldehydes, etc., while cyclic lipopeptides, polyketides, heterocyclic nitrogenous compounds, phenylpyrrole, amino polyols, etc. are non-volatile antibiotics. PGPR releases a biocidal volatile that inhibits the activities of pathogenic fungal, bacterial, and nematode diseases in crops while also increasing plant vigour by activating plant defences Faheem et al. 2015; Raza et al. 2015). In the crop rhizosphere, extracellular emission of biocidal volatile organic compounds (VOC) has been a common feature in a variety of soil microorganisms. Actinobacteria also secrete 70%–80% of the known bioactive natural products. The biological antifungal potential of many antagonistic bacteria is correlated with their ability to produce VOCs (Fialho et al. 2011; Minerdi et al. 2009; Raza et al. 2015).

2.6.1 Myco-Parasitism through Secretion of Cell Wall Lytic Enzymes

Numerous PGPR, capable of secreting extracellular enzymes, are involved in the degradation of a wide range of polymers, such as chitin, protein, cellulose, and hemicelluloses. These traits of bacteria are helpful in myco-parasitism of pathogenic fungi and, hence, in its biocontrol potential (Kumar Jha and Saraf 2015).

PGPR that can produce extracellular lytic enzymes have been found to possess biocontrol activity against a wide range of phytopathogenic fungi, including *Rhizoctonia solani, Fusarium oxysporum Sclerotium rolfsii, Phytophthora spp.*, *Pythium ultimum*, and *Botrytis cinerea* (Bubici et al. 2019).

Siderophore-mediated competitive exclusion of Fe, a biologically important micronutrient, is essential for growth, metabolism, and development activities of all living systems, including plants, bacteria, and fungi. Fe is present in a limited amount in the soil system. Some PGPR secrete extracellular, low molecular weight compounds, called siderophores, to competitively gain ferric ions from the soil solution. Under the Fe-limiting conditions, this compound acts as an Fe chelator by binding firmly and forming a stable complex with the ferric ion and bringing it into the cellular system. Based on the chemical structure, types of ligands, and functional groups, siderophores are grouped into four main classes: hydroxamates, catecholates, carboxylates, and pyoverdines (Hyder et al. 2020; Sharma and Johri 2003; Crowley 2006).

The functional role of microbial siderophores in the rhizosphere can be: (1) competitive sequestration of limited Fe from the medium, (2) prevention of harmful pathogen growth, and (3) making Fe available for plant growth. Usually, bacterial siderophores have more affinity for Fe than do fungal siderophores, hence, a siderophore of PGPR better competes with fungal pathogens for Fe and, thus, obstruct their germination, growth, metabolism, and virulence by making Fe non-accessible for them. Therefore, siderophore-producing PGPR is a major asset in providing the necessary amount of Fe for plant nutrition. Apart from this, siderophore producing microbes can also alleviate heavy metal stress by forming a complex with Fe and enhancing their soluble concentration in the habitat. They play a crucial role in Fe uptake by plants in the presence of other metals, such as nickel and cadmium (Patel et al. 2015; Goswami, Thakker, and Dhandhukia 2016a).

2.7   DETOXIFICATION, INACTIVATION, AND DEGRADATION OF PATHOGENICITY FACTORS

Detoxification, inactivation, and/or degradation of pathogen virulence factors are other mechanisms some antagonistic microbes use to curb pathogen propagation. The detoxification mechanism inactivates pathotoxin and, thereby, reduces its toxicity. These processes strengthen plants' defence mechanism and guard them against invading pathogens. Similarly, agrochemicals that are potentially phytotoxic can also be inactivated by the variety of PGPR (Malathi, Viswanathan, and Padmanaban 2012).

Modes of pathogen virulence detoxification are found in a number of microorganisms. For example, *Burkholderia cepacia* and *Ralstonia solanacearum* are capable of lysis of fusaric acid, a virulent pathotoxin produced by different *Fusarium* strains. These bacteria were able to protect tomato crops from the infection of *Fusarium oxysporum f. sp. lycopersici*, which causes wilt disease. Pathogen detoxification was confirmed by tomato leaf cuttings, which were inoculated with the bacterium before treatment with fusaric acid, protecting the plant from wilt disease. Similarly, pre-inoculation of whole tomato plants with fusaric acid detoxifying bacteria, before pathogen inoculation, also prevented wilt disease infection (Saraf, Pandya, and Thakkar 2014; Kumar Jha and Saraf 2015).

2.8   PLANT–SOIL–MICROBE INTERACTIONS AND THEIR IMPACTS ON SOIL HEALTH AND CROP NUTRITIONAL QUALITY: RHIZO-DEGRADATION AND BIODEGRADATION OF TOXIC SOIL POLLUTANTS

The excessive accumulation of toxic pollutants in soil and water ecosystems is becoming a cause of global concern. Microorganism-mediated bioremediation might be useful to alleviate the ill effects of these pollutants on the ecosystem. In bioremediation techniques, specific living microbes or their enzymes are used to remove, detoxify, or reduce the concentration of toxic contaminants (Wei et al. 2021; Li et al. 2021). The metabolic activity of microorganisms results in the cleaning of the environment via complete mineralization, co-metabolism, transformation, degradation, sequestration, and/or exclusion of the toxic pollutants. Bioremediation is one of the cheapest and eco-friendly approaches to remediating soil and aquifer contamination. The microbe-mediated bioremediation process involves the transformation of pentachlorophenol, which is used in many paper pulp industries and agriculture (Kumar et al. 2018; Kumar, Kaushik, and Saxena 2015; Kumar, Kaushik, and Singh 2016). Rhizo-remediation, a bioremediation approach, involves the combined process of phytoremediation and bioaugmentation to degrade the toxic contaminant in the soil. The phytoremediation process involves plant-mediated extraction of toxic metals from polluted soil (phytoextraction) and their further remediation (Gouda et al. 2018).

The mutual and non-mutualistic plant-microbe interaction in the rhizosphere is crucial for the rhizo-remediation of xenobiotics. Presently, studies on rhizo-degradation using rhizospheric microflora are limited to a certain bacterial species of *Bacillus spp.*, *Pseudomonas aeruginosa*, and genetically engineered *P. fluorescens*. The use of PGPR strains, such as *P. putida* and *P. fluorescens*, have great potential for scavenging dangerous cadmium metal ions from the soil and, thus, are useful in reverting its toxic effect on barley crops (Grobelak, Napora, and Kacprzak 2015).

Kumar et al. (2018) also reported many groups of bacteria are responsible for the degradation of phthalate esters, which are used as plasticizers. All these studies prove the prospects of microorganisms in the removal of organic and inorganic contaminants and other impurities from natural ecosystems.

2.9 RECLAMATION OF PROBLEMATIC SOIL WITH THE APPLICATION OF BENEFICIAL MICROORGANISMS

The increase in soil–salt concentration unfavourably influences plant growth by bringing unfavourable changes in soil reaction, soil exchangeable cations, physical properties, and micronutrient status (Hayat et al. 2010).

Soil microorganisms, with an ability to adapt easily and more rapidly toward changing environments, are vital in maintaining the sustainability of an ecosystem such as extreme pH soil. Specific groups of microorganisms, such as halophiles, cyanobacteria, and mycorrhizal fungi, can be utilized for the reclamation of salt-damaged soil and the escalation of plant growth in such soil. Application of microbial inoculants reclaimed saline-sodic soil as observed by the positive correlation of greater accumulation of P and nitrate-nitrogen (NO_3-N) availability with reduced electrical conductivity (EC) and sodium (Na^+) content in soil solution. Inoculations of EPS-producing organisms have been found to inhibit the apoplastic flow of Na+ ions in salt-stressed soil (Etesami and Glick 2020).

External inoculation of *P. fluorescences* and *Enterobacter aerogenes* lowered Na content in leaves of maize exposed to salt conditions. Certain strains of halophilic bacteria have been found to be effective in the removal of Na ions from liquid media and, thus, could be appropriate for remediation of salt-degraded soils (Arora et al. 2016). Similarly, mycorrhizal fungi and other bacteria that are well adapted to harsh soil conditions support the remediation of such soils along with improvement in crop growth and yield (Zimmer et al. 2009).

2.10   CARBON SEQUESTRATION BY MICROBES

2.10.1   RHIZOSPHERE MICROORGANISMS UNDER CHANGING CLIMATIC CONDITIONS

Soil-inhabiting microorganisms play a significant role in regulating the dynamics of soil carbon pools. On the other hand, photosynthetic foods, formed in higher plants after the reduction of atmospheric carbon dioxide (CO_2), are stored in plant tissue and exuded from roots in the form of rhizodeposits. These rhizodeposits are used by microbiota as a source of nutrients and energy along with the incorporation of soil carbon and organic matter into their cellular components (Philippot et al. 2013; Mirza et al. 2006).

Soil microbiota can be used to enhance carbon sequestration through activities of plant growth support, soil aggregation, and microbial carbon fixation in the soil ecosystem. Thus, partial sequestration of carbon from the terrestrial ecosystem can be attained through the exploitation of these microorganisms. Rhizosphere engineering and manipulation of soil microbes could help in the capture of a considerable amount of atmospheric carbon. Among the soil microorganisms, AMF, which plays a vital role in nutrient dynamics, have great potential to increase the soil carbon pool by enhancing soil aggregation. The alteration of rhizosphere microbial communities and the increasing proportion of AMF through rhizospheric manipulation might be useful in trapping the higher carbon content with an augmented process of soil aggregation. The ability of AMF to produce polysaccharide glomalin and related compounds causes an increase in soil aggregation. For example, AMF, on its own or through co-application of glomalin and glomalin-related soil protein-producing PGPMs or AMF, help to improve the carbon to nitrogen ratio (C: N) storage in the soil system (Walley et al. 2014).

All these studies reveal the scope of soil microbes in capturing a substantial amount of atmospheric carbon. However, a physiological and biochemical adaption of microbes will be crucial in the future to prevent carbon loss from soil and trap more CO_2 in changing scenarios of climate change and warmer weather.

2.11   CONCLUSION AND PROSPECTS

The injudicious use of agrochemicals has caused damaging effects on the environment and humans and has disturbed ecological cycling. As a result, beneficial microbe-based, eco-friendly, and safer alternative technologies, with negligible reliance on chemicals, are being prioritized in sustainable agriculture. The importance of PGPR in plant growth promotion through the supply of mineral nutrition and phytohormone activation and, indirectly, by pathogen inhibition is well recognized. Additionally, PGPR play a vital role in the improvement of plant and soil health via soil structure formation, organic matter decomposition, crop biofortification, rhizo-remediation, abiotic stress mitigation, soil reclamation, and carbon sequestration.

The purposes of sustainable agriculture are being achieved through the application of nanoscience and agricultural biotechnology science that are being explored for strain improvement and precision agricultural input delivery. PGPR have shown a positive impact on crop productivity, plant stress tolerance, balanced nutrient recycling, and soil fertility. An amalgamation of multidisciplinary science involving microbiology, agricultural biotechnology, nanoscience, chemical engineering, and material science will provide an immense opportunity in the development and formulations of PGPR-based bioproducts for agricultural usage. There is a demand for higher crop production and productivity along with maintenance of proper soil health and quality in an eco-friendly manner; therefore, the main focus must include the concept of rhizosphere engineering for crop-yield improvement via the creation of a suitable environment for plant–soil–microbe interactions. The integration of modern scientific tools and techniques can assist with improved management of soil microflora, rhizosphere biology, and increased crop productivity. The exploration of ice-nucleating rhizobacteria to overcome cold-temperature stress in plants is another interesting area of research. An in-depth investigation can be done on potassium and zinc solubilization and PGPR, which are also important components required for balanced crop growth. The PGPR displaying a wide spectrum of growth-promotion potential in different crops, their interaction with the host and native Rhizophora, and an increase in the basic understanding of plant growth promotion also may be further elucidated. The commercial development of effective PGPR-based formulations and their viability, shelf life, success with biological and chemical seed treatments, and delivery system also requires attention. Moreover, factors limiting the commercial success of PGPR products, such as the economics of production, safety and stability, consistent performance, market demand, and mass awareness, should be addressed. Attention must be paid to optimizing growth conditions and enhancement in the shelf life of PGPR-based formulated products, which are not phytotoxic, are cost-effective, and have a broad spectrum of actions, tolerance to adverse environmental stresses, and produce a higher yield. Other issues related to health and safety testing of PGPMs, such as allergenicity, toxicity, pathogenicity, persistence in the environment, and potential for horizontal gene transfer, need critical attention.

FUNDING SOURCE INFORMATION

The research did not receive any specific grants from funding agencies in the public, commercial, or not-for-profit sectors.

CONFLICT OF INTEREST

The authors declare that they have no conflict of interest.

BIBLIOGRAPHY

Ahmadi, Katayoun, Bahar S. Razavi, Menuka Maharjan, Yakov Kuzyakov, Stanley J. Kostka, Andrea Carminati, and Mohsen Zarebanadkouki. 2018. "Effects of Rhizosphere Wettability on Microbial Biomass, Enzyme Activities and Localization to Be Published in Rhizosphere." *Rhizosphere* 7 (September). Elsevier B.V.: 35–42. doi:10.1016/j.rhisph.2018.06.010.

Akinola, Saheed Adekunle, and Olubukola Oluranti Babalola. 2021. "The Fungal and Archaeal Community within Plant Rhizosphere: A Review on Their Contribution to Crop Safety." *Journal of Plant Nutrition* 44 (4). Bellwether Publishing, Ltd.: 600–18. doi:10.1080/01904167.2020.1845376.

Alori, Elizabeth T., Bernard R. Glick, and Olubukola O. Babalola. 2017a. "Microbial Phosphorus Solubilization and It's Potential for Use in Sustainable Agriculture." *Frontiers in Microbiology.* Frontiers Media S.A. doi:10.3389/fmicb.2017.00971.

———. 2017b. "Microbial Phosphorus Solubilization and It's Potential for Use in Sustainable Agriculture." *Frontiers in Microbiology* 8 (JUN). Frontiers Media S.A.: 971. doi:10.3389/fmicb.2017.00971.

Alori, Elizabeth Temitope, and Olubukola Oluranti Babalola. 2018. "Microbial Inoculants for Improving Crop Quality and Human Health in Africa." *Frontiers in Microbiology.* Frontiers Media S.A. doi:10.3389/fmicb.2018.02213.

Armstrong McKay, David I., John A. Dearing, James G. Dyke, Guy M. Poppy, and Les G. Firbank. 2019. "To What Extent Has Sustainable Intensification in England Been Achieved?" *Science of the Total Environment* 648 (January). Elsevier B.V.: 1560–69. doi:10.1016/j.scitotenv.2018.08.207.

Arora, Sanjay, Y.P. Singh, Meghna Vanza, and Divya Sahni. 2016. "Bio-Remediation of Saline and Sodic Soils through Halophilic Bacteria to Enhance Agricultural Production." *Journal of Soil and Water Conservation* 15 (4). Diva Enterprises Private Limited: 302. doi:10.5958/2455-7145.2016.00027.8.

Bai, Z. G., D. L. Dent, L. Olsson, and M. E. Schaepman. 2008. "Proxy Global Assessment of Land Degradation." *Soil Use and Management.* John Wiley & Sons, Ltd. doi:10.1111/j.1475-2743.2008.00169.x.

Bargaz, Adnane, Karim Lyamlouli, Mohamed Chtouki, Youssef Zeroual, and Driss Dhiba. 2018. "Soil Microbial Resources for Improving Fertilizers Efficiency in an Integrated Plant Nutrient Management System." *Frontiers in Microbiology* 9 (July). Frontiers Media SA. doi:10.3389/fmicb.2018.01606.

Beneduzi, Anelise, Adriana Ambrosini, and Luciane M.P. Passaglia. 2012a. "Plant Growth-Promoting Rhizobacteria (PGPR): Their Potential as Antagonists and Biocontrol Agents." *Genetics and Molecular Biology.* doi:10.1590/S1415-47572012000600020.

Beneduzi, Anelise, Adriana Ambrosini, and Luciane M P Passaglia. 2012b. "Plant Growth-Promoting Rhizobacteria (PGPR): Their Potential as Antagonists and Biocontrol Agents." www.sbg.org.br.

Bharti, Nidhi, Shiv Shanker Pandey, Deepti Barnawal, Vikas Kumar Patel, and Alok Kalra. 2016. "Plant Growth Promoting Rhizobacteria Dietzia Natronolimnaea Modulates the Expression of Stress Responsive Genes Providing Protection of Wheat from Salinity Stress." *Scientific Reports* 6 (October). Nature Publishing Group. doi:10.1038/srep34768.

Bhat, Mujtaba Aamir, Vijay Kumar, Mudasir Ahmad Bhat, Ishfaq Ahmad Wani, Farhana Latief Dar, Iqra Farooq, Farha Bhatti, Rubina Koser, Safikur Rahman, and Arif Tasleem Jan. 2020. "Mechanistic Insights of the Interaction of Plant Growth-Promoting Rhizobacteria (PGPR) with Plant Roots toward Enhancing Plant Productivity by Alleviating Salinity Stress." *Frontiers in Microbiology.* Frontiers Media S.A. doi:10.3389/fmicb.2020.01952.

Bhattacharyya, P. N., and D. K. Jha. 2012a. "Plant Growth-Promoting Rhizobacteria (PGPR): Emergence in Agriculture." *World Journal of Microbiology and Biotechnology.* doi:10.1007/s11274-011-0979-9.

———. 2012b. "Plant Growth-Promoting Rhizobacteria (PGPR): Emergence in Agriculture." *World Journal of Microbiology and Biotechnology.* doi:10.1007/s11274-011-0979-9.

Bubici, Giovanni, Manoj Kaushal, Maria Isabella Prigigallo, Carmen Gómez Lama Cabanás, and Jesús Mercado-Blanco. 2019. "Biological Control Agents against Fusarium Wilt of Banana." *Frontiers in Microbiology.* Frontiers Media S.A. doi:10.3389/fmicb.2019.00616.

Calvo, Pamela, Louise Nelson, and Joseph W. Kloepper. 2014. "Agricultural Uses of Plant Biostimulants." *Plant and Soil.* Kluwer Academic Publishers. doi:10.1007/s11104-014-2131-8.

Crowley, David E. 2006. "Microbial Siderophores in the Plant Rhizosphere." In *Iron Nutrition in Plants and Rhizospheric Microorganisms*, 169–98. Springer, the Netherlands. doi:10.1007/1-4020-4743-6_8.

Dey, R., K. K. Pal, D. M. Bhatt, and S. M. Chauhan. 2004. "Growth Promotion and Yield Enhancement of Peanut (Arachis Hypogaea L.) by Application of Plant Growth-Promoting Rhizobacteria." *Microbiological Research* 159 (4). Elsevier GmbH: 371–94. doi:10.1016/j.micres.2004.08.004.

Etesami, Hassan, Hossein Ali Alikhani, and Hossein Mirseyed Hosseini. 2015. "Indole-3-Acetic Acid and 1-Aminocyclopropane-1-Carboxylate Deaminase: Bacterial Traits Required in Rhizosphere, Rhizoplane and/or Endophytic Competence by Beneficial Bacteria." In: Maheshwari D. (ed.) *Bacterial Metabolites in Sustainable Agroecosystem. Sustainable Development and Biodiversity*, Vol. 12. Springer, Cham. doi:10.1007/978-3-319-24654-3_8.

Etesami, Hassan, and Bernard R. Glick. 2020. "Halotolerant Plant Growth–Promoting Bacteria: Prospects for Alleviating Salinity Stress in Plants." *Environmental and Experimental Botany* 178 (October). Elsevier B.V. doi:10.1016/j.envexpbot.2020.104124.

Fasim, Fehmida, Nuzhat Ahmed, Richard Parsons, and Geoffrey M. Gadd. 2002. "Solubilization of Zinc Salts by a Bacterium Isolated from the Air Environment of a Tannery." *FEMS Microbiology Letters* 213 (1). Oxford University Press (OUP): 1–6. doi:10.1111/j.1574-6968.2002.tb11277.x.

Fialho, Mauricio Batista, Luiz Fernando Romanholo Ferreira, Regina Teresa Rosim Monteiro, and Sérgio Florentino Pascholati. 2011. "Antimicrobial Volatile Organic Compounds Affect Morphogenesis-Related Enzymes in Guignardia Citricarpa, Causal Agent of Citrus Black Spot." *Biocontrol Science and Technology* 21 (7): 797–807. doi:10.1080/09583157.2011.580837.

Gaby, John Christian, and Daniel H. Buckley. 2012. "A Comprehensive Evaluation of PCR Primers to Amplify the NifH Gene of Nitrogenase." *PLoS ONE* 7 (7). doi:10.1371/journal.pone.0042149.

Glick, Bernard R. 2012. "Plant Growth-Promoting Bacteria: Mechanisms and Applications." *Scientifica* 2012. Hindawi Limited: 1–15. doi:10.6064/2012/963401.

———. 2014. "Bacteria with ACC Deaminase Can Promote Plant Growth and Help to Feed the World." *Microbiological Research* 169 (1). Urban & Fischer: 30–39. doi:10.1016/j.micres.2013.09.009.

Goswami, Dweipayan, Janki N. Thakker, and Pinakin C. Dhandhukia. 2016a. "Portraying Mechanics of Plant Growth Promoting Rhizobacteria (PGPR): A Review." *Cogent Food & Agriculture* 2 (1). Informa UK Limited. doi:10.1080/23311932.2015.1127500.

Goswami, Dweipayan, Janki N. Thakker, and Pinakin C. Dhandhukia. 2016b. "Portraying Mechanics of Plant Growth Promoting Rhizobacteria (PGPR): A Review-Promoting Rhizobacteria (PGPR); Indole Acetic Acid (IAA); Phos-Phate Solubilization;

Siderophore Production; Antibiotic Production; Induced Systematic Resistance (ISR); ACC Deaminase." *Cogent Food & Agriculture* 19: 1127500. doi:10.1080/23311932.2015.1127500.

Gouda, Sushanto, Rout George Kerry, Gitishree Das, Spiros Paramithiotis, Han Seung Shin, and Jayanta Kumar Patra. 2018. "Revitalization of Plant Growth Promoting Rhizobacteria for Sustainable Development in Agriculture." *Microbiological Research.* Elsevier GmbH. doi:10.1016/j.micres.2017.08.016.

Grobelak, A., A. Napora, and M. Kacprzak. 2015. "Using Plant Growth-Promoting Rhizobacteria (PGPR) to Improve Plant Growth." *Ecological Engineering* 84 (November). Elsevier: 22–28. doi:10.1016/j.ecoleng.2015.07.019.

Hamzah, Amir, Ricky Indri Hapsari, and Erwin Ismu Wisnubroto. 2016. "Phytoremediation of Cadmium-Contaminated Agricultural Land Using Indigenous Plants." *International Journal of Environmental & Agriculture Research* 2: 8–14.

Hayat, Rifat, Safdar Ali, Ummay Amara, Rabia Khalid, and Iftikhar Ahmed. 2010. "Soil Beneficial Bacteria and Their Role in Plant Growth Promotion: A Review." *Annals of Microbiology.* doi:10.1007/s13213-010-0117-1.

Hyder, Sajjad, Amjad Shahzad Gondal, Zarrin Fatima Rizvi, Raees Ahmad, Muhammad Mohsin Alam, Abdul Hannan, Waqas Ahmed, Nida Fatima, and M. Inam-ul-Haq. 2020. "Characterization of Native Plant Growth Promoting Rhizobacteria and Their Anti-Oomycete Potential against Phytophthora Capsici Affecting Chilli Pepper (*Capsicum Annum L.*)." *Scientific Reports* 10 (1). Nature Research. doi:10.1038/s41598-020-69410-3.

Illmer, P., and F. Schinner. 1995. "Solubilization of Inorganic Calcium Phosphates-Solubilization Mechanisms." *Soil Biology and Biochemistry* 27 (3): 257–63. doi:10.1016/0038-0717(94)00190-C.

Istina, Ida Nur, Happy Widiastuti, Benny Joy, and Merry Antralina. 2015. "Phosphate-Solubilizing Microbe from Saprists Peat Soil and Their Potency to Enhance Oil Palm Growth and P Uptake." *Procedia Food Science* 3. Elsevier BV: 426–35. doi:10.1016/j.profoo.2015.01.047.

Kang, Sang Mo, Muhammad Waqas, Abdul Latif Khan, and In Jung Lee. 2014. "Plant-Growth-Promoting Rhizobacteria: Potential Candidates for Gibberellins Production and Crop Growth Promotion." In: *Use of Microbes for the Alleviation of Soil Stresses,* 1: 1–19. Springer, New York. doi:10.1007/978-1-4614-9466-9_1.

Kumar, Dharmendra, Rajeev Kaushik, and Anil K Saxena. 2015. "Biotransformation of Pentachlorophenol to Intermediary Metabolites by Bacteria Isolated from Pulp and Paper Mill Effluent Irrigated Cultivable Land." *Journal of Pure and Applied Microbiology* 9: 201–8.

Kumar, Dharmendra, Rajeev Kaushik, and Surender Singh. 2016. "Molecular and Biochemical Profiling of Pentachlorophenol Utilizing Bacteria from Pulp and Paper Mill Effluent Irrigated Soil in Northern India." *Journal of Pure and Applied Microbiology* 10 (3): 2045–54.

Kumar, Dharmendra, Livleen Shukla, Surender Singh, Shashi Bala Singh, Shalendra Kumar Jha, and G Prakash. 2018. "Assessment of Diverse Phthalate Esters (PAEs) from Irrigated Agriculture Soil under Protected Cultivation in IARI, New Delhi." *International Journal of Chemical Studies* 6 (3): 3432–35.

Kumar Jha, Chaitanya, and Meenu Saraf. 2015. "Plant Growth Promoting Rhizobacteria (PGPR): A Review." *E3 Journal of Agricultural Research and Development* 5 (2): 108–19. http://www.e3journals.org.

Kundan, Rishi, and Garima Pant. 2015. "Plant Growth Promoting Rhizobacteria: Mechanism and Current Prospective." *Journal of Biofertilizers & Biopesticides* 06 (02). OMICS Publishing Group. doi:10.4172/jbfbp.1000155.

Li, Na, Rui Liu, Jianjun Chen, Jian Wang, Liqun Hou, and Yuemei Zhou. 2021. "Enhanced Phytoremediation of PAHs and Cadmium Contaminated Soils by a Mycobacterium." *Science of the Total Environment* 754 (February). Elsevier B.V. doi:10.1016/j. scitotenv.2020.141198.

Lichtfouse, Eric, ed. 2017. *Sustainable Agriculture Reviews*. Vol. 22. Springer International Publishing, Cham. doi:10.1007/978-3-319-48006-0.

Faheem, M, W Raza, Z Jun, S Shabbir, N Sultana. 2015. "Characterization of the Newly Isolated Antimicrobial Strain Streptomyces Goshikiensis YCXU." *Scientific Letters* 3: 94–97.

Maheshwari, Dinesh Kumar, Shrivardhan Dheeman, and Mohit Agarwal. 2015. "Phytohormone-Producing PGPR for Sustainable Agriculture." In: Maheshwari D. (ed.) *Bacterial Metabolites in Sustainable Agroecosystem. Sustainable Development and Biodiversity*, Vol. 12. Springer, Cham. doi:10.1007/978-3-319-24654-3_7.

Maksimov, I. V., R. R. Abizgil'dina, and L. I. Pusenkova. 2011. "Plant Growth Promoting Rhizobacteria as Alternative to Chemical Crop Protectors from Pathogens (Review)." *Applied Biochemistry and Microbiology*. Springer. doi:10.1134/S0003683811040090.

Malathi, P., R. Viswanathan, and P. Padmanaban. 2012. "Identification of Antifungal Proteins from Fungal and Bacterial Antagonists against *Colletotrichum falcatum* Causing Sugarcane Red Rot."*Journal of Biological Control* 26 (1): 49–54.

Martínez, Cristina, Ana Espinosa-Ruiz, and Salomé Prat. 2016. "Gibberellins and Plant Vegetative Growth." *Annual Plant Reviews: The Gibberellins* 49: 285–322. John Wiley and Sons, Ltd. doi:10.1002/9781119210436.ch10.

Mhlongo, Msizi I., Lizelle A. Piater, Paul A. Steenkamp, Nico Labuschagne, and Ian A. Dubery. 2020. "Metabolic Profiling of PGPR-Treated Tomato Plants Reveal Priming-Related Adaptations of Secondary Metabolites and Aromatic Amino Acids." *Metabolites* 10 (5). MDPI AG. doi:10.3390/metabo10050210.

Minerdi, Daniela, Simone Bossi, Maria Lodovica Gullino, and Angelo Garibaldi. 2009. "Volatile Organic Compounds: A Potential Direct Long-Distance Mechanism for Antagonistic Action of Fusarium Oxysporum Strain MSA 35." *Environmental Microbiology* 11 (4): 844–54. doi:10.1111/j.1462-2920.2008.01805.x.

Mirza, M. Sajjad, Samina Mehnaz, Philippe Normand, Claire Prigent-Combaret, Yvan Moënne-Loccoz, René Bally, and Kauser A. Malik. 2006. "Molecular Characterization and PCR Detection of a Nitrogen-Fixing Pseudomonas Strain Promoting Rice Growth." *Biology and Fertility of Soils* 43 (2). Springer Verlag: 163–70. doi:10.1007/ s00374-006-0074-9.

Mushtaq, Zeenat, Shahla Faizan, and Alisha Hussain. 2021. "Role of Microorganisms as Biofertilizers." In: *Microbiota and Biofertilizers*, 83–98. Springer International Publishing. doi:10.1007/978-3-030-48771-3_6.

Olanrewaju, Oluwaseyi Samuel, Ayansina Segun Ayangbenro, Bernard R. Glick, and Olubukola Oluranti Babalola. 2019. "Plant Health: Feedback Effect of Root Exudates-Rhizobiome Interactions." *Applied Microbiology and Biotechnology*. Springer Verlag. doi:10.1007/s00253-018-9556-6.

Olanrewaju, Oluwaseyi Samuel, Bernard R. Glick, and Olubukola Oluranti Babalola. 2017. "Mechanisms of Action of Plant Growth Promoting Bacteria." *World Journal of Microbiology and Biotechnology*. Springer, the Netherlands. doi:10.1007/s11274-017-2364-9.

Ou, Yannan, C. Ryan Penton, Stefan Geisen, Zongzhuan Shen, Yifei Sun, Nana Lv, Beibei Wang, et al. 2019. "Deciphering Underlying Drivers of Disease Suppressiveness against Pathogenic Fusarium Oxysporum." *Frontiers in Microbiology* 10 (November). Frontiers Media S.A. doi:10.3389/fmicb.2019.02535.

Patel, Keyur, Dweipayan Goswami, Pinakin Dhandhukia, and Janki Thakker. 2015. "Techniques to Study Microbial Phytohormones." In: Maheshwari D. (ed.) *Bacterial Metabolites in Sustainable Agroecosystem. Sustainable Development and Biodiversity*, Vol. 12, 1–12. Springer, Cham. doi:10.1007/978-3-319-24654-3_1.

Philippot, Laurent, Jos M. Raaijmakers, Philippe Lemanceau, and Wim H. Van Der Putten. 2013. "Going Back to the Roots: The Microbial Ecology of the Rhizosphere." *Nature Reviews Microbiology*. doi:10.1038/nrmicro3109.

Pretty, Jules. 2008. "Agricultural Sustainability: Concepts, Principles and Evidence." *Philosophical Transactions of the Royal Society B: Biological Sciences*. Royal Society. doi:10.1098/rstb.2007.2163.

Pretty, Jules, and Zareen Pervez Bharucha. 2014. "Sustainable Intensification in Agricultural Systems." *Annals of Botany*. Oxford University Press. doi:10.1093/aob/mcu205.

Raza, Waseem, Jun Yuan, Ning Ling, Qiwei Huang, and Qirong Shen. 2015. "Production of Volatile Organic Compounds by an Antagonistic Strain Paenibacillus Polymyxa WR-2 in the Presence of Root Exudates and Organic Fertilizer and Their Antifungal Activity against *Fusarium Oxysporum f. Sp. Niveum.*" *Biological Control* 80 (January). Academic Press Inc.: 89–95. doi:10.1016/j.biocontrol.2014.09.004.

Riaz, Umair, Ghulam Murtaza, Wajiha Anum, Tayyaba Samreen, Muhammad Sarfraz, and Muhammad Zulqernain Nazir. 2021. "Plant Growth-Promoting Rhizobacteria (PGPR) as Biofertilizers and Biopesticides." In: *Microbiota and Biofertilizers*, 181–96. Springer International Publishing. doi:10.1007/978-3-030-48771-3_11.

Saraf, Meenu, Urja Pandya, and Aarti Thakkar. 2014. "Role of Allelochemicals in Plant Growth Promoting Rhizobacteria for Biocontrol of Phytopathogens." *Microbiological Research* 169 (1): 18–29. doi:10.1016/j.micres.2013.08.009.

Sharma, Alok, and B. N. Johri. 2003. "Growth Promoting Influence of Siderophore-Producing Pseudomonas Strains GRP3A and PRS9 in Maize (Zea Mays L.) under Iron Limiting Conditions." *Microbiological Research* 158 (3). Elsevier GmbH: 243–48. doi:10.1078/0944-5013-00197.

Sharma, Devender, Navin Chander Gahtyari, Rashmi Chhabra, and Dharmendra Kumar. 2020. "Role of Microbes in Improving Plant Growth and Soil Health for Sustainable Agriculture. In: Yadav A., Rastegari A., Yadav N., Kour D. (eds) *Advances in Plant Microbiome and Sustainable Agriculture. Microorganisms for Sustainability*, Vol. 19. Springer, Singapore. doi:10.1007/978-981-15-3208-5_9.

Sharma, Seema B., Riyaz Z. Sayyed, Mrugesh H. Trivedi, and Thivakaran A. Gobi. 2013. "Phosphate Solubilizing Microbes: Sustainable Approach for Managing Phosphorus Deficiency in Agricultural Soils." *SpringerPlus*. SpringerOpen. doi:10.1186/2193-1801-2-587.

Suyal, Deep Chandra, Ravindra Soni, Santosh Sai, and Reeta Goel. 2016. "Microbial Inoculants as Biofertilizer." In: *Microbial Inoculants in Sustainable Agricultural Productivity: Vol. 1: Research Perspectives*, 311–18. Springer India. doi:10.1007/978-81-322-2647-5_18.

Timmusk, Salme. 2012. "Rhizobacterial Application for Sustainable Water Management on the Areas of Limited Water Resources." *Irrigation & Drainage Systems Engineering* 01 (04). OMICS Publishing Group. doi:10.4172/2168-9768.1000e111.

Valette, Marine, Marjolaine Rey, Florence Gerin, Gilles Comte, and Florence Wisniewski-Dyé. 2020. "A Common Metabolomic Signature Is Observed upon Inoculation of Rice Roots with Various Rhizobacteria." *Journal of Integrative Plant Biology* 62 (2). Blackwell Publishing Ltd: 228–46. doi:10.1111/jipb.12810.

Varah, Alexa, Kwadjo Ahodo, Shaun R. Coutts, Helen L. Hicks, David Comont, Laura Crook, Richard Hull, et al. 2020. "The Costs of Human-Induced Evolution in an Agricultural System." *Nature Sustainability* 3 (1). Nature Research: 63–71. doi:10.1038/s41893-019-0450-8.

Vessey, J. Kevin. 2003. "Plant Growth Promoting Rhizobacteria as Biofertilizers." *Plant and Soil.* doi:10.1023/A:1026037216893.

Walker, Travis S., Harsh Pal Bais, Erich Grotewold, and Jorge M. Vivanco. 2003. "Root Exudation and Rhizosphere Biology." *Plant Physiology.* doi:10.1104/pp.102.019661.

Walley, F. L., A. W. Gillespie, Adekunbi B. Adetona, J. J. Germida, and R. E. Farrell. 2014. "Manipulation of Rhizosphere Organisms to Enhance Glomalin Production and C Sequestration: Pitfalls and Promises." *Canadian Journal of Plant Science* 94 (6). Agricultural Institute of Canada: 1025–32. doi:10.4141/CJPS2013-146.

Wang, Wenjing, Qingbin Chen, Shouming Xu, Wen-Cheng Liu, Xiaohong Zhu, and Chun Peng Song. 2020. "Trehalose-6-Phosphate Phosphatase E Modulates ABA-Controlled Root Growth and Stomatal Movement in Arabidopsis." *Journal of Integrative Plant Biology* 62 (10). Blackwell Publishing Ltd: 1518–34. doi:10.1111/jipb.12925.

Wei, Zihan, Quyet Van Le, Wanxi Peng, Yafeng Yang, Han Yang, Haiping Gu, Su Shiung Lam, and Christian Sonne. 2021. "A Review on Phytoremediation of Contaminants in Air, Water and Soil." *Journal of Hazardous Materials* 403 (February). Elsevier B.V.: 123658. doi:10.1016/j.jhazmat.2020.123658.

Yadav, Sunil Kumar, Joyati Das, Rahul Kumar, and Gopaljee Jha. 2020. "Calcium Regulates the Mycophagous Ability of Burkholderia Gladioli Strain NGJ1 in a Type III Secretion System-Dependent Manner." *BMC Microbiology* 20 (1). BioMed Central. doi:10.1186/s12866-020-01897-2.

Ye, Lin, Xia Zhao, Encai Bao, Jianshe Li, Zhirong Zou, and Kai Cao. 2020. "Bio-Organic Fertilizer with Reduced Rates of Chemical Fertilization Improves Soil Fertility and Enhances Tomato Yield and Quality." *Scientific Reports* 10 (1). Nature Research. doi:10.1038/s41598-019-56954-2.

Zhu, Fengling, Lingyun Qu, Xuguang Hong, and Xiuqin Sun. 2011. "Isolation and Characterization of a Phosphate-Solubilizing Halophilic Bacterium Kushneria Sp. YCWA18 from Daqiao Saltern on the Coast of Yellow Sea of China." *Evidence-Based Complementary and Alternative Medicine* 2011. doi:10.1155/2011/615032.

Zimmer, Dana, Christel Baum, Peter Leinweber, Katarzyna Hrynkiewicz, and Ralph Meissner. 2009. "Associated Bacteria Increase the Phytoextraction of Cadmium and Zinc from a Metal-Contaminated Soil by Mycorrhizal Willows." *International Journal of Phytoremediation* 11 (5). Taylor & Francis Inc.: 200–13. doi:10.1080/15226510802378483.

3 Environmental Control of Plant-Microbe Interaction

Young B. Ibiang

CONTENTS

3.1 INTRODUCTION

Plants and microbes interact in different ways in the biosphere. Some interactions result in disease(s) in plants (such as with pathogenic microbes), while others improve their growth and/or fitness (plant growth-promoting microbes). In both scenarios, the prevailing environmental conditions influence the intricate aspects and eventual outcomes of such interactions. Environmental conditions could be chemical, physical, or biological in nature. Chemical conditions include salinity, nutrient levels (low or excess), pH, etc.; physical conditions include temperature, light, moisture, soil compaction, etc.; and biological conditions refer to other organisms, including microbes, plants, and animals. As these factors vary from one place to another, as well as in the extent of their potential impact on the biosphere, understanding their controlling effect on plant-microbe interaction is essential for optimizing the benefits/contributions to agricultural productivity. For example, while arbuscular mycorrhizal (AM) fungi (AMF) are known to improve plant phosphorus (P) uptake and growth, especially under low P soil conditions, such AMF-induced plant growth increases are

DOI: 10.1201/9781003213864-3

less apparent in plants under high P soil conditions (Smith and Read 2008). On the other hand, the impact of a pathogenic microbe (disease incidence/severity/tolerance) could be mitigated by environmental factors as well as genotype by environment interaction (Bocianowski et al. 2020). Thus, for both pathogenic and beneficial microorganisms, it is considered that where the effect of plant-microbe interaction falls along the continuum from negative to positive is largely determined by the prevailing environmental circumstance. Some interactions that are usually mutualistic, such as between leaf inhabiting endophytes and grasses, may become less so (such as with the occurrence of host growth depression) due to change in environmental conditions, e.g., nutrient limitation (Cheplick et al. 1989). Even microorganisms that appear commensal under one set of conditions, may have somewhat different effects under varied prevailing factors. For disease-inducing or growth-promoting tendencies of interactions, there are three important considerations around the environmental control of plant-microbe interaction in relation to agriculture. First is the extent to which agricultural practices such as tillage, monocropping, fertilization, weed control, etc. may affect the microbe community (microbiome) with resultant impact on plant fitness and other ecosystem functions (Lehman et al. 2012; Babalola et al. 2020). Second is the anticipation that projected increase in global temperatures could expand the distribution/impact of pathogens, for example, via an extended habitable area of the vector(s) and/or a direct impact on pathogen characteristics such as reproduction, resilience, and virulence (Velásquez et al. 2018). The possibility of emerging infectious agents in a new area and the way food production is impacted by natural or anthropogenic environmental changes, stemming from effects on populations and the ecological community, merit consideration within the scope of sustainability. Third is the acknowledgement that beneficial effects of microbial inoculants have not had as much efficacy and consistency across the board as would be required for faster adoption and more widespread commercialization (Nadeem et al. 2014) due to "in-situ pressure" from environmental factors such as soil physicochemical properties and competition from native microbes. This realization has led to suggestions such as the use of AM fungal consortium in inoculum production to make it more adaptable to environmental conditions (Trejo-Aguilar and Banuelos 2020). The ability to "design" consistently efficient plant-microbe partnerships for a given range of environmental circumstances will have an enormous impact on food production and environmental sustainability (Figure 3.1).

One of the problems with our grappling of environmental control in plant-microbe interactions, though, is the sheer scope of it. Many discussions around the topic and, indeed, a great many studies are limited to either a gradient of one kind of factor or one sphere (pathogenic or growth-promoting) or mode (biocontrol vs. pathogen) of interaction. In the environment, however, several factors and microbe interactions are in a constant dynamic existence. It is beneficial, therefore, to develop conceptual theories to unify the differential influences of environmental conditions on plant-microbe interactions (Velásquez et al. 2018). A part of this will involve outlining general principles that could provide an overall lens through which things can be viewed, side-by-side discussions of environmental effects on pathogenic and growth-promoting scenarios, broadening (rather than limiting) the discussion beyond a few

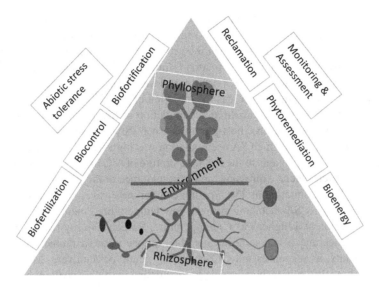

FIGURE 3.1 Plant-microbe interaction plays out in the environment and can be exploited to address some challenges in the agro-environmental sciences.

popular model species, etc. This chapter contributes to furthering the above-stated positions and includes perspectives for the utilization of environmental control of plant-microbe interaction in solving issues in the agro-environmental sciences.

3.1.1 GENERAL PRINCIPLES IN THE ENVIRONMENTAL CONTROL OF PLANT-MICROBE INTERACTION

(i) **Environmental control may stem more from effects on the microbe or the host plant**. During the establishment of plant-microbe interaction, various microbe processes, such as recognition, adherence/attachment, penetration, colonization, and proliferation, (Berg 2009) are subject to possible environmental impact. This situation in the environmental control of the dynamics of plant-microbe interaction can be demonstrated using the interaction between *Agrobacterium tumefaciens* (bacterium that causes crown gall disease) and host plants. This bacterium transfers its virulence (*vir*) genes via an extracellular pilus it attaches to the host cell. However, in some *A. tumefaciens* strains, temperatures greater than 28°C inhibited the microbe's ability to form this pilus (T-pilus) as well as VirB protein accumulation, leading to an inability to induce disease (Baron et al. 2001). Deciphering what entity the environmental control stems more from allows a researcher to decide the primary target of any strategy utilizing plant-microbe interaction to solve an issue. Strategies such as engineering the environment to stimulate in-situ microbe(s), deploying an engineered plant to overcome the burden of pathogen/pathogen control in a given environment, deploying plants and/or competent exotic microbes to extract/contain/remediate pollutants in the environment, etc., take this into account. Plant defense hormones, such as jasmonic acid (JA), salycilic acid (SA), ethylene (ET), and abscisic acid (ABA), are known to be involved in regulating plant response to biotic and/or abiotic stress and play a central role in the growth-defense tradeoffs that characterize plant-microbe–environment interaction (Huot et al. 2014). However, changes in plant hormonal response due to environmental variation, e.g., temperatures, have been reported. For example, whereas the infection of Arabidopsis by *Pseudomonas syringae* at 30°C resulted in impaired accumulation of SA and greater susceptibility of the host to the microbe, host-plant tolerance (SA-induced) was better when infection was at 23°C in which SA accumulation was high (Huot et al. 2017). Again, subjecting Arabidopsis plants to the low temperature of 4°C increased SA accumulation, related defense gene expression, and host tolerance to the pathogen (Kim et al. 2017). Akin to the situation at 23°C, pathogen tolerance was improved in the host, as modulated by its SA metabolism under low temperature conditions, in contrast with the situation at a higher (30°C) temperature. Suffice it to say that an environmental influence may affect some strains of a microbe more than others. The same goes for plants, as responses may differ among genotypes/cultivars. Maize genotypes can show different responsiveness to inoculation with *Azospirillum brasilense*, under nitrogen (N) stress conditions (Vidotti et al. 2019). In general, (ii) **genotypic factors influence plant-microbe interaction response to environmental changes** (Vannier et al. 2015).

(iii) **Environmental control may involve similar or contrasting effects on members of the association**. AMF colonize the roots of many plants, aiding host P nutrition as well as other nutrients like copper, zinc, and ammonium (NH_4^+) in addition to improving host tolerance to abiotic stresses (Smith and Read, 2008). However, it is believed that the sufficiency of P in the soil reduces the host dependency on AMF, and AMF root colonization indices, such as new arbuscule development (Kobae et al. 2016), may be *inhibited* under high soil P, even when the plant vegetative growth is apparently *increased*. AMF-induced plant tolerance to excess soil trace elements is widely recognized, but this is a *bidirectional* situation. While excess trace elements may tend to reduce plant growth, the AMF root colonization of the host may *also* be reduced (Ibiang et al. 2018). Excess soil Zn may increase *or* decrease AMF colonization of the host plant, depending on other conditions (Ibiang et al. 2017; Ibiang and Sakamoto 2018). There exists some diversity in the AMF response to excess heavy metals, with some species being more tolerant to metal toxicity than others. For example, *Claroideoglomus claroideum* is considered more tolerant of metal toxicity, as it does not show as much inhibition of extraradical hyphal growth as some other *Glomus* species (delVal et al. 1999). Drought is another condition that can increase *or* decreasing AMF colonization (Auge 2001; Klironomos et al. 2001). In general, host response to AMF is a complex issue involving the plant-microbe–environment interaction in any given case (Smith and Smith 2011). Under different N levels (limiting vs. sufficiency), root-associated bacteria (*Bacillus* sp. or *A. brasilense*) influence maize root anatomy, growth, and physiology in similar as well as in distinct ways (Calzavara et al. 2018).

(iv) **Environmental control assumes a direct and/or an indirect nature**. Environmental control may be characterized by *direct* effects on the member(s) in a plant-microbe interaction, as can be seen from the illustration above involving temperature and some strains of *A. tumefaciens*. In the Arabidopsis–*Colletotrichum tofieldiae* association, the host growth-promoting effect of microsymbiont kicks in under P deficiency in the host but fades away under P sufficiency (Hiruma et al. 2016). However, *indirect* effects, such as those that involve other entities in the biosphere or nebulous effects pervading the microhabitat, are also perceptible. For example, herbicides applied on a field to kill weeds can cause changes in the microclimate around crops that ultimately lessen disease development (Agrios 2005). Another illustration of this can be drawn from the broad-based bioprotective effects of some species in the microbiome against the potential pathogenesis induced by others in the host rhizosphere. The plant-associated microbe community (microbiome) can reduce disease severity not only by direct pathogen antagonism but also by broadly priming host defenses (Conrath et al. 2002). The selection processes for root-associated microbes is characterized by enrichment and/or exclusion of microbes during their interaction in the rhizosphere (Mine et al. 2014). These processes, as well as the microbiome itself, is affected by environmental factors, such as temperature, that can strongly affect soil properties, such as carbon mineralization rate, which affects interacting plants and microbes (Tang et al. 2018). Environmental variation, leading to changes in plant tolerance to an opportunistic pathogen, can result in increased susceptibility to disease due to a change in the host defense priming that may or may not be inclusive of a direct impact on the pathogen. In a 3-year study, long-term warm and humid weather in some growing areas was reported to favor the development of pests and fungal disease of rice, illustrating connectedness between a direct effect on the pathogen itself and effects on other entities (pests) that may weaken a plant and tilt the interaction more in favor of the pathogen (Hu et al. 2016). The fungus, *Trichoderma harzianum* helps plants by producing antibiotics with direct action against the growth of other pathogenic fungi, but it also acts (via chitinases) against insect pests that trouble plants and aggravate susceptibility to disease (Shakeri and Foster 2007).

(v) **Although one or a few may discernibly predominate, environmental control stems from the aggregate effect of several factors**. Plant-microbe interaction always takes place within an environmental context. In plant–pathogen interaction, it is widely understood that both host and pathogen have optimum environmental conditions for their maximum possible fitness or virulence, but the final outcomes of that interaction (say, disease severity) depend on where the balance exists between these two environmental endpoints. In other words, if the aggregate of prevailing environmental factors favors pathogen virulence more than plant fitness, disease severity is likely to be high. But, if plant fitness is favored by the environmental conditions rather than pathogen virulence, disease severity is likely to be lower. The same argument holds for an interaction between plants and beneficial microbes, with respect to the host dependency on the beneficial microbe. The effect of grazing on AM fungal symbiosis is moderated by diversity and biomass of plants, nutrient (e.g., N and P) availability, and soil edaphic properties, such as bulk density, moisture levels, pH, and organic carbon (Faghihinia et al. 2020). These factors aggregate to shape and give context to the symbiosis. Randall et al. (2020) observed that climate-induced increase in soil water availability affects soil bacterial community assemblages exposed to long-term differential phosphorus fertilization. But, it is often valuable to delineate one or a few obvious predominating environmental factor(s) around plant-microbe interactions. This has led scientists to sometimes connect two factors, e.g., temperature and moisture as the major controlling factors in many plant–pathogen interactions (Manstretta and Rossi 2015; Velásquez et al. 2018) or nutrient availability and in situ microbe community in many interactions between plants and beneficial microbes (Fuentes-Ramírez et al. 1999; Davitt et al. 2011).

3.1.2 INFLUENCE OF TEMPERATURE ON PLANT-MICROBE INTERACTION

The effect of temperature on plant–pathogen interaction is important with respect to disease tolerance because disease development naturally progresses at certain optimal temperature ranges, amid the fluctuations between day and nighttime values prevailing at any place and/or time (Velásquez et al. 2018). Generally, with conditions favorable for disease, pathogenic microbes that associate with a potential host plant can overcome the basal immune response that is elicited in the plant upon recognition of microbe-associated molecular patterns (MAMPs) present on the microbe surface (Boller and He 2009). Such basal immune responses, referred to as pattern-triggered immunity (PTI), can be induced by xenobiotic entities, such as flagellin and chitin of bacterial and fungal cells, respectively (Segonac and Zipfel 2011). As the microbe transmits unique proteins (effectors) to the host cells, a more serious second-tier plant defense response, called effector-triggered immunity (ETI), is activated through the action of host immune receptors known as nucleotide-binding leucine-rich repeats (NLR). However, elevation in temperature is known to potentially suppress ETI, and this may have to do with reduced expression of ETI-related genes (Cheng et al. 2013). It is thought that such effects may be due to changes in nucleo-cytoplasmic localization of associated NLR receptor proteins (Zhu et al. 2010) and transcription factors (Wang et al. 2019). However, not all NLRs show nuclear localization, and there is not yet a full understanding of how elevated temperature inhibits ETI in plants (Cheng et al. 2019). Because of plant dependence on ETI to counter crop pathogens, the effects of a warmer climate on the balance between crop production and pathogen tolerance/control may be imagined. For example, soft rot and blackleg disease of potatoes is caused by opportunistic microbes that exhibit widespread latent infection of roots and stem of the host but cause serious disease when host resistance is breached (Pérombelon 2002). According to Hasegawa et al. (2005), elevated temperature enhances virulence of *Pectobacterium astrosepticum*, a soft rot bacterium, to the potato plant, as it increases the production of the quorum-sensing signals that modulate the production of plant cell wall degrading enzymes that soften the host tissues. Host resistance to pathogens may be reduced by higher temperatures, such as with *Pseudomonas syringae* in *Arabidopsis thaliana* for which both basal and resistance (R) gene-mediated defense responses were reduced by increased temperature (Wang et al. 2009). At 35°C, high temperature-induced adult resistance to stripe rust caused by *Puccinia striiformis* in wheat (*Triticum aestivum*) was reported by Fu et al. (2009), but this resistance was not observed at the lower temperature of 20°C.

This trait, believed to be controlled by a resistance-conferring quantitative trait loci (QTL), conferred temperature-dependent resistance to several races of *P. striiformis* at high temperatures (Carter et al. 2009; St Clair, 2010). This phenomenon shows that the expectation of greater disease susceptibility under higher temperatures is not a one-way street, as there are potentially many genes with unknown functions under varying environmental conditions.

The effect of temperature on plant–virus interaction is also important and recently gaining increased attention, as many plant diseases are due to viral infections. Obrępalska-Stęplowska et al. (2015) studied the effect of temperature on pathogenesis due to peanut stunt virus (PSV) (genus: *Cucumovirus*) in *Nicotiana benthamiana* grown at 21°C or 27°C and found that pathogenesis was faster at the latter than the former. In plants grown at 27°C, disease symptoms were already well developed at 21 days post infection (dpi), with a significant drop in photosynthesis and carbohydrate metabolism proteins. Higher temperatures increased the replication and survival of the turnip crinkle virus in Arabidopsis (Zhang et al. 2012), while the accelerated spread of the tobacco mosaic virus (TMV) due to temperature-dependent suppression of host resistance was reported at temperatures above 28°C (Király et al. 2008). Higher temperatures also weakened host defense against tomato spotted wilt virus (TSWV) in pepper plants (Moury et al. 1998). But other authors, such as Szittya et al. (2003), have reported increased host defenses at higher temperatures due to a more efficient RNA silencing-mediated host defense responses in which higher temperature (21°C – 27°C) enhanced RNAi-based defenses against *Cymbidium* ringspot virus (CymRSV) as compared with lower temperatures (15°C) in *N. benthamiana*. In cassava, DNA geminiviruses, such as the cassava mosaic virus, tend to show greater virulence at 25°C compared to 30°C (Chellappan et al. 2005), while geminivirus-mediated gene silencing from cotton leaf crumple virus was enhanced by low temperatures in cotton (Tuttle et al. 2008). Thus, the effect of temperature on viral pathogenesis may be contradictory, depending on plant and virus identity.

In the case of host interaction with beneficial microorganisms, several reports show that elevation in temperatures could also have an effect, with some organisms conferring tolerance to temperature stress on the host. In the case of popular examples such as AMF, increased allocation of host photosynthates to rhizosphere due to moderately increased temperatures may stimulate higher AM fungal root colonization and external hyphal length and respiration (Heinemeyer et al. 2006; Compant et al. 2010). Warmer temperatures promoted higher AMF colonization by *Ambispora leptoticha* than either *Claroideoglomus claroideum* or *Funneliformis mosseae*, but this pattern was reversed under colder conditions (Antunes et al. 2011). There are also reports of minimal or negative effects of increased temperature on AMF (see Compant et al. 2010), indicating a case-by-case situation in the light of other prevailing conditions. Interaction between the tropical plant *Dichanthelium lanuginosum* and *Curvularia protuberate* enables both organisms to survive under elevated soil temperatures (up to 55°C), with the fungus also requiring infection by a dsRNA virus to confer heat-stress tolerance (Márquez et al. 2007). In the absence of this interaction, neither plant nor fungus may survive soil temperatures above

38°C–40°C, but symbiotically, a soil temperature of 65°C was tolerated (Redman et al. 2002; Rodriguez et al. 2009). Such three-way symbioses involving a virus in a fungus infecting a plant host, illustrates the complexity of plant-microbe interactions in nature. Research to understand more precisely how this interaction works may enable its use for protecting other plants from temperature stress, especially as the thermotolerance conferred by this fungus may not be exclusive to *D. lanuginosum* (Rodriguez et al. 2008). Thermotolerance has also been conferred on host plants by bacteria such as by *Parabukholderia phytofirmans* (PsJN) strain on tomato (Issa et al. 2018).

3.1.3 INFLUENCE OF ATMOSPHERIC CO$_2$ ON PLANT-MICROBE INTERACTION

The elevation of atmospheric gases, such as CO$_2$ (and others), can induce changes in the plant pool of flavonoid-related compounds that are considered key in plant defense and signaling mechanisms (Kretzschma et al. 2009). This is one of the means through which atmospheric gases can broadly impact plant-microbe interactions involving both pathogenic and beneficial microorganisms. Often, studies on atmospheric gases are imbued with temperature and/or climate change related perspectives. As global human population, lifestyle needs, and carbon footprints are ever-increasing, this has potential ramifications for agricultural production in this century. There are reports that an increase in atmospheric CO$_2$ levels may increase disease severity in some important crops, including cereals such as wheat and rice (Kobayashi et al. 2006; Velásquez et al. 2018). According to Váry et al. (2015), the severity of wheat diseases increased when plants and pathogens were acclimatized to elevated levels of CO$_2$. In their study, the virulence of isolates of the fungal pathogen (*Fusarium graminearum*) and the susceptibility of wheat varieties increased because of elevated CO$_2$ levels. *Epichloë bromicola* is an endophytic fungus that causes choke disease in its grass host, *Bromus erectus*. In the interaction between *E. bromicola* and *B. erectus*, elevated CO$_2$ levels favored fungal over host reproductive vigor (Groppe et al. 1999). It is thought that elevated CO$_2$ reduces plants' resistance to herbivorous insect pests and enhances water-use efficiency of C$_3$ plants, allowing for improved feeding efficiency of aphids (Guo et al. 2017). Because aphids are important vectors of many infectious agents (such as viruses) in plants, the (indirect) impact of elevated CO$_2$ levels on plant disease they spread should receive more attention. In contrast, reduction in disease severity due to elevated CO$_2$ levels was reported by Eastburn et al. (2010) in the interaction between soybean and the oomycete, *Peronospora manshurica*. Atmospheric CO$_2$ levels influence the resistance of Arabidopsis to infection by *P. syringae* via effects on abscisic acid accumulation and stomatal responsiveness to coronatine (Zhou et al. 2017). Because stomatal responsiveness to coronatine is also influenced by light (Panchal et al. 2016), we can deduce (as previously pointed out) that atmospheric gases act in concert with other environmental factors to foist an outcome on plant-microbe interactions. N fixation by rhizobia may be altered under changing atmospheric conditions in the form of elevated CO$_2$ and temperature between two seasons (summer and autumn) (Sanz-Sáez et al. 2012). Their study reported that Alfalfa plants in symbiosis with

Sinorhizobium meliloti generally produced more dry weight in autumn than in summer, with an indication of a lowered N_2-fixation during the hotter summer months than in autumn. They also observed that higher Alfalfa yield with *S. meliloti* strain 102F78 was more related to a lower carbon consumption (C cost) for N_2-fixation in summer. This points to environment-driven variations in the Alfalfa-*S. meliloti* symbiosis, that affects different rhizobium strains and harkens back to the general principles highlighted previously.

3.1.4 INFLUENCE OF NUTRIENTS ON PLANT-MICROBE INTERACTION

Plants and microbes rely on nutrients for survival. The nutritional status of hosts is likely to influence the physiology of associated bacteria and fungi. There is evidence that the transcriptome of associated root colonizing bacteria is affected by the nutritional status of maize hosts due to changes in the composition of host root exudates (Carvalhais et al. 2013). One instance in which the influence of nutrients on plant-microbe interaction has been extensively studied is in the plant–AM fungal symbiosis. In this interaction, plants provide photosynthates and a home/anchor (its roots) to AMF, while the fungus mines and supplies nutrients, mainly P, to the host (Smith and Read 2008). Obviously, this is construed to be a mutualistic interaction because both partners gain in the association, but it is known that the host P status plays a huge role in how things play out. When P status is high, initiation of the symbiosis is inhibited, but when P status is deficient/low, the symbiosis assumes greater (mutualistic) value and is enhanced (Koide 1991). In an already colonized root/established symbiosis, increased P supply/nutrient levels inhibits the development of new AMF structures (arbuscules) critical to P exchange between the fungus and host (Kobae et al. 2016), with an overall tendency to lower the hosts' dependency on AMF. Another popular plant-microbe interaction influenced by nutrient status is the legume–rhizobium symbiosis in which N fixation by root nodule rhizobacteria nourish the host plant. In this symbiosis, which also influences the trace element homeostasis of the host (González-Guerrero et al. 2016) and has some similarities with AM symbiosis, N availability/status in the shoots plays a role in the autoregulation of nodulation, helping to determine the number of root nodules to be maintained (Polacco and Todd 2011). In a non-legume, high levels of N fertilization inhibited the colonization of sugarcane roots by *Acetobacter diazotrophicus*, an endophytic species (Fuentes-Ramírez et al. 1999). Aside from P and N, the plant-microbe symbiosis is also affected by the status of trace elements. AM fungal symbiosis tends to enhance the host tolerance to excessive trace elements in polluted soils and is potentially attractive for aiding phytoremediation (Smith and Read 2008). Mycorrhizal response to excess nutrients is affected by the nature of the fungus, plant, trace element, levels of contamination, etc. For example, in studies using similar levels of Zn treatments (0, 200, and 400 mg Zn kg^{-1} soil) but different soil storage times and AMF species, mycorrhizal colonization of the soybean host root increased (Ibiang et al. 2017) or decreased (Ibiang and Sakamoto 2018) under excess soil Zn. The nutrient status in host plants is fine-tuned by mycorrhizal and rhizobial symbionts, and there is evidence of synergized action of both symbionts on the modulation of host trace element concentrations under excess soil elements (Ibiang et al. 2017). Whereas microsymbiont-derived benefits depend on the nutrient status of the host, it has been observed that deficiency of the nutrient pushes the association between host and microbe more toward the mutualistic end of the spectrum. For example, in the association between Arabidopsis and *Colletotrichum tofieldiae*, the plant growth-promoting effect of the endophyte is apparent under phosphate deficient growth conditions, but less so, under P sufficiency (Hiruma et al. 2016). In the case of phosphorus, a phosphate starvation response (PSR) under genetic control is generally part of the signaling network for sensing phosphorus availability in plants, and these genes are responsive to microbe and environmental influences (Chiou and Lin 2011). Excess arsenic (As) could be toxic to plants and enhance their susceptibility to pathogens. Lakshmanan et al. (2016) reported that rice exposure to arsenic enhanced their susceptibility to the rice blast pathogen, *Magnaporthe oryzae*. In their study, the inoculation of a plant growth-promoting rhizobacteria, *Pantoea* sp. EA106, reduced arsenic uptake and the incidence of blast in the host plant, highlighting a link between the As status in the host plant and tolerance to blast disease. Changes in the forms (e.g., from inorganic to organic) of nutrients supplied to the field is also capable of inducing changes in the microbial community, which could impact plant fitness. Bi et al. (2020) reported that the partial replacement of inorganic P by organic manure reshaped the phosphate solubilizing bacterial community and promoted P bioavailability in a paddy soil, leading to some improvements in plant growth.

3.1.5 INFLUENCE OF MOISTURE ON PLANT-MICROBE INTERACTION

The cell is made up of a large percentage of water; therefore, the life cycles of plants and microbes depend on moisture. Moisture levels in the soil, plant, and air affects plant-microbe interactions. Scarcity as well as excessive moisture conditions have significant effects on these associations. Early reports suggest that scarcity of moisture may increase the harmful impact of pathogens infecting plants, such as the enhancement of bacterial leaf scorch symptoms in *Parthenocissus quinquefolia* infected by *Xylella fastidiosa* (McElrone et al. 2001). Many microbes, though, rely on wetness of host organs for infection, and it is generally thought that moisture propels microbe activity, including the dispersal and germination of spores and motility of bacterial cells, amongst others. Fungi and oomycetes require wind and high humidity for sporulation and dispersal to establish their association with hosts. Some bacteria establish an aqueous space/microenvironment in plant tissue that is essential for their virulence. Xin et al. (2016) reported that bacteria release effectors into plant cells that stimulate water-soaking of tissues during pathogenesis and that high atmosphere humidity bolsters the function/virulence of the effector proteins (AvrE and HopM1) involved. Recent studies using *Xanthomonas gardneri* suggest that bacterial effectors may modulate the expression of genes involved in the production of enzymes, such as pectate lyases, that affect the host cell wall integrity to create water-soaked lesions in bacterial spot of tomato (Schwartz et al. 2017). A water-soaked host tissue may favor microbes in several ways, including in nutrient acquisition from tissues, dilution of host defense molecules, and microbe multiplication and spread (Cheng et al. 2019), all of which allow the microbe to gain the upper hand in the plant-microbe interaction. High atmospheric humidity (acting synergistically with elevated temperature) may also reduce hypersensitivity-related plant defense response, such as the repression of Cf-4/Avr4- and Cf-9/Avr9-dependent hypersensitive cell death and defense gene expression in tomatoes infected by the leaf mold fungal pathogen, *Cladosporium fulvum* (Wang et al. 2005). In canola, pathogenesis of blackleg disease caused by *Leptosphaeria* spp. (*L. maculans* and *L. biglobosa*) is modulated by relative humidity (El-Hadrami et al. 2009). While *L. maculans* and *L. biglobosa* are considered highly and weakly aggressive, respectively, elevations in relative humidity enhanced disease progression by *L. biglobosa* and coincided with a reduced accumulation of lignin. High rhizosphere moisture also affects the interaction between plants and pathogens. In ginger (*Zingiber officinale*), high soil moisture elevates the plant susceptibility to *Ralstonia solanacearum*, which causes bacterial wilt – one of the most important crop production constraints (Prasath et al. 2014). Although the exact mechanisms involved are not fully understood, the down regulation of *WAK16* and *WAK3-2* (wall-associated receptor kinase genes) as well as *PRX* (peroxidase), *CPY* (cytochrome P450), and *XET* (xyloglucan endotransglucosylase) genes indicate that disturbance to cell wall metabolism and inhibition of hypersensitive response, which are heightened under high rhizosphere moisture, are involved (Jiang et al. 2018).

AM diversity and biomass were affected by AMF inoculation and water limitation, with inoculated plants showing increased performance only under water limitation/stress (Omirou et al. 2013). Water limitation appeared to boost the host dependency on the mycobiont. Many studies show that mycorrhizal symbiosis is intensified under drought stress in a way that provides some mitigation of the negative effects in the host. Both fungal and host physiological responses are modulated by the environmental pressure of moisture deficit (Smith and Read 2008). Legumes show a significant reduction in N fixation under drought stress, as nodule development and function are affected adversely (Sinclair et al. 1987). The perennial ryegrass (*Lolium perenne*) is often infected by the leaf-inhabiting endophyte, *Neotyphodium lolii*. Although this asexual endophyte has the potential to impact host growth, survival, and reproduction, it was reported that this interaction aids the tolerance to drought stress and may confer a selective advantage in the Mediterranean regions (Kane 2011). Aside from being maternally transferred because of its ovule colonization tendency, the ability of *N. lolii* to confer benefits to the host under arid conditions has likely assisted in its ecological foothold/perpetuity in this region of the globe. Leuchtmann et al. (2014) proposed the realignment of *Neotyphodium* into the genus *Epichloë*.

3.1.6 INFLUENCE OF LIGHT ON PLANT-MICROBE INTERACTION

Light is widely known to affect the growth/activity of plants and microbes individually. While it is essential for photosynthesis, microbial processes, such as spore production and dispersal, hyphal growth, bacterial phototaxis, adhesion, etc., are directly influenced by light. The establishment and sustenance of plant-microbe interaction is, therefore, closely linked to light and plant circadian rhythms. For example, the stomata of plant leaves serve as entry points for many microbes invading plants, and stomatal opening and closing is under circadian influence. While there is evidence that plant immune responses to pathogen infection may follow a circadian rhythm (Wang et al. 2011), pathogen infection may also modulate host circadian activity (Lu et al. 2017). The circadian rhythm of microbes may also influence their interaction with hosts, as reported by Hevia et al. (2015), in which a circadian oscillator in the fungus *Botrytis cinerea* regulates virulence during infection of *Arabidopsis thaliana*. Santamaría-Hernando et al. (2018) showed that, by integrating monochromatic light signals during the day and regulating gene expression, *Pseudomonas syringae* exploits the absence of light signals to optimize virulence and colonization of leaves. The coronatine biosynthesis genes, *cmaA* and *cfa5*, which, respectively, encode coronamic acid synthetase and coronafacic acid synthetase, are modulated by light, with lights of various wavelengths repressing (and darkness activating) coronatine biosynthesis, which facilitates stomatal opening at night and entry by *P. syringae* during infection (Panchal et al. 2016).

An interesting effect of the environment on plant-microbe symbiosis can be seen between the fungal leaf endophyte, *Neotyphodium lolii*, and its host, *Lolium perenne*. According to Newsham et al. (1998), plants grown symbiotically with *N. lolii* had 7% thicker leaves, 4% thicker stem bases, and 7% fewer tillers than those grown non-symbiotically. However, the fertility of plants grown symbiotically with *N. lolii* was significantly reduced by outdoor elevated UV-B+A exposure, with 70% fewer spikes, 75% fewer seeds, and 71% lower total seed weight compared to symbiotic plants exposed to ambient radiation. In non-symbiotic plants, elevated UV-B+A exposure had no significant effect on these host parameters when compared with ambient radiation. This showed that a change in environmental factors led to a loss of host fitness that did not occur in non-symbiotic state. Light waves, such as UV, can inhibit/eliminate microbe activity and have been used as a sterilant. *Podosphaera pannosa* causes powdery mildew disease in roses. Kobayashi et al. (2013) showed that supplemental UV radiation with low levels of UV light (via a UV-B fluorescent lamp) suppress *P. pannosa* disease development and that UV light supplied at night was more effective than in the day. The authors identified inhibition of fungal growth and upregulation of host secondary metabolite gene pathways (such as phenylalanine ammonia lyase and chalcone synthase) as the related mechanisms involved. The tomato spotted wilt virus (TSWV) has a worldwide distribution and a wide host range, including tomato, lettuce, eggplant, pepper, broad beans, and celery (Sherwood et al. 2000). It can lead to huge economic losses due to crop disease and is transmitted exclusively by thrips, such as *Frankliniella occidentalis*, which is believed to be the most efficient insect vector (Whitfield et al. 2005). However, a recent study by Escobar-Bravo et al. (2019) showed that UV radiation exposure time and intensity modulate tomato resistance to herbivory by *F. occidentalis* with SA and JA signaling playing a role in the enhancement of resistance to thrips. Thus, the possibility for (environmental) control of TSWV disease via the exploitation of UV radiation regimes and JA biochemistry can be contemplated.

3.1.7   INFLUENCE OF SALINITY ON PLANT-MICROBE INTERACTION

Salinity is an important factor in plant production. In many countries, it reduces the utility of otherwise cultivable land for crop production, such as rice. In many parts of West Africa, including Nigeria, large areas of mangrove swamps are affected by salinity (Zaidi et al. 2014). High salinity is toxic to plants as well as microbes due to the toxicity of ions and osmotic stress. It reduces microbial activity, leading to effects such as reduced N fixation and crop yields (Kumar and Verma 2018). Salinity inhibits nodulation and may induce the premature senescence of legume nodules (Swaraj and Bishnoi 1999). Reductions in rhizobial infectivity, nodule development and nitrogenase biosynthesis have been observed as negative effects of high salinity on plant–rhizobial symbiosis (Tripathi et al. 2002; Bouhmouch et al. 2005). Salinity may change the abundance of soil bacteria and fungi and their functions in a saline ecosystem (Zhang et al. 2019). Salt stress caused changes in the microbial diversity of rice rhizosphere, and the presence of *Bacillus amyloliquefaciens* stimulated the population of certain (osmoprotectant-utilizing) bacteria that confer salinity tolerance to the host (Nautiyal et al. 2013).

The endophyte *Fusarium culmorum* colonizes non-embryonic tissues of coastal dune grass (*Leymus mollis*), and this interaction allows both organisms to tolerate high levels of salinity (reaching 300–500 mM NaCl) associated with their native habitat (Rodriguez et al. 2009). However, if exposed to this level of salinity stress when grown non-symbiotically, the coastal dune grass does not thrive while fungal growth is also retarded by such high salinity levels. In a study by Haque and Matsubara (2018), AMF (*Gigaspora margarita*) improved growth under salt stress (200mM NaCl), *Fusarium* stress (*Fusarium oxysporum* f. sp. *fragariae*), *Fusarium* + salt dual stress, and no-stress controls compared with non-mycorrhizal control. Disease incidence, index, and pathogen population were lower due to AMF under both salt stress and no-salt stress. However, compared with no-salt stress treatment, plant dry weight increases by AMF were reduced in salt stress treatment, although mycorrhizal plants in salt stressed groups had higher shoot dry weights than non-mycorrhizal. This reduction in AMF-induced dry weight gain was higher in dual-stress compared to either single-stress treatments, although AM plants did better than non-AM ones under dual stress. Their study showed the attenuating effect of elevated salinity against the growth-promoting potential of *G. margarita* and its aggravating effects on the severity of disease caused by *Fusarium*. Salinity may enhance susceptibility to *Fusarium* disease, and high salinity may even reduce the tolerance of *Fusarium* resistant tomato cultivars (Besri 1993). *Verticilium dahlia* is affected by salinity, as increasing salinity may increase its mycelial growth and number of conidia and microsclerotia. Thus, tomato seedlings irrigated with saline water were more susceptible to *V. dahlia* than those with non-saline water (Besri 1993). Salinity enhanced the disease severity of *Phytophthora parasitica* in tomato, partly due to reduced root growth rate and enhanced root death rate (Snapp et al. 1991). Salinity and alkalinity stress affect plant performance, and some soils sometimes contain overlapping areas of high salt and alkalinity. Combined alkaline and salinity stress is more deleterious than either alone (Bui et al. 2014). Salinity–alkalinity stress induced changes in microbial population and enzyme activity of the common bean, with increases in the number of bacteria compared to fungi and actinomycetes (Guo et al. 2019).

3.1.8 INFLUENCE OF BIOLOGICAL AGENTS ON PLANT-MICROBE INTERACTION

AMF may systemically induce resistance to diseases in plants (Vos et al. 2012). According to Chandanie et al. (2006), the induction of systemic resistance to anthracnose disease in cucumber caused by *Colletotrichum orbiculare* was influenced by inoculation with two plant growth-promoting fungi – *Phoma sp.* and *Penicillium simplicissimum* – and AMF. Plants inoculated with single *Phoma sp.* isolate showed considerable protection against *C. orbiculare*, but when co-inoculated with AMF, there was a reduction in the extent of disease protection/tolerance. The study observed that there was a suppression of *Phoma sp.* colonization by the AMF. However, there was no reduction of disease tolerance during co-inoculation of *P. simplicissimum* with AMF. Mycorrhiza mediated the tolerance to anthracnose disease caused by *Colletotrichum gloeosporioides*, with reductions in disease incidence and severity and increases in dry weights of strawberry plants (Li et al. 2006). Mycorrhizas are influenced by biological entities, such as insects, interacting with the above-ground tissues of the host. Herbivory by aphids, for example, may influence the root colonization of the host by mycorrhiza, with reports of both increases and decreases in levels of root colonization (Barto and Rillig 2010; Babikova et al. 2014). In wheat, aphid herbivory reduced the allocation of plant carbon to AM fungal symbiont, but mycorrhizal uptake of P was maintained regardless of aphid herbivory or elevated CO_2 (Charters et al. 2020). The plant type and nature of herbivory influence the eventual impact on mycorrhizal colonization, but it is thought that host carbon reallocation, as well as other mechanisms that affect root hormone profiles and host emissions, fine tune the herbivory-induced effects on mycorrhizal colonization of host roots (Pozo et al. 2015).

Interactions between a fungal endophyte, *Epichloë amarillans*, and its grass host, *Agrostis hyemalis*, was influenced by changing water availability gradients and the presence or absence of soil microbes. According to Davitt et al. (2011), the benefits of this symbiosis was strongest when water was limiting, as the symbiotic host had greater fitness than the non-symbiotic host, highlighting the existence of endophyte-mediated drought tolerance. However, this symbiotic benefit was reduced in the presence of soil microbes, indicating that the symbiosis became more costly in the presence of other microbes. Tripartite AMF–rhizobia–plant interactions often produce better results in terms of host fitness, than the single microbe–plant associations (see Meena et al. 2018). Such reports generally point to dual inoculations as likely more versatile and beneficial. Nevertheless, dual inoculation outcomes could depend on other factors including host and fungal consortium identity, inoculum material, and soil condition (Ibiang et al. 2020). Negative outcomes, such as leaf wilting and mortality, were reported during co-infection with *Penicillium pinophilum* EU0013 and *Claroideoglomus etunicatum*, in a lettuce–AMF–*Penicillium* interaction, depending on the soil and inoculum substrate material (Ibiang et al. 2020). In a study on endophyte–grass–mycorrhiza interaction, Vignale et al. (2020) reported that *Epichloë* endophytes of a wild grass promoted mycorrhizal colonization of neighbor grasses. Compared to AMF-only treatment, co-inoculation of *R. intraradices* and *Massilia* sp. RK4 benefits AM fungal colonization and the host's nutrient uptake under salt stress (Krishnamoorthy et al. 2016). There have been reports that associated microorganisms may mitigate or enhance pathogenicity of some plant pathogens. Pink disease in pineapple is characterized by dark discoloration and caused by *Tatumela ptyseos*. However, antagonism of *Burkholderia gladioli* on *T. ptyseos* detected in vitro and *in planta* led to a decline in disease severity in pineapple hosts (Marín-Cevada et al. 2012). Again, dual application of some bacterial isolates and a fungal pathogen, *Stagonospora* (*Septoria*) *nodorum*, which causes glume blotch of wheat, was reported to significantly increase the pathogenicity of the fungus. The bacterial isolates that showed this ability included *Xanthomonas maltophilia*, *Sphingobacterium multivorum*, and *Enterobacter agglomerans* (Dewey et al. 1999). On the other hand, some associated bacteria (called mycorrhizal helper bacteria) may enhance mycorrhizal symbiosis (Deveau and Labbé 2016). In an example of insect-induced effect on foliar fungal community, differences in the foliar fungal endophyte community inhabiting the leaves of the chestnut (*Castanea sativa*) exist between galls (caused by the invasive gall wasp, *Dryocosmus kuriphilus*) and the surrounding leaf tissue (Fernandez-Conradi et al. 2019). The reduction (in both richness and composition) observed in galls compared to surrounding tissue indicated the stifling of some endophytes that would normally maintain association with the host due to a less convenient gall microhabitat.

3.1.9 Influence of Pesticides on Plant-Microbe Interaction

Widely applied to the environment, pesticides are important entities with respect to plant-microbe interaction. Obviously, a lot of pesticides are directly harmful to microbial pathogens of plants and, thus, disrupt plant–pathogen interaction in favor of hosts. There is already plentiful information on how, for example, application of fungicides to crops in the fields affect fungal pathogens. Consequently, a focus on non-target effects involving plant interaction with beneficial microbes and microbiome-wide/community structure changes (Ratcliff et al. 2006), perhaps, may be more apt here. Pesticides have the potential to reduce rhizobial nodulation in legumes (Ibiang et al. 2015) and may also induce reductions in microbial populations in the field, although this effect subsides with the passage of time (Banerjee and Dey 1992). The most popular herbicide, glyphosate, inhibits the enzyme 5-enolpyruvyl-shikimic acid-3-phosphate synthase (EPSPS), which enables plants to synthesize aromatic amino acids (Duke 2018). Genetically modified glyphosate-tolerant crops can withstand glyphosate application to kill weeds, thus, causing its extensive use for weed control on transgenic crop fields and potential effects on microbe associations with plants. Because plants utilize phenols and other compounds derived from the shikimic acid pathway to combat disease, glyphosate may tend to enhance disease susceptibility in plants because of its mode of action interfering with normal plant defense mechanisms (Hammerschmidt 2017). Plant lignin (which presents a physical barrier to microbe infection and disease development) and phytoalexins, produced using aromatic amino acids, are reduced by glyphosate (Lévesque and Rahe 1992). However, there are also reports of glyphosate synergizing microbial bioherbicides for greater impact, such as with *Gloeocercospora sorghi* and *Colletotrichum gramini-cola* in the control of shattercane, *Sorghum bicolor* (Mitchell et al. 2008). While many bacteria and fungi contain EPSPS enzymes that are potentially sensitive to this herbicide, other microbes are competent in degrading it. Thus, there is a wide range of results on effects of glyphosate on microbial communities, including detrimental (Cherni et al. 2015), minimal/negligible (Zabaloy et al. 2016), or stimulatory impacts (Imparato et al. 2016) on bacterial abundance, activity, and diversity (Schlatter et al. 2017). Sometimes, both increases and decreases are reported for different groups of microbes. In maize and soybean rhizosphere, *Proteobacteria* increased in rela-tive abundance, while *Acidobacteria* decreased, in response to glyphosate exposure (Newman et al. 2016). Sensitivity of rhizobia and reductions in N fixation due to glyphosate has been reported (dos Santos et al. 2005). Effects, ranging from negative (Savin et al. 2009) to minimal (Pasaribu et al. 2013), on AM fungal colonization/attributes have also been ascribed to glyphosate. In general, while inhibitory effects of pesticides have been reported on plant pathogens, biocontrol agents, and plant growth-promoting microorganisms (Duke 2018), pesticide effects on biota depend on many factors, including identity, formulation, dosage, etc., and with moderate appropriate use, look to retain their importance in the agroecosystem.

3.1.10 INFLUENCE OF FARMING PRACTICES ON PLANT-MICROBE INTERACTION

Farming practices differ from place to place and in scale. But there is a broad cat-egorization based on the full use of synthetic means of pest control and fertilization (conventional system) or an integrated system approach with the objective of avoid-ing/minimalizing synthetic chemicals input (organic system) (Ishaq 2017). The prac-tices within farming systems, such as pesticide application, irrigation, composting/manuring, fertilizer addition, cropping system, bush burning, tillage, etc., modify the environment and potentially impact plant-microbe interaction. Features such as soil organic matter content, which is generally higher in organic than conventional farming systems, influence microbial diversity (Pershina et al. 2015) and help to set the scope of interactions in the host's rhizosphere. Local farmers in many countries are aware that bush fallowing and mixed cropping tend to support healthier plant growth. The mineralization of cover crop/legume residues – rich in N due to the activity of associated N-fixing bacteria replenish soil fertility, while, during fallow periods, microbiome reconstitution limits pathogen buildup. Cereal–green manure rotations shape the rhizosphere microbial community and enrich beneficial bacte-ria (Zhang et al. 2017). According to Tiemann et al. (2015), crop rotation diversity enhances below-ground communities and functions in an agroecosystem. In contrast, in a field where 15 years of continuous cropping was practiced, pathogen buildup in the last 8 years allowed a rapid surge in *Fusarium* infection of tobacco plants, leading to crop failure (Santhanam et al. 2015). Field trials conducted later showed that inoculation with a mixture of native bacterial isolates significantly reduced the disease incidence and mortality in infected plots. A core consortium of five bacteria (which are normally recruited from seed germination) were later shown to be essential for disease reduction (Santhanam et al. 2015). AM fungal community structure differed significantly under organic versus conventional farm management. The abundance of individual AMF could also be affected by farm management. In one study, *Funneliformis mosseae* had greater abundance under a conventional management system, while an organic system had a greater abundance of *Claroideoglomus claroideum* (Schneider et al. 2015).

Soil tillage mixes up litter, soil moisture, oxygen, etc., creating conditions that influence microbial activity and diversity, which is generally beneficial. But fungi are also physically affected by tillage, as it mechanically beaks their hyphae in topsoil. Tillage-induced disruption of mycorrhiza are well documented, and reports of reduced mycorrhizal colonization in disturbed soils abound, but instances in which reduced colonization was not seen have been reported (McGonigle et al. 1990). In such cases, soil-disturbance effects on extraradical mycelia are likely to be more important than on the internal root colonization. Soil tillage affects AM fungal communities colonizing roots of plants as compared to no-till versus conventional tillage (Mirás-Avalos et al. 2011). Effects of tillage may also be shaded by other conditions. For example, intensive tillage restricted species composition and reduced AMF attributes, but root colonization may increase if there is a high soil organic carbon and decrease with elevated levels of inorganic N (dela Cruz-Ortiz et al. 2020). AM spore populations and root colonization of winter wheat were higher under low-input than conventional systems, with spore populations of *Glomus occultum*-type AMF being more numerous in non-tilled versus tilled soil (Galvez et al. 2001). Via the production of glomalin, AMF improve the stability of soil structure, thus, aiding overall soil quality. A study in Chile showed that, under no-tillage and conventional-tillage plots over a 6- and 10-year period, the number of mycorrhizal propagules, total soil carbon, and glomalin-related soil protein was higher in no-tillage for 6 years compared to conventional-tillage and no-tillage for 10 years (Curaqueo et al. 2011). Their study suggested a potential benefit of occasional ploughing in long-term (a decade) no-tillage systems, even though parameters were improved during a comparatively short-term (6 years), no-tillage system. Plant disease is also affected by tillage and other farming practices. According to Guo et al. (2005), tillage reduced disease when it was performed with a single-crop rotation, but the effect was reduced with a two-crop rotation. Their study concluded that tillage and crop rotation could be combined to reduce the severity of blackleg disease of canola.

Fires affect plants and microbes and, thus, can be expected to affect their interaction. However, the impact of fires in agroecosystems is shaped by several factors, including duration, frequency, and intensity of the fire as well as the season (summer fires are more severe/destructive than spring), amongst others. Fire affects AM fungal diversity, more likely via direct effects of fire on AMF rather than via indirect effects based on changes in soil abiotic properties (Longo et al. 2014). Bellgard et al. (1994) studied the effect of wildfire on infectivity and abundance of AMF and observed that post-fire soil had less infectivity than pre-fire soil, but there was, ultimately, no difference in root colonization of plants, although spore numbers were significantly lower post-fire for the most abundant spore type. Spring burning of prairie field plots significantly reduced AMF species diversity, but there was no significant effect on root colonization and extraradical hyphae development (Eom et al. 1999). Studies pointing to possible fire-induced reduction in mycorrhizal colonization exist, though. According to Dove and Hart (2017), fire reduced mycorrhizal colonization in situ, however, when ex situ assessment was used, the effect was not significant. In addition, in situ reductions in mycorrhizal colonization were alleviated with time. Hewitt et al. (2016) reported that, as fire severity increased, seedling biomass decreased while an increase in proportion of pathogenic soil fungi was also observed. Foliar microbes are also affected by fires. Huang et al. (2016) studied the abundance, diversity, and composition of endophytes in foliage of *Juniperus deppeana* and *Quercus* spp. collected contemporaneously from areas affected by wildfire vs. no-fire areas. They observed significant effects, including shifts in community structure and taxonomic composition, indicating that some host–endophyte associations were likely more affected than others. Recently, Semenova-Nelson et al. (2019) reported that frequent fires reorganize fungal communities across a heterogeneous pine savanna landscape, suggesting that the persistence of fire-adapted ecosystems is aided by fire-induced fungal community reorganizations. It is worth pointing out that the soil fungal community also changed in response to long-term fire cessation and N fertilization in tallgrass prairies (Carson et al. 2019).

3.2 CONCLUSION

The efforts at mainstreaming the exploitation of plant-microbe interaction for the benefit of humankind and the environment will be boosted by an increased understanding of plant-microbe interaction response to environmental variability. Some general principles that hold for both pathogenic and beneficial microbe interactions were highlighted and illustrated in the subsequent sections discussing the influence of some physical, chemical, and biological factors. Obviously, the environmental control of plant-microbe interaction is a vast area of discussion with multidisciplinary bearings, and, in covering its importance, general principles, and applications, it is acknowledged that mentions of articles were omitted in the interest of space for this chapter.

ACKNOWLEDGEMENTS

The author thanks the University of Calabar, Nigeria, and Chiba University, Japan, for the award of a postdoctoral fellowship (2019–2020), during which portions of this chapter were developed.

REFERENCES

Agrios G.N. 2005. *Plant Pathology*, 5th edn. London: Elsevier Academic Press.
Antunes P.M., Koch A.M., Morton J.B., Rillig M.C., and J.N. Klironomos 2011. Evidence for functional divergence in arbuscular mycorrhizal fungi from contrasting climatic origins. *New. Phytol.*, 189: 507–514.

Auge R.M. 2001. Water relations, drought and vesicular-arbuscular mycorrhizal symbiosis. *Mycorrhiza*, 11: 3–42.

Babalola O.O., Fadiji A.E., Enagbonma B.J., Alori E.T., Ayilara M.S., and A.S. Ayangbenro 2020. The nexus between plant and plant microbiome: revelation of the networking strategies. *Front. Microbiol.*, 11: 548037. https://doi.org/10.3389/fmicb.2020.548037

Babikova Z., Gilbert L., Bruce T., Dewhirst S.Y., Pickett J.A., and D. Johnson 2014. Arbuscular mycorrhizal fungi and aphids interact by changing host plant quality and volatile emission. *Funct. Ecol.*, 28: 375–385.

Banerjee M.R., and B.K. Dey 1992. Effects of different pesticides on microbial populations, nitrogen mineralisation, and thiosulphate oxidation in the rhizosphere of jute (*Corchoruscapsularis* L. cv.). *Biol. Fert. Soils.*, 14: 213–218.

Baron C., Domke N., Beinhofer M., and S. Hapfelmeier 2001. Elevated temperature differentially affects virulence, VirB protein accumulation, and T-pilus formation in different Agrobacterium tumefaciens and Agrobacterium vitis strains. *J. Bacteriol.*, 183: 6852–6861.

Barto E.K., and M.C. Rillig 2010. Does herbivory really suppress mycorrhiza? A meta-analysis. *J. Ecol.*, 98: 745–753.

Bellgard S.E., Whelan R.J., and R.M. Muston 1994. The impact of wildlife on vesicular-arbuscular mycorrhizal fungi and their potential to influence the re-establishment of post-fire plant communities. *Mycorrhiza*, 4: 139–146.

Berg G. 2009. Plant-microbe interactions promoting plant growth and health: perspectives for controlled use of microorganisms in agriculture. *Appl. Microbiol. Biotechnol.*, 84: 11–18.

Besri M. 1993. Effects of salinity on plant diseases development. In: *Towards the Rational Use of High Salinity Tolerant Plants*, ed. H. Lieth and A.A. AlMasoom, 67–74. *Tasks for Vegetation Science, Vol 28*. Springer, Dordrecht. https://doi.org/10.1007/978-94-011-1860-6_8

Bi Q-F., Li K.J., Zheng B.X., et al. 2020. Partial replacement of inorganic phosphorus (P) by organic manure reshapes phosphate mobilizing bacterial community and promotes P bioavailability in a paddy soil. *Sci. Total Environ.* https://doi.org/10.1016/j.scitotenv.2019.134977

Bocianowski J., Tratwal A., and K. Nowosad 2020. Genotype by environment interaction for area under the disease-progress curve (AUDPC) value in spring barley using additive main effects and multiplicative interaction model. *Australas. Plant Pathol.*, 49: 525–529. https://doi.org/10.1007/s13313-020-00723-7

Boller T., and S. Y. He 2009. Innate immunity in plants: an arms race between pattern recognition receptors in plants and effectors in microbial pathogens. *Science*, 324, 742–744. https://doi.org/10.1126/science.1171647

Bouhmouch I., Souad-Mouhsine B., Brhada F., and J. Aurag 2005. Influence of host cultivars and rhizobium species on the growth and symbiotic performance of Phaseolus vulgaris under salt stress. *J. Plant Physiol.*, 162: 1103–1113.

Bui E.N., Thornhill A., and J.T. Miller 2014. Salt-and alkaline-tolerance are linked in Acacia. *Biol. Lett.*, 10: 20140278.

Calzavara A.K., Paiva P.H.G., Gabriel L.C. et al. 2018. Associative bacteria influence maize (Zea mays L.) growth, physiology and root anatomy under different nitrogen levels. *Plant Biol. (Stuttg)*, 20(5): 870–878. https://doi.org/10.1111/plb.12841

Carson C.M., Jumpponen A., Blair J.M., and L.H. Zeglin 2019. Soil fungal community changes in response to long-term fire cessation and N fertilization in tallgrass prairie. *Fungal Ecol.*, 41: 45–55.

Carter A.H., Chen X.M., Garland-Campbell K., and K.K. Kidwell 2009. Identifying QTL for high-temperature adult-plant resistance to stripe rust (*Puccinia striiformis* f. sp. *tritici*) in the spring wheat (*Triticum aestivum* L.) cultivar 'Louise'. *Theor. Appl. Genet.*, 119: 1119–1128.

Carvalhais L.C., Dennis P.G., Fan B., et al. 2013. Linking plant nutritional status to plant-microbe interactions. PLoS One 8(7): e68555. https://doi.org/10.1371/journal.pone.006 8555

Chandanie W.A., Kubota M., and M. Hyakumachi 2006. Interactions between plant growth promoting fungi and arbuscular mycorrhizal fungus Glomus mosseae and induction of systemic resistance to anthracnose disease in cucumber. *Plant Soil.*, 286: 209–217.

Charters M.D., Sait S.M., and K.J. Field 2020. Aphid herbivory drives asymmetry in carbon for nutrient exchange between plants and an arbuscular mycorrhizal fungus. *Curr. Biol.*, 30: 1801–1808.

Chellappan P., Vanitharani R., Ogbe F., and C.M. Fauquet 2005. Effect of temperature on geminivirus-induced RNA silencing in plants. *Plant Physiol.*, 138: 1828–1841.

Cheng C., Gao X., Feng B., Sheen J., Shan L., and P. He 2013. Plant immune response to pathogens differs with changing temperatures. *Nat. Commun.*,4: 2530.

Cheng Y.T., Zhang L., and S.Y. He 2019. Plant-microbe interactions facing environmental challenge. *Cell Host Microbe*, 26: 183–192. https://doi.org/10.1016/j.chom.2019.07.009

Cheplick G.P., Clay K., and S. Marks 1989. Interactions between infection by endophytic fungi and nutrient limitation in the grasses Lolium perenne and Festuca arundinacea. *NewPhytol.*, 111: 89–97.

Cherni A.E., Trabelsi D., Chebil S., Barhoumi F., Rodríguez-Llorente I.D., and K. Zribi 2015. Effect of glyphosate on enzymatic activities, Rhizobiaceae and total bacterial communities in agricultural Tunisian soil. *Water Air Soil Pollut.*, 226: 145. https://doi.org/10.1 007/s11270-014-2263-8

Chiou T.J., and S.I. Lin 2011. Signaling network in sensing phosphate availability in plants. *Annu. Rev. Plant Biol.*, 62: 185–206.

Compant S., van der Heijden M.G.A., and A. Sessitsch 2010. Climate change effects on beneficial plant-microorganism interactions. *FEMS Microbiol. Ecol.*, 73: 197–214. https:// doi.org/10.1111/j.1574-6941.2010.00900.x

Conrath U., Pieterse C.M.J., and B. Mauch-Mani 2002. Priming in plant-pathogen interactions. *Trends Plants* Sci., 7: 210–216.

Curaqueo G., Barea J.M., Acevedo E., Rubio R., and P. Cornejo 2011. Effects of different tillage system on arbuscular mycorrhizal fungal propagules and physical properties in a Mediterranean agroecosystem in central Chile. *Soil Tillage Res.*, 113: 11–18.

Davitt A.J., Chen C., and J.A. Rudgers 2011. Understanding the context-dependency in plant-microbe symbiosis: the influence of abiotic and biotic contexts on host fitness and the rate of symbiont transmission. *Environ. Exp. Bot.*, 71(2): 137–145.

delVal C., Barea J.M., and C. Azcón-Aguilar 1999. Assessing the tolerance to heavy metals of arbuscular mycorrhizal fungi isolated from sewage sludge-contaminated soils. *Appl. Soil Ecol.*, 11: 261–269.

Dela Cruz-Ortiz Á.V., Álvarez-Lopeztello J., Robles C., and L. Hernández-Cuevas 2020. Tillage intensity reduces the arbuscular mycorrhizal fungi attributes associated with Solanum lycopersicum, in the Tehuantepec Isthmus (Oaxaca) Mexico. *App. Soil Ecol.*, 149: 103519. https://doi.org/10.1016/j.apsoil.2020.103519

Deveau A., and J. Labbé 2016. Mycorrhizal helper bacteria. In: *Molecular mycorrhizal symbiosis*, ed. F. Martin, 437–450. John Wiley and Sons. https://doi.org/10.1002/978111 8951446.ch24

Dewey F.M., Wong Y.L., Seery R., Hollins T.W., and S.J. Gurr 1999. Bacteria associated with *Stegonospora* (*Septoria*) *nodorum* increase pathogenicity of the fungus. *New Phytol.*, 144: 489–497.

Dos Santos J.B., Ferreira E.A., Kasuya M.C.M., da Silva A.A., and P.S. de Oliveira 2005. Tolerance of Bradyrhizobium strains to glyphosate formulations. *Crop Prot.*, 24: 543–547.

Dove N.C., and S.C. Hart 2017. Fire reduces fungal species richness and in-situ mycorrhizal colonization: a meta-analysis. *Fire Ecol.*, 13: 37–65.

Duke S.O. 2018. Interaction of chemical pesticides and their formulation ingredients with microbes associated with plants and plant pests. *J. Agric. Food Chem.*, 66: 7553–7561.

Eastburn D.M., Degennaro M.M., Delucia E.H., Dermody O., and A.J. McElrone 2010. Elevated atmospheric carbondioxide and ozone alter soybean diseases at SoyFACE. *Glob. Chang. Biol.*, 16: 320–330.

El-Hadrami A., Dilantha-Fernando W.G., and F. Daayf 2009. Variations in relative humidity modulate Leptosphaeria spp. pathogenicity and interfere with canola mechanisms of defense. *Eur. J. Plant Pathol.*, 126: 187–202.

Eom A-H., Hartnett, D.C., Wilson G.W.T., and A.A.H. Figge 1999. The effect of fire, mowing and fertilizer amendment on arbuscular mycorrhizas in tallgrass prairie. *Am. Midl. Nat.*, 142: 55–70.

Escobar-Bravo R., Chen G., Kim H.K., et al. 2019. Ultraviolet radiation exposure time and intensity modulate tomato resistance to herbivory through activation of jasmonic acid signaling. *J. Exp. Bot.*, 70: 315–327.

Faghihinia M., Zou Y., Chen Z., et al. 2020. Environmental drivers of grazing effects on arbuscular mycorrhizal fungi in grasslands. *Appl. Soil Ecol.*, 153. https://doi.org/10.1 016/j.apsoil.2020.103591

Fernandez-Conradi P., Castagneyrol B., Jactel H., and C. Robin 2019. Fungal endophyte communities differ between chestnut galls and surrounding foliar tissues. *Fungal Ecol.*, 42: 100876.

Fu D., Uauy C., Distelfeld A., et al. 2009. A novel kinase-START gene confers temperature-dependent resistance to wheat stripe rust. *Science*, 323: 1357–1360.

Fuentes-Ramírez L.E., Caballero-Melado J., Sepúlveda J., and E. Martínez-Romero 1999. Colonization of sugarcane by *Acetobacterdiazotrophicus* is inhibited by high N-fertilization. *FEMS Microbiol Ecol.*, 29: 117–128.

Galvez L., Douds D.D., and P. Wagoner 2001. Tillage and farming system affect AM fungus populations, mycorrhizal formation, and nutrient uptake by winter wheat in a high-P soil. *Am. J. Alter. Agric.*, 16(4): 152–160.

González-Guerrero M., Escudero V., Saéz Á., and M. Tejada-Jiménez 2016. Transition metal transport in plants and associated endosymbionts: arbuscular mycorrhizal fungi and rhizobia. *Front. Plant Sci.*,7: 1088. https://doi.org/10.3389/fpls.2016.01088

Groppe K., Steinger I., Schmid B., Wiemken A., and T. Boller 1999. Interaction between the endophytic fungus *Epichloëbromicola* and the grass *Bromuserectus*: effects of endophytic infection, fungal concentration and environment on grass growth and flowering. *Mol. Ecol.*, 8: 1827–1835.

Guo H., Peng X., Gu L., Wu J., Ge F., and Y. Sun 2017. Up-regulation of MPK4 increases the feeding efficiency of the green peach aphid under elevated CO_2 in *Nicotiana attenuata*. *J. Exp. Bot.*, 68 (21–22): 5923–5935. https://doi.org/10.1093/jxb/erx394

Guo X., Wang X., Liang H., et al. 2019. Effects of salinity-alkalinity stress on rhizosphere soil microbial quantity and enzyme activity of common bean. *Acta Agric. Boreali-Sinica*, 34(4): 148–157.

Guo X.W., Fernando W.G.D., and M. Entz 2005. Effect of crop rotation and tillage on blackleg disease of canola. *Can. J. Plant Pathol.*, 27: 53–57.

Hammerschmidt R. 2017. How glyphosate affects plant disease development: it is more than enhanced susceptibility. *Pest Mgt. Sci.*, 74: 1054–1063. 10.1002/ps.4521

Haque S.I., and Y-I Matsubara 2018. Cross-protection to salt stress and Fusarium wilt with the alleviation of oxidative stress in mycorrhizal strawberry plants. *Environ. Control Biol.*, 56(4): 187–192.

Hasegawa H., Chatterjee A., Cui Y., and A.K. Chatterjee 2005. Elevated temperature enhances virulence of Erwinia carotovora subs. Carotovora strain EC153 to plants and stimulates production of the quorum sensing signal, N-acetyl homoserine lactone, and extracellular proteins. *Appl. Environ. Microbiol.*, 71: 4655–4663.

Heinemeyer A., Ineson P., Ostle N., and A.H. Fitter 2006. Respiration of the external mycelium in the arbuscular mycorrhizal symbiosis shows strong dependence on recent photosynthates and acclimation to temperature. *New Phytol.* 171: 159–170.

Hevia M.A., Canessa P., Müller-Esparza H., and L.F. Larrondo 2015. A circadian oscillator in the fungus *Botrytiscinereal* regulates virulence when infecting of *Arabidopsisthaliana*. *Proc. Natl. Acad. Sci.*, 112: 8744–8749.

Hewitt R.E., Hollingsworth T.N., Chapin III F.S., and D.L. Taylor 2016. Fire-severity effects on plant-fungal interactions after a novel tundra wildfire disturbance: implications for arctic shrub and tree migration. *BMC Ecol.*, 16: 25. https://doi.org/10.1186/s12898-01 6-0075-y

Hiruma K., Gerlach N., Sacristán S., et al. 2016. Root endophyte *Colletotrichumtofieldiae* confers plant fitness benefits that are phosphate status dependent. *Cell*, 165: 464–474.

Hu X-F., Cheng C., Luo F., et al. 2016. Effects of different fertilization practices on the incidence of rice pests and diseases: a three-year case study in Shanghai, in subtropical southeastern China. *Fields Crops Res.*, 196: 33–50.

Huang Y-L., Devan M.M.N., U'Ren J.M., Furr S.H., and A.E. Arnold 2016. Pervasive effects of wildfire on foliar endophyte communities in montane forest trees. *Microb. Ecol.*, 71: 452–468. https://doi.org/10.1007/s00248-015-0664-x

Huot B., Castroverde C.D.M., Velásquez A.C., et al. 2017. Dual impact of elevated temperature on plant defence and bacterial virulence in Arabidopsis. *Nat.Commun.*, 8: 1808.

Huot B., Yao J., Montgomery B.L., and S.Y. He 2014. Growth-defense tradeoffs in plants: a balancing act to optimize fitness. *Mol. Plant.*, 7: 1267–1287.

Ibiang Y.B., Ekanem B.E., Usanga D.A., and U.I. Williams 2015. Germination and root nodule formation of soybean (Glycine max (L.) Merr.) in ridomil and chlorpyriphos treated soil. *Am. J. Environ. Prot.*, 4: 17–22.

Ibiang Y.B., Innami H., and K. Sakamoto 2018. Effect of excess zinc and arbuscular mycorrhizal fungus on bioproduction and trace element nutrition of tomato (*Solanum lycopersicum* L. cv. Micro-Tom). *Soil Sci. Plant Nutr.*, 64: 342–351.

Ibiang Y.B., Mitsumoto H., and K. Sakamoto 2017. Bradyrhizobia and arbuscular mycorrhizal fungi modulate manganese, iron, phosphorus, and polyphenols in soybean (*Glycinemax* (L.) Merr.) under excess zinc. *Environ. Exp. Bot.*, 137: 1–13.

Ibiang Y.B., and K. Sakamoto 2018. Synergic effect of arbuscular mycorrhizal fungi and bradyrhizobia on biomass response, element partitioning and metallothionein gene expression of soybean-host under excess soil zinc. *Rhizosphere*, 6: 56–66.

Ibiang S.R., Sakamoto K., and N. Kuwahara 2020. Performance of tomato and lettuce to arbuscular mycorrhizal fungi and *Penicilliumpinophilum* EU0013 inoculation varies with soil, culture media of inoculum, and fungal consortium composition. *Rhizosphere*, 16. https://doi.org/10.1016/j.rhisph.2020.100246

Imparato V., Santos S.S., Johansen A., Geisen S., and A. Winding 2016. Stimulation of bacteria and protists in rhizosphere of glyphosate-treated barley. *Appl. Soil Ecol.*, 98: 47–55. 10.1016/j.apsoil.2015.09.007

Ishaq S.L. 2017. Plant-microbial interactions in agriculture and the use of farming systems to improve diversity and productivity. *AIMS Microbiol.*, 3: 335–353.

Issa A., Esmaeel Q., Sanchez L., et al. 2018. Impacts of *Paraburkholderiaphytofirmans* strain PsJN on tomato (*Lycopersicumesculentum* L.) under high temperature. *Front. Plant Sci.*, 9: 1397.

Jiang Y., Huang M., Zhang M., et al. 2018. Transcriptome analysis provides novel insights into high-soil-moisture-elevated susceptibility to *Ralstoniasolanacearum* infection in ginger (*Zingiberofficinale* Roscoe cv. Southwest). *Plant Physiol. Biochem.*, 132: 547–556.

Kane K.H. 2011. Effects of endophyte infection on drought stress tolerance of *Loliumperenne* accessions from the Mediterranean region. *Environ. Exp. Bot.*, 71(3): 337–344.

Kim Y.S., An C., Park S., et al. 2017. CAMTA-mediated regulation of salicylic acid immunity pathway genes in Arabidopsis exposed to low temperature and pathogen infection. *Plant Cell*, 29: 2465–2477.

Király L., Hafez Y. M., Fodor J., and Z. Király 2008. Suppression of tobacco mosaic virus-induced hypersensitive-type necrotization in tobacco at high temperature is associated with downregulation of NADPH oxidase and superoxide and stimulation of dehydro-ascorbate reductase. *J. Gen. Virol.* 89: 799–808. https://doi.org/10.1099/vir.0.83328-0

Klironomos J.N., Hart M.M., Gurney J.E., and P. Moutoglis 2001. Interspecific differences in the tolerance of arbuscular mycorrhizal fungi to freezing and drying. *Can. J. Bot.*, 79: 1161–1166.

Kobae Y., Ohmori Y., Saito C., Yano K., Ohtomo R., and T. Fujiwara 2016. Phosphate treatment strongly inhibits new arbuscule development but not the maintenance of arbuscule in mycorrhizal rice roots. *Plant Physiol.*, 171: 566–579.

Kobayashi M., Kanto T., Fujikawa T., et al. 2013. Supplemental UV radiation controls rose powdery mildew disease under the greenhouse conditions. *Environ. Control Biol.*, 51(4): 157–163.

Kobayashi T., Ishiguro K., Nakajima T., Kim H.Y., Okada M., and K. Kobayashi 2006. Effects of elevated atmospheric CO_2 concentration on the infection of rice blast and sheath blight. *Phytopathology*, 96: 425–431.

Koide R.T. 1991. Nutrient supply, nutrient demand and plant response to mycorrhizal infection. *New Phytol.*, 117: 365–386.

Kretzschma F.D., Aidar M.P.M., Salgado I., and M.R. Braga 2009. Elevated CO_2 atmosphere enhances procuction of defense-related flavonoids in soybean elicited by NO and a fungal elicitor. *Env. Exp. Bot.*, 65(2–3): 319–329. https://doi.org/10.1016/j.envexpbot.2008.10.001

Krishnamoorthy R., Kim K., Subramanian P., Senthilkumar M., Anandham R., and T. Sa 2016. Mycorrhizal fungi and associated bacteria isolated from salt-affected soil enhance the tolerance of maize to salinity in coastal reclamation soil. *Agric. Ecosyst. Environ.*, 231: 233–239.

Kumar A. and J.P. Verma 2018. Does plant-microbe interaction confer stress tolerance in plants: a review? *Microbiol. Res.*, 207: 41–52.

Lakshmanan V., Cottone J., and H.P. Bais 2016. Killing two birds with one stone: Natural rice rhizospheric microbes reduce arsenic uptake and blast infections in rice. *Front. Plant Sci.*, https://doi.org/10.3389/fpls.2016.01514

Lehman R.M., Taheri W.I., Osborne S.L., Buyer J.S., and D.D. Douds Jr. 2012. Fall cover cropping can increase arbuscular mycorrhizae in soils supporting intensive agricultural production. *Appl. Soil Ecol.*, 61: 300–304.

Leuchtmann, A., Bacon, C. W., Schardl, C. L., White, J. F., and M. Tadych 2014. Nomenclatural realignment of Neotyphodium species with genus Epichloë. *Mycologia*, 106(2): 202–215. https://doi.org/10.3852/13-251

Lévesque C.A., and J.E. Rahe 1992. Herbicide interactions with fungal root pathogens, with special reference to glyphosate. *Annu. Rev. Phytopathol.*, 30: 579–602.

Li Y., Miyawaki Y., Matsubara Y-I., and K. Koshikawa 2006. Tolerance to anthracnose in mycorrhizal strawberry plants grown by capillary watering method. *Environ. Control Biol.*, 44(4): 301–307.

Longo S., Nouhra E., Goto B.T., Berbara R.L., and C. Urcelay 2014. Effects of fire on arbuscular mycorrhizal fungi in the Mountain Chaco forest. *For. Ecol. Manage.*, 315: 86–94.

Lu H., McClung C.R., and C. Zhang 2017. Tick tock: circadian regulation of plant innate immunity. *Annu. Rev. Phytopathol.*, 55: 287–311.

Manstretta V.,and V. Rossi 2015. Effects of temperature and moisture on development of Fusarium graminearum perithecia in maize stalk residues. *Appl. Environ. Microbiol.*, 82: 184–191.

Marín-Cevada V., Muñoz-Rojas J., Caballero-Mellado J. et al. 2012. Antagonistic interactions among bacteria inhabiting pineapple. *Appl. Soil Ecol.*, 61: 230–235.

Márquez L.M., Redman R.S., Rodriguez R.J., and M.J. Roossinck 2007. A virus in a fungus in a plant: three-way symbiosis required for thermal tolerance. *Science*, 315: 513–515. https://doi.org/10.1126/science.1136237

McElrone A.J., Sherald J.L., and I.N. Forseth 2001. Effects of water stress on symptomatology and growth of *Parthenocissus quinquefolia* infected by *Xylella fastidiosa*. *Plant Dis.*, **85**: 1160–4.

McGonigle T.P., Evans D.G., and M.H. Miller 1990. Effect of degree of soil disturbance on mycorrhizal colonization and phosphorus absorption by maize in growth chamber and field experiments. *New Phytol.*, 116: 629–636.

Meena R.S., Vijayakumar V., Yadav G.S., and T. Mitran 2018. Response and interaction of Bradyrhizobium japonicum and arbuscular mycorrhizal fungi in the soybean rhizosphere. *Plant Growth Regul.*, 84: 207–223.

Mine A., Sato M., and K. Tsuda 2014. Towards a systems understanding of plant-microbe interactions. *Front. Plant Sci.*, 5: 423.

Mirás-Avalos, J.M., Antunes P.M., Koch A., Khosla K., Klironomos J.N., and K.E. Dunfield 2011. The influence of tillage on the structure of rhizosphere and root-associated arbuscular mycorrhizal fungal communities. *Pedobiologia*, 54(4): 235–241.

Mitchell J.K., Yerkes C.N., Racine S.R., and E.H. Lewis 2008. The interaction of two fungal bioherbicides and a sub-lethal rate of glyphosate for the control of shattercane. *Biol. Control.*, 46: 391–399.

Moury B, Selassie KG, Marchoux G, Daubèze A, and A. Palloix 1998. High temperature effects on hypersensitive resistance to *Tomato spotted wilt tospovirus* (TSWV) in pepper (*Capsicum chinense* Jacq.). *Eur. J. Plant Pathol.*,104: 489–498

Nadeem S.M., Ahmad M., Zahir Z.A., Javaid A., and M. Ashraf 2014. The role of mycorrhizae and plant growth promoting rhizobacteria (PGPR) in improving crop productivity under stressful environments. *Biotechnol. Adv.*, 32: 429–448.

Nautiyal C.S., Srivastava S., Chauhan P.S., et al. 2013. Plant growth-promoting bacteria Bacillus amyloliquefaciens NBRISN13 modulates gene expression profile of leaf and rhizosphere community in rice during salt stress. *Plant Physiol. Biochem.*, 66: 1–9.

Newman M.M., Hoilett N., Lorenz N., et al. 2016. Glyphosate effects on soil rhizosphere associated bacterial communities. *Sci. Tot. Env.*, 543: 155–160.

Newsham K.K., Lewis G.C., Greenslade P.D., and A.R. Mcleod 1998. *Neotyphodiumlolii*, a fungal leaf endophyte, reduces fertility of *Loliumperenne* exposed to elevated UV-B radiation. *Ann. Bot.*, 81(3): 397–403. https://doi.org/10.1006/anbo.1997.0572

Obrępalska-Stęplowska A., Renaut J., Planchon S., et al. 2015. Effect of temperature on the pathogenesis, accumulation of viral and satellite RNAs and on plant proteome in peanut stunt virus and satellite RNA-infected plants. *Front. Plant Sci.* 6:903. https://doi.org/10.3389/fpls.2015.00903

Omirou M., Ioannides I.M., and C. Ehaliotis 2013. Mycorrhizal inoculation affects arbuscular mycorrhizal diversity in watermelon roots, but leads to improved colonization and plant response under water stress only. *Appl. Soil. Ecol.*, 63: 112–119.

Panchal, S., Roy, D., Chitrakar, R., et al. 2016.Coronatine facilitates *Pseudomonas syringae* infection of *Arabidopsis* leaves at night. *Front. Plant Sci.*, **7**: 880. https://doi.org/10. 3389/fpls.2016.00880

Pasaribu A., Mohamad R.B., Hashim A., et al. 2013. Effect of herbicide on sporulation and infectivity of vesicular arbuscular mycorrhizal (Glomus mosseae) symbiosis with peanut. *Plant J. Anim. Plant Sci.*, 23: 1671–1678.

Pérombelon M.C.M. 2002. Potato diseases caused by soft rot erwinias: an overview of pathogenesis. *Plant Pathol.*, 51: 1–12.

Pershina E., Valkonen J., Kurki P., et al. 2015. Comparative analysis of prokaryotic communities associated with organic and conventional farming systems. *PLOS One*, 10: e0145072.

Polacco J.C., and C.D. Todd (ed.) 2011. *Ecological Aspects of Nitrogen Metabolism in Plants.* West Sussex: John Wiley and Sons.

Pozo M.J., López-Ráez J.A., Azcón-Aguilar C., and J.M. García-Garrido 2015. Phytohormones as integrators of environmental signals in the regulation of mycorrhizal symbioses. *New Phytol.*, https://doi.org/10.1111/nph.13252

Prasath D., Karthika, R., Habeeba N.T., et al. 2014. Comparison of the transcriptomes of ginger (Zingiber officinale Rosc.) and mango ginger (Curcuma amada Roxb.) in response to the bacterial wilt infection. *PloS One*, 9: e99731.

Randall K.C., Brennan F., Clipson N. et al. 2020. An assessment of climate induced increase in soil water availability for soil bacterial communities expose to long-term differential phosphorus fertilization. *Front. Microbiol.*, 11: 682. https://doi.org/10.3389/fmicb.2020. 00682

Ratcliff A.W., Busse M.D., and C.J. Shestak 2006. Changes in microbial community structure following herbicide (glyphosate) additions to forest soil. *Appl. Soil Ecol.*, 34: 114–124. 10.1016/j.apsoil.2006.03.002

Redman R.S., Sheehan K.B., Stout R.G., Rodriguez R.J., and J.M. Henderson 2002. Thermotolerance generated by plant/fungal symbiosis. *Science*, 298: 1581.

Rodriguez R.J., Henson J., van Volkenburgh E., et al. 2008. Stress tolerance in plants via habitat-adapted symbiosis. *ISMEJ.*, 2: 404–416. https://doi.org/10.1038/ismej.2007.106

Rodriguez R.J., White Jr J.F., Arnold A.E., and R.S. Redman 2009. Fungal endophytes: diversity and functional roles. *New Phytol.*, 182: 314–330. http://doi.org/10.1111/j.1469-8137. 2009.02773.x

Santamaría-Hernando S., Rodríguez-Herva J.J., Martínez-García P.M., et al. 2018. Pseudomonas syringae pv. tomato exploits light signals to optimize virulence and colonization of leaves. *Environ. Microbiol.*, 20: 4261–4280. https://doi.org/10.1111/1462-2920.14331

Santhanam R., Luu VT, Weinhold A., Goldberg J., Oh Y., and I.T. Baldwin 2015. Native root-associated bacteria rescue a plant from sudden-wilt disease that emerged during continuous cropping. *Proc. Natl. Acad. Sci.*, 112: E5013–E5020. https://doi.org/10.1073/ pnas.1505765112

Sanz-Sáez Á., Erice G., Aguirreolea J., Irigoyen J.J., and M. Sánchez-Díaz 2012. Alfalfa yield under elevated CO_2 and temperature depends on the Sinorhizobium strain and growth season. *Env. Exp. Bot.*, 77: 267–273.

Savin M.C., Purcell L.C., Daigh A., and A. Manfredini 2009. Response of mycorrhizal infection to glyphosate applications and P fertilization in glyphosate-tolerant soybean, mean, and cotton. *J. Plant Nutr.*, 32: 1702–1717.

Schlatter D.C., Yin C., Hulbert S., Burke I., and T. Paulitz 2017. Impacts of repeated glyphosate use on wheat-associated bacteria are small and depend on glyphosate use history. *Appl. Env. Microbiol.*, https://doi.org/10.1128/AEM.01354-17

Schneider K.D., Lynch D.H., Dunfield K., Khosla K., Jansa J., and R.P. Voroney 2015. Farm system management affects community structure of arbuscular mycorrhizal fungi. *Appl. Soil Ecol.*, 96: 192–200.

Schwartz A.R., Morbitzer R., Lahaye T.,and B.J. Staskawicz 2017. TALE-induced bHLH transcription factors that activate a pectate lyase contribute to water soaking in bacterial spot of tomato. *Proc. Natl. Acad. Sci.*, 114: E897–E903.

Segonac C.and C. Zipfel 2011. Activation of plant pattern-recognition receptors by bacteria. *Curr. Opin. Microbiol.*, 14: 54–61.

Semenova-Nelson T.A., Platt W.J., Patterson T.R., Huffman J.,and B.A. Sikes 2019. Frequent fire reorganizes fungal communities and slows decomposition across a heterogeneous pine savanna landscape. *New Phytol.*, 222: 916–927.

Shakeri J., Foster H.A. (2007). Proteolytic activity and antibiotic production by *Trichodermaharzianum* in relation to pathogenicity to insects. *Enzyme Microb. Technol.*, 40: 961–968.

Sherwood J., German T. L., Moyer J. W., Ullman D. E., and A. E. Whitfield 2000. Tomato spotted wilt. In: *Encyclopedia of Plant Pathology*, ed. Maloy O. C., and T. D. Murray, 1030–1031. New York: John Wiley and Sons.

Sinclair T.R., Muchow R.C., Bennett J.M., and L.C. Hammond 1987. Relative sensitivity of nitrogen and biomass accumulation to drought in field-grown soybean. *Agron. J.*, 79: 986–991.

Smith F.A., and S.E. Smith 2011. What is the significance of the arbuscular mycorrhizal colonization of many economically important crop plants? *Plant Soil.*, 348: 63–79.

Smith S.E., and D.J. Read 2008. *Mycorrhizal Symbiosis* 3rd edn. San Diego: Academic Press.

Snapp S.S., Shennan C., and A.H.C. Van Bruggen 1991. Effects of salinity on severity of infection by Phytophthora parasitica Dast., ion concentrations and growth of tomato, *Lycopersicumesculentum* Mill. *NewPhytol.*, 119: 275–284.

St Clair D.A. 2010. Quantitative disease resistance and quantitative resistance loci in breeding. *Annu Rev. Phytopathol.*, 48: 247–268.

Swaraj K., and N.R. Bishnoi 1999. Effect of salt stress on nodulation and nitrogen fixation in legumes. *Indian J. Exp. Biol.*, 37: 843–848.

Szittya G., Silhavy D., Molnár A., et al. 2003. Low temperature inhibits RNA silencing-mediated defence by the control of siRNA generation. *EMBO J.* 22: 633–640. https://doi.org/10.1093/emboj/cdg74

Tang Z., Sun X., Luo Z., He N., and O.J. Sun 2018. Effects of temperature, soil substrate, and microbial community on carbon mineralization across three climatically contrasting forest sites. *Ecol. Evol.*, 8: 879–891. https://doi.org/10.1002/ece3.3708

Tiemann L.K., Grandy A.S., Atkinson E.E., Marin-Spiotta E., and M.D. McDaniel 2015. Crop rotational diversity enhances belowground communities and functions in an agroecosystem. *Ecol. Lett.*, 18: 761–771.

Trejo-Aguilar D., and J. Banuelos 2020. Isolation and culture of arbuscular mycorrhizal fungi from field samples. In: *Arbuscular Mycorrhizal Fungi: Methods and Protocols*, ed. Ferrol N. and L. Lanfranco, 1–16, *Methods in Molecular Biology*, vol 2146. https://doi.org/10.1007/978-1-0716-0603-2_1

Tripathi A.K., Nagarajan T., Verma S.C., and D. LeRudulier 2002. Inhibition of biosynthesis and activity of nitrogenase in Azospirillum brasilense Sp7 under salinity stress. *Curr. Microbiol.*, 44: 363–367.

Tuttle J. R., Idris A. M., Brown J. K., Haigler C. H., and D. Robertson 2008. Geminivirus-mediated gene silencing from Cotton leaf crumple virus is enhanced by low temperature in cotton. *Plant Physiol.*148: 41–50. 10.1104/pp.108.123869

Vannier N., Mony C., Bittebière A.K., and P. Vandenkoornhuyse 2015. Epigenetic mechanisms and microbiota as a toolbox for plant phenotypic adjustment to environment. *Front. Plant Sci.*, 6: 1159.

Váry Z., Mullins E., McElwain J.C., and F.M. Doohan 2015. The severity of wheat diseases increases when plants and pathogens are acclimatized to elevated carbon dioxide. *Glob. Chang. Biol.*, 21: 2661–2669.

Velásquez A.C., Castroverde C.D.M., and S.Y. He 2018. Plant and pathogen warfare under changing climate conditions. *Curr. Biol.*, 28: R619–R634. https://doi.org/10.1016/j.cub.2018.03.054

Vidotti M.S., Lyra D.H., and J.S. Morosini (2019). Additive and heterozygous (dis)advantage GWAS models reveal candidate genes involved in the genotypic variation of maize hybrids to Azospirillum brasilense. *PLoS One*, 14(9): e0222788.

Vignale M.V., Iannone L.J., and M.V. Novas 2020. *Epichloë* endophytes of a wild grass promote mycorrhizal colonization of neighbor grasses. *Fungal Ecol.*, 45. https://doi.org/10.1016/j.funeco.2020.100916

Vos C.M., Tesfahun A.N., Panis B., Waele D.D., and A. Elsen 2012. Arbuscular mycorrhizal fungi induce systemic resistance in tomato against sedentary nematode *Meloidogyneincognita* and the migratory nematode *Pratylenchuspenetrans*. *Appl. Soil Ecol.*, 61: 1–6.

Wang C., Cai X., and Z. Zheng 2005. High humidity represses Cf-4/Avr4- and Cf-9/Avr9-dependent hypersensitive cell death and defense gene expression. *Planta*, 222: 947–956.

Wang W., Barnaby J.Y., Tada Y., et al. 2011. Timing of plant immune responses by a central circadian regulator. *Nature*, 470:110–114.

Wang Y., Bao Z., Zhu Y., and J. Hua 2009. Analysis of temperature modulation of plant defense against biotrophic microbes. *Mol. Plant Microbe Interact.*, 22: 498–506.

Wang Z., Cui D., Liu C., et al. 2019. TCP transcription factors interact with ZED1-realted kinases as components of the temperature-regulated immunity. *Plant Cell Environ.*, 42: 2045–2056.

Whitfield A. E., Ullman D. E., and T. L. German 2005. Tospovirus-thrips interactions. *Annu. Rev. Phytopathol.* 43: 459–489.

Xin XF, Nomura K, Aung K, et al. 2016. Bacteria establish an aqueous living space in plants crucial for virulence. *Nature*, 539: 524–529.

Zabaloy M.C., Carné I., Viassolo R., Gómez M.A., and E. Gomez 2016. Soil ecotoxicity assessment of glyphosate use under field conditions: microbial activity and community structure of Eubacteria and ammonia-oxidizing bacteria. *Pest. Manag. Sci.*, 72: 684–691. 10.1002/ps.4037

Zaidi N.W., Dar M.H., Singh S., and U.S. Singh 2014. Trichoderma species as abiotic stress relievers in plants. In: *Biotechnology and Biology of Trichoderma*, ed. Gupta V.K. et al., 515–525. https://doi.org/10.1016/B978-0-444-59576-8.00038-2

Zhang W-W., Wang C., Xue R., and L-J. Wang 2019. Effects of salinity on the soil microbial community and soil fertility. *J. Integr. Agric.*, 18(6): 1360–1368.

Zhang X., Zhang R., Gao J., et al. 2017. Thity-one years of rice-rice-green manure rotations shape the rhizosphere microbial community and enrich beneficial bacteria. *Soil Biol. Biochem.*, 104: 208–217.

Zhang X., Zhang X., Singh J., Li D., and F. Qu 2012. Temperature-dependent survival of Turnip crinkle virus-infected arabidopsis plants relies on an RNA silencing-based defense that requires dcl2, AGO2, and HEN1. *J. Virol.*, 86: 6847–6854.

Zhou Y., Vroegop-Vos I., Schuurink R.C., Pieterse C.M.J., and S.C.M. Van Wees 2017. Atmospheric CO_2 alters resistance of Arabidopsis to Pseudomonas syringae by affecting abscisic acid accumulation and stomatal responsiveness to coronatine. *Front. Plant Sci.*, 8: 700.

Zhu Y., Qian W., and Hua J. 2010. Temperature modulates plant defense responses through NB-LRR proteins. *PLOS Pathog.*, 6: e1000844.

4 Vocabulary of Volatile Compounds (VCs) Mediating Plant-Microbe Interactions

Ekta Khare

CONTENTS

4.1 INTRODUCTION

The ubiquitous occurrence of volatile compounds (VCs) in ecosystems influences biological interactions among organisms. Many biogenic as well as anthropogenic sources are responsible for the emission of VCs (Wagner and Kuttler 2014; Liu et al. 2018; Yang et al. 2018). In 2014, the National Emissions Inventory (NEI) reported that vegetation and soil emitted biogenic VCs, among all other sources, contributed around 70% in the US (U.S. EPA 2018). In soil decomposing litter, organic waste materials, plant roots, associated mycorrhiza, and microorganisms are responsible

DOI: 10.1201/9781003213864-4

for the generation of VCs (Insam and Seewald 2010). The presence of VCs in the phytosphere is important to plants, as it stimulates production of organic nitrates (Monson and Holland 2001). In the atmosphere of soil, several microbial processes, including nitrification, nitrogen mineralization, denitrification, and methane oxidation are influenced by the presence of precise VCs (Bending and Lincoln 2000; Smolander et al. 2006).

Biogenic VCs are of immense value for the development of ecosystems. Volatile metabolites produced by the microbial and/or plant metabolism can be differentiated into inorganic and organic molecules. The most relevant inorganic volatile molecules include CO, CO_2, H_2, N_2, O_2, NH_3, H_2S, NO_2^-, SO_2, SO_3, and HCN now known for a wide variety of biological functions, such as defence compound, interspecies communication (e.g., quorum sensing/quenching), and antibiotic resistance (Avalos et al. 2019; Tilocca et al. 2020). Volatile organic compounds (VOCs) are characterized by a lipophilic moiety, odour, low molecular weight, low boiling point, and high vapour pressure. From a chemical point of view, VOCs are comprised of several molecular classes, including hydrocarbons, alcohols, thioalcohols, aldehydes, ketones, thioesters, cyclohexanes, heterocyclic compounds, phenols, and benzene derivatives (Chiron and Micherlot 2005; Morath et al., 2012). Signal communications mediated by VOCs are responsible for maturity of microbial communities (Kai et al. 2009). It is considered that VOCs as info-chemicals mediate communications within species, between species, and even between genus or kingdoms. These VCs act as a signal of environmental conditions, increasing ecological competence in plants and microorganisms. Adeptness in sensing and responding to VOCs increases microbial/plants survival rate in competitive and stressed environments. Bacterial VOCs are known for the effect on laccase activity of fungi (Mackie and Wheatley 1999). Antagonistic interactions are another imperative goal of VOCs. In 2003, Ryu et al. reported, for the first time, that bacterial VOCs can modulate the growth and health processes in plants even in stressed environment. This experiment with *Bacillus subtilis* GB03 found that VOCs induce the growth of *Arabidopsis thaliana* (Ryu et al. 2003).

Accordingly, microbial VCs can influence plant growth by affecting the microflora and their processes, increasing availability of nutrients, antagonistic interactions with pathogens, and induction of systemic resistance (Lee et al. 2012; Park et al. 2015; Tahir et al. 2017). One benefit VCs have over other biological preparations is that their mode of action does not require any physical contact with plant or microbes (Xie et al. 2014; Yunus et al. 2016). With the past studies on VCs, it became easy to understand the interaction between plant–plant or plant–microorganisms. The modern era requires such a valuable vocabulary of VCs that can be used in multiple ways for effective and sustainable agriculture. As smart technologies have improved, it has become possible to study, derive accurate information, and implement VCs in agriculture. This chapter aims to provide summarized information with all the facts known so far about the biogenic VCs useful for successful implementation of this treasure in contemporary agriculture.

4.2 PLANT VCs

4.2.1 CONCEPT AND CHEMICAL PROPERTIES OF PLANT VCs

The emission of VCs from various plant parts, viz., leaves, flowers, and roots, is a constitutive process. Exposure of plants to abiotic as well as biotic stresses enhances this emission of VCs (Kigathi et al. 2019). The rate of release of VCs differs between community levels genotypes, with the age of the plant and season, and even from leaf to leaf (Niinemets et al. 2010; Alves et al. 2014). For example; isoprenoid is constitutively emitted several shrubs and green tress however its rate of emission and concentration varies in response to various environment factors, such as light intensity and temperature (Monson 2013). Methanol emission is associated with the synthesis and breakdown of the pectin material of cell walls and, thus, is linked to the various plant stages from growth to senescence (Hüve et al. 2007).

Several abiotic stresses associated with water availability, flooding, high temperature, salt, and oxidative damage caused by ozone can induce VCs emissions (Vuorinen et al. 2004; Vallat et al. 2005; Wahid et al. 2007). The emission of VCs from plants can also be induced by several biotic factors from beneficial to pathogenic microbes and pollinators to herbivores. Upon mechanical injury by herbivores, monoterterpenes and green leaf volatiles (GLVs) are released instantly, and a subsequent systemic response leads to production of several VOCs (β-pinene, (E)-β-ocimene, α-humulene, (E,E)-αfarnesene, aldoxime phenylacetaldoxime, nitrile benzyl cyanide, etc.) from damaged as well as undamaged leaves with a delay of minutes, hours, days, or seasons (De Moraes et al. 2000; Irmisch et al. 2014). The invasion of fungal and bacterial pathogens is well known for induction of 3-octanone, methyl salicylate, α-farnesene, and 4,8-dimethylnona-1,3,7-triene (DMNT) emission (Yi et al. 2009; Niederbacher et al. 2015).

Around 1,700 VCs produced by plants can be grouped according to chemical structure into terpenoids, oxygenated VOCs (the common GLVs (methanol (CH_4O), acetone (C_3H_6O), acetaldehyde (C_2H_4O), their acetates), few sulphur compounds, and furanocoumarins (Dicke and Loreto 2010; Vivaldo et al. 2017). Terpenoids are synthesized by polymerization of C5 isoprenoid units and contribute to different scents in plants. Monoterpenes, diterpenes, and tetraterpenes are synthesized in plastids by polymerization of two, four, and eight isoprene units, respectively, through the methylerythritol pathway (MEP) (Rajabi Memari et al. 2013). Cytosol and endoplasmic reticulum are the sites of sesquiterpenes (three isoprene units) production by the mevalonic acid pathway (Rosenkranz and Schnitzler, 2013). GLVs can be of several configurational isomeric forms produced via the lipoxygenase (LOX) pathway from damaged plant tissues (Brilli et al. 2019). The shikimic acid pathway is used for biosynthesis of aromatic volatiles, such as methyl salicylate (Paré and Tumlinson 1996). Plants use VCs to achieve a variety of goals – ranging from intra- and interspecific plant communication to protection against abiotic stress factors – as a signal for interaction with pollinators and microorganisms to direct or indirect protection against pathogens or natural enemies (Karban et al. 2013; Kigathi et al. 2019; Mumm and Dicke 2010) (Table 4.1).

TABLE 4.1

Some Selected Examples of the Origin of Plant VOCs with Biological Activities

Origin	Volatile compound	Biological activity	References
Plants (all parts)	Ethylene	Induces shade avoidance response	Kegge and Pierik 2010
Damaged plants	(Z)-3-hexenol, (Z)- 3-hexenyl acetate	Induces the defence response in neighbour plants	Shiojiri et al. 2012
Plants (essential oil)	Citral, carvacrol, trans-2-hexenal	Fungistatic activity against *Monilinialaxa*	Neri et al., 2007
Plants (Secretory tissues, herbivore wounded tissues)	Methyl jasmonate	Induces SAR in plants against fungi, bacteria, and viruses	Shulaev et al. 1997; Song and Ryu 2018
Plants (wounded tissues)	Methyl salicylate	Required for SAR signal perception in systemic tissues	Webster et al. 2008; Schmidt-Busser et al. 2009
Damaged leaf	Indole	Primes SAR	Ameye et al. 2015
Damaged leaf	Z-3-hexenyl acetate	Primes SAR, reduces tissue damage due to cold stress	Cofer et al. 2018
Most plants (mesophyll cells)	Isoprene and monoterpenes	Protects plants against abiotic stresses	Bertamini et al. 2019
Thymus vulgaris	α-phellandrene, O-cymene, γ-terpinene and β-caryophyelene	Determines compounds involved in drought stress adaptation	Mahdavi et al. 2020
Necrotic lesions of leaf tissues and flowers	β-ionone	Repels *Phyllotreta cruciferae*, inhibits the sporulation and growth of the fungus *Peronospora tabacina,* and attracts pollinators	Bouvier et al. 2005; Gruber et al. 2009
Root secretory Cells	Vetiverol	Produced in root cells upon bacterial transformation, toxic to insects and mammals (rats, mice, rabbits)	Zhu et al. 2001; Bhatia et al. 2008

4.2.2 PLANT VCs IN PLANT–PLANT INTERACTIONS

In natural situations, plants do not grow as individuals but in community with conspecific and heterospecific plants. So, the studies on plant VCs should be based on communities, where plants interact and compete for light as well as nutrients (Grace and Tilman 1990) (Table 2.1). The release of VCs in the surrounding environment can be perceived by the different parts of same plant or by proximate neighbours (Frost et al. 2007). As light is the primary requirement, plants sense the competitors before becoming shaded through the perception of far-red (FR) light relative to red (R) light reflected by leaves of adjacent plants. Phytochrome photoreceptors (especially phytochrome B) play a prime role in FR light perception and detecting neighbouring plants (Smith 2000; Franklin 2008). Several reports indicate the role of ethylene as a volatile cue in shade avoidance. The release of ethylene levels increase in conditions of low R:FR and induce shade avoidance response as analysed using shade-insensitive tobacco cultivar (Pierik et al. 2004; Kegge and Pierik 2010). Below-ground competition between plants also uses volatile cues emitted by roots in the rhizosphere to identify neighbours and to avoid competitors' roots (Peñuelas et al. 2014, Schmidt et al. 2015).

Biotic stress-induced emission of VCs also plays an important role in plants' cross talk. Emitted VCs have a priming effect on parts of the same plant and neighbour plants and make them respond quickly to encounters with stressors, such as herbivores (Frost et al. 2008; Li et al. 2012; Erb et al. 2015). Shiojiri et al. (2012) reported the induction of defence responses in neighbour plants that were exposed to trace amounts of (Z)-3-hexenol and (Z)- 3-hexenyl acetate that were emitted by damaged plants. Another important point associated with VCs-induced defence response is that conspecific neighbours are more responsive than less genetically related plants (Karban et al. 2003). An experiment with *Artemisia tridentata* Nutt. showed that plants exposed to VCs released by genetically alike ramets or by conspecifics of same chemotype were less susceptible to damage by herbivores (Karban and Shiojiri2009,Karban et al. 2014). Damaged leaves emitted volatile cues that not only induce plant defences but can also inhibit seed germination in neighbourhood, thus, impacting the structure of the plant community (Karban 2007). Below ground, plant–insect interactions are also influenced by the presence of specific neighbours. *Taraxacum officinale* becomes more susceptible to the *Melolontha melolontha* larvae once it has been exposed to the root-emitted VCs from *Centaurea stoebe* (Schmid et al. 2015).

After VCs have been detected, receiver plants or plant parts start gene expression and synthesis of plant defence proteins and phytohormones as a part of their direct defence responses (Engelberth et al. 2004; Heil and Kost 2006; Sugimoto et al. 2014). It has been documented that jasmonic acid (JA) controls the emission of VCs associated with herbivore defence (Kost and Heil 2008). However, in competitive environments, low R-FR light checks the emission of JA inducible volatiles and in turn it attenuates the herbivores attraction mechanisms. (Kegge et al. 2013). Indirect defence mechanisms involve the volatile-cues-induced emission of VCs and the secretion of extrafloral nectar to attract hyperparasites of the third trophic level (Engelberth et al. 2004; Heil and Silva Bueno 2007; Li et al. 2012). Light quality also affects indirect defences controlled by JA and herbivore-induced synthesis of extrafloral nectar (Izaguirre et al. 2013). Perception of volatile cues emitted from neighbour plants is an important strategy for plants to modify their growth pattern in order to face ecological challenges.

4.2.3 PLANT VCs IN GROWTH REGULATORY AND DEFENSIVE RESPONSE AGAINST MICROORGANISMS

Plant-emitted volatiles can protect plants from pathogenic diseases either by direct antimicrobial action or as a signal for induction of defence responses against microbes (Table 4.1). Several reports confirm the antagonistic role of GLVs on germination and growth of plant pathogens. Neri et al. (2007) found that in vitro growth of *Monilinia laxa* (the causal agent of brown rot in stone fruit) was hampered by citral, carvacrol, and trans-2-hexenal. While (+)-limonene, a monoterpene, is quite effective in the control of necrotrophic fungi *Botrytis cinerea*, it stimulates growth of *Penicillium digitatum*. The same *P. digitatum* strain was found to be sensitive to citral (Simas et al. 2017). Plant volatiles are known for their action on fungi and bacteria by effecting membrane integrity and permeability (Sikkema et al. 1995). Lipophilic volatiles, such as GLVs and indole derivatives, change the membrane potential. Cytoskeleton integrity is also disturbed by indole volatiles (Mei et al. 2019). Terpenes integrate between acyl chains of membrane phospholipids and, therefore, are responsible for leakage of ions and some metabolites (Lambert et al. 2001). Joshi et al. (2016) reported that plant volatiles modulate the bacterial quorum sensing and consequently disrupt microbe–microbe communications. Though less information is available regarding antagonistic activity of VCs against plant pathogens, several studies reported plants emitting volatiles in response to infection by fungal and bacterial pathogens (Attaran et al. 2008; Sharifi et al. 2018).

Plants are able to prepare their defence systems for the forthcoming stress based on the memory of previous experiences (Crisp et al. 2016; Hilker et al. 2016). The contour of stress memory in plants is created by several factors known as "priming stimuli". VCs, being volatile in nature, can easily reach to the neighbouring environment and, thus, function as a critical priming agent (Heil and Kost 2006; Mauch-Mani et al. 2017). Methyl salicylate and methyl jasmonate are two volatile plant defence hormones that induce systemic defence response (SAR) in plants. Song and Ryu (2018) checked the priming effect of methyl salicylate by repeated application to uninfected *Nicotiana benthamiana* seedlings against *Pseudomonas syringae* and *Pectobacterium carotovorum*. Greater protection against pathogens is associated with an increase in expression of pathogenesis-related 1a (NbPR1a) and NbPR2 genes every time methyl salicylate treatment confirms its priming role for the enhancement in SAR capacity. Tobacco plants infected with tobacco mosaic virus emit methyl salicylate, which is reported to reduce the mosaic symptoms in neighbour plants (Shulaev et al. 1997). Induction of the defence response in plants follows the similar rhythm of methyl jasmonate treatment (Lundborg et al. 2019). Similar to their non-volatile analogues, methyl salicylate induces defence responses against biotrophic pathogens, while methyl jasmonate tends toward necrotrophic pathogens.

Indole volatiles emitted by grasses on herbivore damage also act as priming stimuli triggers in a cascade of events that start with production of reactive oxygen species to enhanced expression of defence genes on exposure to hemibiotrophic or necrotrophic pathogens. Another example of volatile is Z-3-hexenyl acetate, which is released from leaf tissue after mechanical injury primes the SAR in wheat plants against *Fusarium graminearum* (Ameye et al. 2015). VCs, such as benzothiadiazole, are also know to prime defence responses against bacterial pathogens (Yi et al. 2009). Reports on plant VCs priming defences against phytopathogens explain the possible implementation of VCs as green vaccines in agriculture (Luna-Diez 2016). VCs mediated priming for induction of SAR is a favourable process, as it does not require activation of metabolic pathways (van Hulten et al. 2006; Martinez-Medina et al. 2016) and, thus, is a sustainable approach for future agriculture production.

4.2.4 PLANT VCs IN DEFENCE AGAINST ENVIRONMENTAL STRESSES

An array of environmental conditions, viz., temperature, drought, humidity, salinity, ozone, light intensity, etc., comes under the category of abiotic stress factors modulating the emission of constitutive volatiles in positive or negative ways (Staudt and Lhoutellier 2011; Blande et al. 2014). Moreover, the brutality of abiotic stress determines the intensity and bouquet of volatile emission. Constitutively released volatiles act as antioxidants and can directly help plants to withstand abiotic stress by making their membranes more stable (Vickers et al. 2009; Possell and Loreto 2013). Indirectly, VCs act as signal molecules for volatile emitters and even neighbouring plants, reducing the effect of abiotic stress on plants (Gommers et al. 2013) (Table 4.1). Isoprene and monoterpenes are especially nominated for their role in protection of plants against abiotic stresses (Bertamini et al. 2019). Various stress factors induce defence pathways that combine at the stage of oxidative signalling, thus, inducing resistance to a diversity of agents (Fujita et al. 2006; Koornneef and Pieterse 2008).

Heat stress, especially in low water availability, leads to harmful effects on plants due to physical damage of tissue and changes in metabolism (Bita and Gerats 2013; Carvalho et al. 2015). Terpenes are known to protect plants until heat stress reaches up to as high as 40°C –45°C, which is the optimum temperature for enzyme activity in the MEP pathway involved in terpenes synthesis. Isoprenes and monoterpenes stabilize the thylakoid membrane lipid and protein interactions in high temperatures and, thus, increase photosynthetic thermotolerance. *In vitro* application of terpenes relieves thermal stress as evidenced by Copolovici et al. (2005). Generation of reactive oxygen species (ROS) due to the presence of strong solar radiation during heat stress ultimately leads to oxidative stress (Hasanuzzaman et al. 2013). Volatiles emitted because of high light intensity perform a crucial role as scavenger of ROS, thus, stabilizing membrane structure and functions (Zuo et al. 2019). For example: Isoprene is a volatile which is known for antioxidant activity due to its ability to react with hydroxyl radicals a form of ROS and ozone (Loreto et al. 2001). Several GLVs, including Z-3-hexenyl acetate, are known to release quickly in response to mechanical damage of leaf tissues, reducing tissue damage due to cold stress in maize plants (Cofer et al. 2018).

Generation of ROS and free radicals increases during drought conditions and negatively affect membrane integrity (Ionenko and Anisimov 2001; Munns 2002). According to the report by Timmusk et al. (2014), emission of several VCs, such as monoterpenes, geranyl acetone, and benzaldehyde, are increased in wheat plants when they are under drought stress. An experiment on isoprene-emitting transgenic *Nicotiana tabacum* showed no significant increase in ROS and lipid peroxidation levels as compared to plants negative for isoprene emission, which are severely affected by generation of ROS under drought conditions (Ryan et al. 2014). A study on drought tolerant *Thymus vulgaris* plant revealed that α-phellandrene, O-cymene, γ-terpinene, and β-caryophyelene are among the most determinant compounds involved in drought stress adaptation (Mahdavi et al. 2020). On the other hand, water logging or flooding creates hypoxia that induces plants to emit ethylene. Increased levels of ethylene lead to elongation of submerged plant parts and development of new adventitious roots, leaf epinasty, aerenchyma tissues, etc. (Rani et al. 2017). Though plants have defence system to fight for survival under different abiotic stress conditions, understanding the role VCs play in this complicated process requires multi-directional thinking because plants always face more than one stress at a time.

4.3   MICROBIAL VCs (mVCs)

4.3.1   CONCEPT AND CHEMICAL PROPERTIES OF MICROBIAL VCs

VCs are produced during the primary and secondary metabolism of microorganisms. Until recently, these compounds remained largely unexplored. Some VCs produced as secondary metabolite (by-products) of the primary metabolic process of an organism (plant/bacteria/fungi etc) such as nucleic acids, amino acids, and fatty acids synthesis (Schulz-Bohm et al. 2017). Living systems make use of fundamentally the same catalysed enzyme reactions in primary metabolism, hence, the majority of volatiles released by organisms are essentially similar. The deficiency of primary carbon and nitrogen sources, together with other disorders, induces a secondary metabolism that leads to the synthesis of various microbial volatile compounds (mVCs) (Bjurman 1999; Korpi 2001). Production of volatiles depends on several aspects, such as physiological condition of the microbe, availability of nutrients, pH, temperature, humidity etc. (Korpi et al. 2009; Insam and Seewald 2010). mVCs can participate in chemical reactions in the environment and can be converted to other compounds, viz., alcohols to aldehydes to carboxylic acids and ketones to aldehydes (Wilkins et al. 1997; Atkinson et al. 2000).

VCs produced by microorganisms belong to chemical classes of alkanes, alkenes, alcohols, ketones, esters, benzenoids, pyrazines, sulphur compounds and terpenes (Schulz and Dickschat 2007; Kanchiswamy et al. 2015a) (Figure 4.1). Characteristics or specific volatile compound profiles depend on the precise metabolism type of that microorganism (Kai et al. 2009). Specific groups of phylogenetically related genus or species produce similar volatile compound characteristics (Insam and Seewald 2010). Species-specific differences in production of VCs, especially dimethyldisulfide, dimethyltrisulfide, and isoprene, were reported for *Pseudomonas* spp., *Serratia* spp., and *Enterobacter* spp. (Schöller et al. 1997). Several studies reported the involvement of the GacS/GacA two-component regulatory system in production of certain VCs by bacteria (Cheng et al. 2016; Ossowicki et al. 2017). As is the case with bacteria, fungi also produce several common VCs among phylogenetic groups in addition to some specific or unique volatiles (Schnürer et al. 1999).

Though produced as by-products, recent research has supported the biological activities of mVCs (Schmidt et al. 2015; Tyc et al. 2017a). A comparative study on the role of volatile and non-volatile metabolites in microbial interactions revealed higher performance of mVCs (Tirranen and Gitelson 2006). Microbial volatiles are also

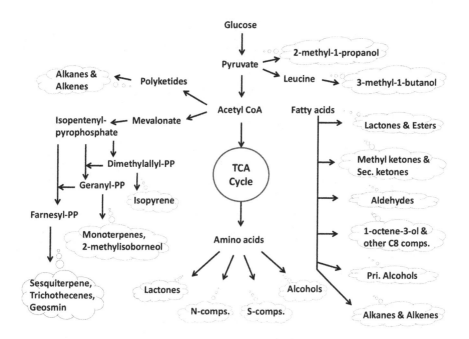

FIGURE 4.1 Metabolic pathways for the production of some microbial volatile organic compounds (mVOCs) (Gadd 1988; Sunesson 1995; Korpi 2001; Simon et al. 2017). Abbreviations: CoA = coenzyme A, N = nitrogen, PP = pyrophosphate, S = sulphur, TCA = tricarboxylic acid.

critical for various processes related to plant health through the induction of proper response to neighbouring organism and environmental conditions. mVCs released in the rhizosphere modulate interactions between phytopathogens and their host plants and are reported to exhibit antagonistic properties (Vespermann et al. 2007; Martín-Sánchez et al. 2020) (Table 4.2). Different organisms respond to mVCs in multipartite ways and setup complex trophic interactions. Thorough and systematic investigations into mVCs will provide methods to address agricultural as well as environmental issues.

4.3.2 MICROBIAL VCs IN MICROBE–MICROBE INTERACTIONS

Bacterial volatile metabolites are known mainly for their inhibitory effect on bacteria's vital functions that sometimes reach a bactericidal level (Table 4.2). Far fewer studies have been conducted on mVCs ability to stimulate than on their ability to inhibit (Tirranen and Gitelson 2006). Several reports showed that bacterial volatiles inhibited the growth activity of most of the *Burkholderia cepacia* complex strains (Papaleo et al. 2012; Orlandini et al. 2014). Dimethyl disulphide is a principal volatile produced by certain rhizospheric bacteria, viz., *Pseudomonas fluorescens* and *Serratia phymuthica* known for their bacteriostatic effect on the phytopathogens *Agrobacterium tumefaciens* and *A. vitis* (Dandurishvili et al. 2011). Gürtler et al. (1994) reported the antagonistic

TABLE 4.2

Selected Examples of the Origin of Microbial VOCs with Biological Activities

Origin	Volatile compound	Biological activity	References
Rhizospheric bacteria, viz., *Pseudomonas fluorescens* and *Serratia phymuthica*	Dimethyl disulphide	Bacteriostatic effect on *Agrobacterium tumefaciens* and *A. vitis*, provides sulphur to plants, inhibits growth of *Sclerotinia minor* by damaging its membrane, and induces systemic resistance in host plants	Dandurishvili et al. 2011; Takahashi et al. 2011; Tyagi et al. 2020
Streptomyces albidoflavus	Albaflavenone, Trimethylamine	Antagonistic role against *Bacillus subtilis*	Gürtler et al. 1994; Jones et al. 2017
Collimonas partensis	β-pinene	Antimicrobial (antifungal, antibacterial)	Song et al. 2015
Bacillus pumilus and *Paenibacillus* sp.	2,5-dimethyl pyrazine, 1-octen-3-ol	Inhibits growth of *Phaeomoniella chlamydospora*	Enebe and Babalola 2019
Burkholderia ambifaria	3-Hexanone	Promotes plant growth	Groenhagen et al. 2013
Bacillus subtilis, *B. amyloliquefaciens* and *Pseudomonas chlororaphis*	2,3-butanediol	Synthesizes virulence factors in *Pseudomonas aeruginosa* and *Pectobacterium carotovorum*, triggers ISR in plants and tolerance to drought	Cho et al. 2008; Lee et al. 2012; Audrain et al. 2015
Bacillus subtilis and *Bacillus amyloliquefaciens*	Acetoin	Enhances total leaf surface area, triggers ISR	Ryu et al. 2003; Rudrappa et al. 2010
Plant growth-promoting bacteria	Dimethylhexadecyl-amine (DMHDA)	Involved in development of root and root hair	Castulo-Rubio et al. 2015
Plant growth promoting bacteria	Indole	Modulates auxin signalling pathway	Bailly et al. 2014
Bacillus subtilis FB17	3-hydroxy-2-butanone	Triggers ISR in *Arabidopsis thaliana*	Rudrappa et al. 2010
Bacillus amyloliquefaciens GB03	Acetoin, cyclohexan, dodecane, undecane, hexadecane, benzaldehyde, 2-butanone-3metioxy-3 methyl	Enhances biosynthesis of secondary metabolites and the antioxidant level in *Mentha piperita* L. grown under salinity stress conditions	Cappellari et al. 2020
		Fungal VOCs	
Trichoderma viride and *T. harzianum*	6-Pentyl-α-pyrone	Suppresses seedling blight and phytotoxicity during seedling formation and triggers ISR	Hung et al. 2015; Kottb et al. 2015

(Continued)

TABLE 4.2 (CONTINUED)

Selected Examples of the Origin of Microbial VOCs with Biological Activities

Origin	Volatile compound	Biological activity	References
Trichoderma spp.	2,3-hexanal, 2,3-hexenol, E-2-hexenal	Systemic acquired resistance	Nawrocka et al. 2018
Fusarium oxysporum	β-Caryophyllene	Antimicrobial, plant growth promotion	Minerdi et al. 2011
Hypoxylon sp.	1,8-Cineole	Antifungal	Tomsheck et al. 2010
Most fungi	2-Methyl-1-propanol	Fungivore attractant	Schalchli et al. 2014
Irpexlacteus	5-Pentyl-2-furaldehyde	Anti-fungal against *Bulmeria graminis, Fusarium oxysporum, Colletotrichum fragarie, Botrytis cinerea*	Hung et al. 2015
Botrytis cinerea, Penicillium expansum	Geosmin	Microbe detection in *Drosophila*	Stensmyr et al. 2012
Candida albicans	Farnesol	Apoptosis in *Aspergillus nidulans*, altered morphology and reduced fitness in *Fusarium graminearum*	Semighini et al. 2006; Semighini et al. 2008
Most fungi	1-octen-3-ol	Fungal spore production, inhibition and induction, insect attractant, and repellent, ISR	Kishimoto et al. 2007; Berendsen et al. 2013
Epichloe sp.	Chokol K.	Insect attractant	Steinebrunner et al. 2008
Ampelomyces sp. and *Cladosporium* sp.	m-cresol and methyl benzoate	Induced systemic resistance	Naznin et al. 2014
Pichia kudriavzevii, Pichia occidentalis, and *Meyerozyma guilliermondii/Meyerozyma caribbica*	Ethyl esters of medium-chain fatty acids, phenylethyl alcohol, and its acetate ester	Antagonistic activity against *Mucor* spp., *Penicillium chrysogenum, Penicillium expansum, Aspergillus flavus, Fusarium cereals, Fusarium poae, Botrytis cinerea*	Choińska et al. 2020

role of albaflavenone, a sesquiterpene synthesized by *Streptomyces albidoflavus* against *Bacillus subtilis*. Trimethylamine released by *Streptomyces* exhibits antibacterial action against *B. subtilis* and *Micrococcus luteus*, which might be due to an increase in the pH of the growth medium (Jones et al. 2017). Exposure to volatiles produced by *Veillonella* sp. and *Bacteroides fragilis* inhibited the growth of enteropathogenic bacteria (Hinton and Hume 1995; Wrigley 2004). The compounds 2,5-dimethyl pyrazine and 1-octen-3-ol produced by *Bacillus pumilus* and *Paenibacillus* sp. inhibit growth of *Phaeomoniella chlamydospora*, which is the causal agent of grapevine trunk disease (Enebe and Babalola 2019).

In contrast to the antagonistic activity of bacterial VCs, *Collimonas pratensis* and *Serratia plymuthica* emitted volatiles that induce the growth of *Pseudomonas fluorescens* Pf0-1 (Garbeva et al. 2014). VCs produced by microorganisms affect the motility and pathogenicity of bacteria. The compounds 2,3 butanediol and acetoin are necessary for the synthesis of virulence factors in *Pseudomonas aeruginosa* and *Pectobacterium carotovorum* (Audrain et al. 2015). Exposure to mVCs modulate the expression of genes, e.g., global virulence regulator PhcA, the type III secretion system, and extracellular polysaccharide (EPS) production known for bacterial virulence (Schulz-Bohm et al. 2017). mVCs also modulate the resistance of bacteria to antibiotics. Ammonia, trimethylamine, hydrogen sulphide, nitric oxide, and 2-amino-acetophenone affect the biofilm formation/dispersal and motility of bacteria, thus, altering the resistance to antibiotics (Audrain et al. 2015; Raza et al. 2016b). VCs produced by several bacteria inhibit the germination or growth of fungal propagules (Garbeva et al. 2011). Cordovez et al. (2015) reported the antifungal activity of volatiles from *Streptomyces* spp. against *Rhizoctonia solani*. Recently, antifungal activity of VCs from endophytic *B. subtilis* isolated from *Eucommia ulmoides* was reported against *Curvularia lunata* (Xie et al. 2020).

Fungi also produce volatile metabolites that positively or negatively modulate various activities of bacteria, such as biofilm formation, motility (both swimming as well as swarming), and growth alteration. In this way, fungi, through production of VOCs, employ a strategy to attract mutualistic bacteria and repel competitors (Piechulla et al. 2017). For instance, *Trichoderma atroviride* produces volatiles that increase the synthesis of 2,4-diacetylphloroglucinol by *Pseudomonas fluorescens* by increasing the expression of the *phlA* gene (Lutz et al. 2004). Fungal volatiles are known for their inhibitory effect against fungi. *Trichoderma* spp. produced VCs having antagonistic activity against *Fusarium oxysporum*, *Rhizoctonia solani*, *Sclerotium rolfsii*, *S. sclerotiorum*, and *Alternaria brassicicola* (Amin et al. 2010). Recently, Li et al. (2018) reported that volatiles released by *F. oxysporum* induce *Trichoderma* spp. to produce a significantly higher amount of antifungal metabolites. Choińska et al. (2020) reported that the VCs released by the yeast strains isolated from organic grapes and rye grains acted antagonistically against plant pathogenic moulds. They reported that fungal VOCs acted as quorum sensing signal molecules to regulate growth and germination of conspecific mycelia and spores (Nemcovič et al. 2008; Stoppacher et al. 2010) (Table 4.2). In their proteomic study, Fialho et al. (2016) reported the obstructing role of fungal VOCs on vital metabolic pathways responsible for inhibitory effect on fungal growth.

4.3.3 MICROBIAL VCs IN STIMULATION OF PLANT GROWTH

Plants' exposure to biogenic VCs from microorganisms is a ubiquitous phenomenon. After the first report by Ryu et al. (2003), several researchers reported VOCs' ability to increase plant cell size, leaf size and number, fruit and seeds yield, lateral roots, and root hair development (Table 4.2). Hormonal signalling is one factor responsible for plant growth and health improvement under the influence of VOCs. Several physiological activities of plants, viz., nutrients uptake, photosynthesis, and storage of sugar, are also enhanced under the influence of microbial volatiles (Ryu et al. 2004; Minerdi et al. 2011; Sánchez-López et al. 2016; Tahir et al. 2017). VCs emitted by both bacteria and fungi are known for their modulating effect on the expression of several genes, including expansin, that deal with cell wall extension and rigidity, which increase cell size (Zhang et al. 2007, Minerdi et al. 2011).

Dimethylhexadecylamine (DMHDA) and indole are the common volatiles produced by plant growth-promoting rhizobacteria (PGPR) involved in the development of primary roots, the increase in length and number of lateral roots, and the enhanced density of root hairs (Bailly et al. 2014; Castulo-Rubio et al. 2015). Pathogenic *Fusarium oxysporum* produces VOCs that affect auxin transport and signalling in *A. thaliana* and *Nicotiana tabacum* (Bitas et al. 2015). The auxin signalling pathway modulated by volatile indole controls the secondary root development in *A. thaliana* (Bailly et al. 2014). However, in contrast to the auxin signalling pathway, root hair length enhancement is under the control of the reactive oxygen species (ROS)-dependent mechanism. An alteration in the sesquiterpenes profile for the below-ground portion and synthesis of superoxide anion radicals (O_2^-) in the roots elicit the enlargement of root hairs in *A. thaliana* (Ditengou et al. 2015).

Bacterial VCs improve the iron uptake by plants that required chlorophyll for synthesis and, thus, chlorophyll content. The photosynthesis rate is also enhanced via modulation of the photosystem and electron transport chain activity (Briat 2007). Iron solubility increases because volatiles induce the release of three times more protons and, thus, reducing the pH of the rhizosphere. Expression of Fe-deficiency-inducing transcription factor 1 (FIT1) is also upregulated by volatiles, inducing the expression of ferric reduction oxidase 2 (FRO2) and iron regulated transporter 1 (IRT1) genes essential for iron uptake from soil (Sharifi and Ryu 2018). *Alternaria alternata*, a plant pathogen, emits VOCs that are known to stimulate growth, flowering, and photosynthesis by increasing plastidic cytokinin in *Arabidopsis*, maize, and pepper plants (Sanchez-Lopez et al. 2016).

Microbial volatiles not only promote plant growth but are also known for their inhibitory activity (Vespermann et al. 2007; Kai et al. 2008; Kai et al. 2010; Liu and Zhang 2015). Several fungi, for example endophyte *Muscodoryu catanensis*, produce VOCs that are toxic for amaranth, tomato, and barnyard grass roots and inhibit seed germination. However, these VOCs might help the host plant to compete with other plants in close vicinity (Macias-Rubalcava et al. 2010). As reported by several researchers, VOCs from several members of *Burkholderia*, *Chromobacterium*, *Pseudomonas*, *Serratia*, and *Stenotrophomonas* inhibit plant growth or may extend to phytotoxic levels (Kai et al. 2009; Bailly and Weisskopf 2012). During a study on the effect of volatiles emitted from *Serratia plymuthica* and *Stenotrophomonas maltophilia* on *Arabidopsis thaliana*, Wenke et al. (2012) found growth inhibition, mediated by WRKY18, a member of WRKY transcription factor family, that regulates several developmental processes of plants. Soil-borne bacterial pathogens might produce certain VOCs that act as effectors and suppress host plant defence responses (Blom et al. 2011). Interestingly, the accumulation of growth-inducing volatiles above a certain threshold, or change in plant growth conditions, might cause the opposite effect of VOCs on plants (Liu and Zhang 2015). VC indole is one such example for which low concentrations induce growth, but high levels kill plants (Blom et al. 2011). Dimethyl disulphide (DMDS) is also known for both growth promotion and inhibitory activity (Kai et al. 2010; Meldau et al. 2013; Tyagi et al. 2020). Thus, the outcome for plants exposed to microbial VOCs depends on the components of volatile blend and duration of exposure in particular environmental conditions and stage of plant development (Lee et al. 2016).

4.3.4 Microbial VCs in Elicitation of the Plant's Defence System

Microbial VCs indirectly support plant growth by eliciting a defence response against biotic stressors (Ryu et al. 2003) (Table 4.2). It is well known that several bacterial VOCs, such as DMDS and 2-methylpentanoate, are toxic to phytopathogens (Groenhagen et al. 2013; Cordovez et al. 2015; Raza et al. 2016a; Ossowicki et al. 2017), while volatiles, such as acetoin, 2,3-butanediol, and tridecane, trigger induced systemic resistance (ISR) in plants against numerous pathogens (Lee et al. 2012). In the natural environment, the volatiles' elicitation of ISR appears to be an imperative mechanism for disease repression (Sharifi and Ryu 2016). The compound 2,3-butanediol is a major VC produced by *Bacillus subtilis* and *B. amyloliquefaciens* but follows different pathways to elicit the defensive system in *Arabidopsis thaliana* against *Erwinia carotovora*. Volatiles from *B. subtilis* induce ethylene dependent pathways against *E. carotovora*, while VOCs from *B. amyloliquefaciens* does not require this signalling pathway (Ryu et al. 2004). *Pseudomonas chlororaphis* also releases 2,3-butanediol and induces resistance in tobacco plants against same pathogen *E. carotovora* (Han et al. 2006). Several rhizospheric plant growth-promoting microbes produce 2,3-butanediol of three possible stereoisomers with two enantiomers 2R,3R and 2S,3S, and a meso-type 2R,3S. Out of the three isomers, 2R,3R and 2R,3S butanediol most effectively induce systemic resistance in pepper plants against cucumber mosaic virus (CMV) and tobacco mosaic virus (TMV) by priming salicylic acid, jasmonic acid, and ethylene-dependent signalling pathways (Kong et al. 2018). *B. subtilis* FB17 releases 3-hydroxy-2-butanone that elicits systemic resistance in *A. thaliana* against *Pseudomnas syringae* pv. *tomato* DC3000(Pst) that involves salicylic acid and ethylene pathways (Rudrappa et al. 2010).

VCs produced by soil fungi also prime plant defence systems against pathogens' attack (Table 4.2). *Ampelomyces* sp. and *Cladosporium* sp. emit m-cresol and methyl benzoate (MeBA) that also protect *A. thaliana* against Pst (*Pseudomonas syringae*). The molecular defence mechanism was determined using mutants and transgenic

Arabidopsis plants impaired in various signalling pathways and revealed combined SA-and JA-dependent pathways induced by m-cresol while MeBA elicits mainly JA-signalling pathways with SA-signals playing a minor role (Naznin et al. 2014). Exposure of *A. thaliana* to 1-octen-3-ol (also known as mushroom alcohol) upregulates defence genes and protects plant from *Botrytis cinerea* (Kishimoto et al. 2007). Volatile 6-pentyl-α-pyrone produced by *Trichoderma* spp. activates SA-dependent defence pathways in *A. thaliana* against *B. cinerea* and *Alternaria brassicicola* (Kottb et al. 2015). Fungal VOCs 2,3-hexanal, 2,3-hexenol, and E-2-hexenal from *Trichoderma* spp. mainly contribute to the generation of systemic acquired resistance in plants through the enhanced expression of SA defence genes PR1 and PR5 (Nawrocka et al. 2018).

Endophytic fungi also produce VCs that indirectly help plants defend against various pathogens (Morath et al. 2012, Kaddes et al. 2019). *Oxyporuslate marginatus* produces volatiles antagonistic to phytopathogenic fungi *Alternaria alternata*, *Colletotrichum gloeosporioides*, and *Fusarium oxysporum* f. sp. *lycopersici* that infect a broad range of plants (Lee et al. 2009). Mercier and Manker (2005) reported that VCs from endophyte *Muscodor albus* controls *Rhizoctonia solani* (causal agent of damping-off disease in broccoli and root rot of bell pepper caused by *Phytophthora capsici*). In the natural environment, plants are exposed to a blend of VCs from numerous microorganisms, so the result of their interactions with plants is quite complicated. Tahir et al. (2017) provide a report on interaction of VOCs from *B. subtilis* and the pathogen *Ralstonia solanacearum* and their individual and combined effects on tobacco plants. Individual exposure of tobacco plants with *B. sublilis* VOCs induce systemic resistance in plants and promote growth through increased activities of polyphenol oxidase (PPO) and phenylalanine ammonia lyases (PAL), while the ethylene pathway causes downregulation in the expression of genes. Volatiles produced by *R. solanacearum* do not show significant effects on plant growth. Plant exposure to VOCs of both bacteria reduces the growth-promoting potential of *B. subtilis* volatiles. Both bacteria inhibit growth were found to inhibit the growth of each other when grown in co-culture condition; however, *B. subtilis* superimposes the inhibitory activity of pathogen *R. solanacearum*. Still more studies are required to understand the fate of applied volatiles on plants for agriculture production.

4.3.5 MICROBIAL VOCs IN INDUCTION OF PLANT TOLERANCE TO ABIOTIC STRESSES

Certain rhizospheric microorganisms produce volatiles that induce physicochemical changes in plants to generate tolerance to abiotic stresses, termed as induced systemic tolerance (IST) (Farag et al. 2013) (Table 4.2). Beneficial microorganisms modulate proline content, antioxidants level, synthesis of hormones and reduced accumulation of Na^+ in plants, which are all common factors of tolerance to abiotic stresses (Liu and Zhang 2015; Ngumbi and Kloepper 2016; Sharifi and Ryu 2017). Bacterial volatiles are known to induce plant tolerance for salinity and drought.

Volatile metabolites from *Bacillus subtilis* GB03 increase the accumulation of choline and glycine betaine in *Arabidopsis* that protect plants against abiotic stresses through osmoregulation (Zhang et al. 2010). Salinity conditions cause osmotic stress because of excessive buildup of Na^+ concentration in plants. Exposing soybean to VOCs emitted by *Pseudomonas simiae* leads to a decrease in Na^+ level in the roots and augments the accumulation of proline that protect plants from osmotic stress (Vaishnav et al. 2015). The *Arabidopsis* HKT1 gene encodes for high affinity xylem parenchyma expressed Na^+ transporter (Sunarpi et al. 2005; Horie et al. 2009). Exposing *Arabidopsis* to *B. subtilis* VOCs modulates the expression of HKT1 in a tissue-specific manner that is downregulated in roots and upregulated in shoots to lessen the overall Na^+ accumulation, resulting in increased tolerance to salinity stress (Zhang et al. 2008). A recent study reported the impact of mVOCs emitted by *B. amyloliquefaciens* GB03 on enhanced biosynthesis of secondary metabolites and the antioxidant level in *Mentha piperita* L. grown under salinity stress conditions (Cappellari et al. 2020).

Tolerance to drought in *A. thaliana* was observed after exposing it to 2,3-butanediol emitted by *Pseudomonas chlororaphis*. The isomeric form 2R,3R of 2,3-butanediol induces systemic tolerance to drought through induction of salicylic acid, jasmonic acid, and ethylene signalling pathways (Cho et al. 2008). Dehydration is the outcome of osmotically stressed plants exposed to salinity, drought, or cold environments. Dehydration stress conditions induce plants to accumulate osmoprotectants to increase osmotic pressure of cells that lower the free-water potential and, hence, reduce water loss required for the stability of the structure of proteins and membranes (Yancey 1994). Choline and glycine betaine are important solutes required for tolerance of dehydration induced osmotic stress. Volatiles released by *Bacillus subtilis* GB03 induce the gene expression for enzyme phosphoethanolamine N-methyltransferase (PEAMT) that are required to synthesize choline and glycine betaine (Mou et al. 2002; Zhang et al. 2010). To combat the low water condition, plant utilize the abscisic acid signalling pathway. However, this case was not found for *B. subtilis* GB03 volatiles-induced tolerance to dehydration in *Arabidopsis* (Zhang et al. 2010, Kang et al. 2014). Bacterial VCs enhance the ability of bacteria to synthesize exopolysaccharide, as it has been reported for *P. aeruginosa*. Exopolysaccharides increase the water-holding capacity of soil, thus, VCs can indirectly enhance tolerance of plants to dehydration conditions (Naseem and Bano 2014, Chen et al. 2015).

Mineral nutrients are essential for plant growth and development, thus, their deficiency is deleterious for plants. As discussed in Section 4.3.3, Fe uptake by plants enhances exposure to VCs emitted by several microorganisms. Fe transporter IRT1, in addition to Fe, uptakes other metal ions, such as cadmium (Cd), from the soil (Nishida et al. 2011), and Cd stress conditions may worsen the plant's toxicity (Liu and Zhang 2015). Plants absorb sulphur from the soil in the form of SO_4^- that reduces the energy consuming process for assimilation. Plant growth reduction due to sulphur deprivation is also ameliorated by volatiles such as DMDS (Meldau et al. 2013; Sharifi and Ryu 2018). DMDS provides sulphur to plants without needing to expend any energy, unlike the case of SO_4^- absorption (Takahashi et al. 2011).

4.4 CONCLUSION AND FUTURE PROSPECTS

Plants and microbes interact with each other using VCs as a chemical language. VCs, either of plant or microbial origin, modulate the physiology and hormonal signalling pathways of plants to combat biotic and abiotic stresses. Plants exposed to microbial VCs grow healthier in terms of growth parameters such as root volume, leaf number and size, flower number, and fruit and seed production. This picture of VCs attracts attention for application in bio-farming. However, our knowledge still mainly depends upon in vitro trials. To be applicable for comprehensive agriculture and horticulture, VCs must be effective in open field conditions. So far, the use of plant VCs follows the strategy of intercropping in which VC-emitting plants species that repel herbivores are surrounded by plants whose volatiles attract herbivores, thus, keeping them away from the field (Picket and Khan 2016; Brilli et al. 2019). Song and Ryu (2013) conducted a study in an open field environment that showed the effectiveness of microbial VCs in triggering the defensive system of plants against pathogens and herbivores. Volatiles can be applied in the field through foliar spray or soil dumping (Kanchiswamy et al. 2015b), but it still requires an optimized method of field application. Mycofumigation is also an effective strategy that makes use of antimicrobial VCs for the control of post-harvest vegetables and fruits diseases (Gomes et al. 2015).

Despite of outstanding success in understanding volatiles' origin, properties, and possible applications, there is still much to explore before implementation as an agro-production strategy. It is an unavoidable fact that little information is available about the synthesis pathways of VCs, though omic studies are on the way to correlate gene expression to volatile synthesis. The next upcoming question is on the molecular mechanism of the perception of volatiles by plants. An enormous number of reports confirms that VCs modulate plant physiological and defence pathways. However, studies need to be conducted to determine whether microbial VCs provoke alteration on volatile biosynthetic pathways of plants. This possibility cannot be ruled out, as there are several reports on alteration in VCs of plants due to plant-associated bacteria and fungi (Sharifi et al. 2018). Most of the studies conducted so far are based on one-sided effects of VCs produced by a single microorganism on other organisms on the perceiving hand, neglecting the fact that interaction is a two-sided event. When we talk about utilization of VCs in agriculture, we must concern rhizosphere microbiota and macrobiota (e.g., nematodes, earthworms) that may also emit volatiles, which is an abandoned part of this approach. Hence, there is a need for a holistic approach that considers the fact that the soil community can alter the predictable outcome of applied volatiles, or VCs treatment might affect the natural soil community structure.

Furthermore, most of the studies describe the effect of a particular volatile compound and its effective concentration for getting a desirable result. However, living organisms produce a blend of VCs, and their individual concentrations may vary depending on the environment conditions. So, for maximum benefit, a mixture of volatiles can be used with the condition that the volatiles in the blend must be compatible and work synergistically. Before selection of volatiles for their beneficial features to plants, we must evaluate side effects on non-target organisms, including humans.

For instance, DMDS promotes development of plant roots and uptake of sulphur from soil (Meldau et al. 2013), but it adversely effects nematodes and *Drosophila melanogaster* (Popova et al. 2014). Another volatile, 1-octen-3-ol, is important for induction of systemic resistance to phytopathogens and promotion of plant growth (Kishimoto et al. 2007); however, it negatively affects the eyes and respiratory system of humans (Araki et al. 2010). Microbial VCs can indirectly affect human health by enhancing the virulence of human pathogenic bacteria or by increasing pathogen resistance to antibiotics. Though, several studies have showed the potential of VCs in plant growth promotion and defense, till date there is no report for its practical application in field. Advancement in technology to analyse the profile of volatile organisms, genomics, and metabolomics and tools to study changes at the molecular and physiological level will greatly increase our knowledge on this chemical ecology frontier. This will make possible the restoration and fortification of plant-microbes interactions and lessen the requirement of agrochemicals.

ACKNOWLEDGEMENTS

The author is grateful for the support from the Vice Chancellor of Chhatrapati Shahu Ji Maharaj University, Kanpur, UP, India.

REFERENCES

Alves, E. G., P. Harley, J. F. D. Goncalves, C. E. D. Moura, and K. Jardine. 2014. Effects of light and temperature on isoprene emission at different leaf developmental stages of *Eschweilera coriacea* in central Amazon. *Acta Amazonica* 44:9–18.

Ameye, M., K. Audenaert, N. De Zutter, K. Steppe, L. Van Meulebroek, L. Vanhaecke, D. De Vleesschauwer, G. Haesaert, and G. Smagghe. 2015. Priming of wheat with the green leaf volatile Z-3-hexenyl acetate enhances defense against *Fusarium graminearum* but boosts deoxynivalenol production. *Plant Physiol* 167:1671–84.

Amin, F., V. Razdan, F. Mohiddin, K. Bhat, and P. Sheikh. 2010. Effect of volatile metabolites of *Trichoderma* species against seven fungal plant pathogens in-vitro. *J Phytol* 2:34–7.

Araki, A., T. Kawai, Y. Eitaki, A. Kanazawa, K. Morimoto, K. Nakayama, E. Shibata, M. Tanaka, T. Takigawa, T. Yoshimura, H. Chikara, Y. Saijo, and R. Kishi. 2010. Relationship between selected indoor volatile organic compounds, so-called microbial VOC, and the prevalence of mucous membrane symptoms in single family homes. *Sci Total Environ* 408:2208–15.

Atkinson, R., E. C. Tuazon, and S. M. Aschmann. 2000. Atmospheric chemistry of 2-pentanone and 2-heptanone. *Environ Sci Technol* 34(4):623–31.

Attaran, E., M. Rostás, and J. Zeier. 2008. *Pseudomonas syringae* elicits emission of the terpenoid (E, E)-4, 8, 12-trimethyl-1, 3, 7, 11-tridecatetraene in *Arabidopsis* leaves via jasmonate signaling and expression of the terpene synthase TPS4. *Mol Plant Microbe Interact* 21(11):1482–97.

Audrain, B., M. A. Farag, C. M. Ryu, and J. M. Ghigo. 2015. Role of bacterial volatile compounds in bacterial biology. *FEMS Microbiol Rev* 39:222–33.

Avalos, M., P. Garbeva, J. M. Raaijmakers, and G. P. van Wezel. 2019. Production of ammonia as a low-cost and long-distance antibiotic strategy by *Streptomyces* species. *ISME J* 2019:1–15.

Bailly, A., U. Groenhagen, S. Schulz, M. Geisler, L. Eberl, and L. Weisskopf. 2014. The inter-kingdom volatile signal indole promotes root development by interfering with auxin signalling. *Plant J* 80:758–71.

Bailly, A., and L. Weisskopf. 2012. The modulating effect of bacterial volatiles on plant growth: current knowledge and future challenges. *Plant Signal Behav* 7:79–85.

Bending, G. D., and S. D. Lincoln. 2000. Inhibition of soil nitrifying bacteria communities and their activities by glucosinolate hydrolysis products. *Soil Biol Biochem* 32(8–9):1261–69.

Berendsen, R. L., S. I. C. Kalkhove, L. G. Lugones, J. J. P. Baars, H. A. B. Wösten, and P. A. H. M. Bakker. 2013. Effects of the mushroom-volatile 1-octen-3-ol on dry bubble disease. *Appl Microbiol Biotechnol* 97(12):5535–43.

Bertamini, M., M. S. Grando, P. Zocca, M. Pedrotti, S. Lorenzi, and L. Cappellin. 2019. Linking monoterpenes and abiotic stress resistance in grapevines. *BIO Web of Conferences* 13:01003.

Bhatia, S. P., D. McGinty, C. S. Letizia, and A. M. Api. 2008. Fragrance material review on vetiverol. *Food Chem Toxicol* 46 Supplement 11:S297–S301.

Bita, C. E., and T. Gerats. 2013. Plant tolerance to high temperature in a changing environment: scientific fundamentals and production of heat stress-tolerant crops. *Front Plant Sci* 4:273.

Bitas, V., N. McCartney, N. Li, J. Demers, J.-E. Kim, H.-S. Kim, K. M. Brown, and S. Kang. 2015. *Fusarium oxysporum* volatiles enhance plant growth via affecting auxin transport and signaling. *Front Microbiol* 6:1248.

Bjurman, J. 1999. Release of MVOCs from microorganisms. In *Organic Indoor Air Pollutants: Occurrence, Measurement, Evaluation*, ed. T. Salthammer, 259–73. Wiley–VCH, Weinheim.

Blande, J. D., J. K. Holopainen, and Ü. Niinemets. 2014. Plant volatiles in polluted atmospheres: stress responses and signal degradation. *Plant Cell Environ* 37:1892–904.

Blom, D., C. Fabbri, E. C. Connor, F. P. Schiestl, D. R. Klauser, T. Boller, L. Eberl, and L. Weisskopf. 2011. Production of plant growth modulating volatiles is widespread among rhizosphere bacteria and strongly depends on culture conditions. *Environ Microbiol* 13:3047–58.

Bouvier, F., J. C. Isner, O. Dogbo, and B. Camara. 2005. Oxidative tailoring of carotenoids: a prospect towards novel functions in plants. *Trends Plant Sci* 10:187–94.

Briat, J. F. 2007. Iron dynamics in plants. *Adv Bot Res* 46:37–180.

Brilli, F., F. Loreto, and I. Baccelli. 2019. Exploiting plant volatile organic compounds (VOCs) in agriculture to improve sustainable defense strategies and productivity of crops. *Front Plant Sci* 10:264.

Cappellari, L. del R., J. Chiappero, T. B. Palermo, W. Giordano, and E. Banchio. 2020. Volatile organic compounds from rhizobacteria increase the biosynthesis of secondary metabolites and improve the antioxidant status in *Mentha piperita* L. grown under salt stress. *Agronomy* 10(8):1094.

Carvalho, L. C., J. L. Coito, S. Colaço, M. Sangiogo, and S. Amâncio. 2015. Heat stress in grapevine: the pros and cons of acclimation. *Plant Cell Environ* 38(4):777–89.

Castulo-Rubio, D. Y., N. A. Alejandre-Ramírez, M. del Carmen Orozco-Mosqueda, G. Santoyo, L. I. Macías-Rodríguez, and E. Valencia-Cantero. 2015. Volatile organic compounds produced by the rhizobacterium *Arthrobacter agilis* UMCV2 modulate *Sorghum bicolor* (strategy II plant) morphogenesis and SbFRO1 transcription in vitro. *J Plant Growth Regul* 34:611–623.

Chen, Y., K. Gozzi, F. Yan, and Y. Chai. 2015. Acetic acid acts as a volatile signal to stimulate bacterial biofilm formation. *MBio* 6:e00392.

Cheng, X., V. Cordovez, D. W. Etalo, M. van der Voort, and J. M. Raaijmakers. 2016. Role of the GacS sensor kinase in the regulation of volatile production by plant growth-promoting *Pseudomonas fluorescens* SBW25. *Front Plant Sci* 7:1706.

Chiron, N., and D. Micherlot. 2005. Odeurs des champignons: chimie et rôle dans les interactions biotiques - une revue. *Cryptogam Mycol* 26:299–364.

Choińska, R., K. Piasecka-Jóźwiak, B. Chabłowska, J. Dumka, and A. Łukaszewicz. 2020. Biocontrol ability and volatile organic compounds production as a putative mode of action of yeast strains isolated from organic grapes and rye grains. *Antonie van Leeuwenhoek*. doi:10.1007/s10482-020-01420-7

Cho, S. M., B. R. Kang, S. H. Han, A. J. Anderson, J. Y. Park, Y. H. Lee, B. H. Cho, K. Y. Yang, C. M. Ryu, and Y. C. Kim. 2008. 2R,3R-butanediol, a bacterial volatile produced by *Pseudomonas chlororaphis* O6, is involved in induction of systemic tolerance to drought in *Arabidopsis thaliana*. *Mol Plant Microbe Interact* 21:1067–75.

Cofer, T. M., M. Engelberth, and J. Engelberth. 2018. Green leaf volatiles protect maize (*Zea mays*) seedlings against damage from cold stress. *Plant Cell Environ* 41:1673–82.

Copolovici, L. O., and Ü. Niinemets. 2005. Temperature dependencies of Henry's law constants and octanol/water partition coefficients for key plant volatile monoterpenoids. *Chemosphere* 61:1390–400.

Cordovez, V., V. J. Carrion, D. W. Etalo, R. Mumm, H. Zhu, G. P. van Wezel, and J. M. Raaijmakers. 2015. Diversity and functions of volatile organic compounds produced by *Streptomyces* from a disease-suppressive soil. *Front Microbiol* 6:1081.

Crisp, P. A., D. Ganguly, S. R. Eichten, J. O. Borevitz, and B. J. Pogson. 2016. Reconsidering plant memory: intersections between stress recovery, RNA turnover, and epigenetics. *Sci Adv* 2:e1501340. doi: 10.1126/ sciadv.1501340

Dandurishvili, N., N. Toklikishvili, M. Ovadis, P. Eliashvili, N. Giorgobiani, R. Keshelava, M. Tediashvili, A. Vainstein, I. Khmel, E. Szegedi, and L. Chernin. 2011. Broad-range antagonistic rhizobacteria *Pseudomonas fluorescens* and *Serratia plymuthica* suppress *Agrobacterium* crown gall tumours on tomato plants. *J Appl Microbiol* 110:341–352.

De Moraes, C. M., W. J. Lewis, and J. H. Tumlinson. 2000. Examining plant–parasitoid interactions in tritrophic systems. *Anais da Sociedade Entomológica do Brasil* 29:189–203.

Dicke, M., and F. Loreto. 2010. Induced plant volatiles: from genes to climate change. *Trends Plant Sci* 15:115–17.

Ditengou, F. A., A. Muller, M. Rosenkranz, J. Felten, H. Lasok, M. M. van Doorn, V. Legue, K. Palme, J. P. Schnitzler, and A. Polle. 2015. Volatile signalling by sesquiterpenes from ectomycorrhizal fungi reprogrammes root architecture. *Nat Commun* 6:6279.

Enebe, M. C., and O. O. Babalola. 2019. The impact of microbes in the orchestration of plants' resistance to biotic stress: a disease management approach. *Appl Microbiol Biotechnol* 103:9–25. doi:10.1007/s00253-018-9433-3

Engelberth, J., H. T. Alborn, E. A. Schmelz, and J. H. Tumlinson. 2004. Airborne signals prime plants against insect herbivore attack. *Proc Natl Acad Sci USA* 101:1781–85.

Erb, M., N. Veyrat, C. A. M. Robert, H. Xu, M. Frey, J. Ton, and T. C. J. Turlings. 2015. Indole is an essential herbivore-induced volatile priming signal in maize. *Nat Commun* 6:6273.

Farag, M. A., H. Zhang, and C. M. Ryu. 2013. Dynamic chemical communication between plants and bacteria through airborne signals: induced resistance by bacterial volatiles. *J Chem Ecol* 39:1007–18.

Fialho, M. B., A. de Andrade, J. M. C. Bonatto, F. Salvato, C. A. Labate, and S. F. Pascholati. 2016. Proteomic response of the phytopathogen *Phyllosticta citricarpa* to antimicrobial volatile organic compounds from *Saccharomyces cerevisiae*. *Microbiol Res* 183(Suppl. C):1–7.

Franklin, K. A. 2008. Shade avoidance. *New Phytol* 179:930–44.

Frost, C. J., M. Appel, J. E. Carlson, C. M. De Moraes, M. C. Mescher, and J. C. Schultz. 2007. Within-plant signalling via volatiles overcomes vascular constraints on systemic signalling and primes responses against herbivores. *Ecol Lett* 10:490–98.

Frost, C. J., M. C. Mescher, C. Dervinis, J. M. Davis, J. E. Carlson, and C. M. De Moraes. 2008. Priming defense genes and metabolites in hybrid poplar by the green leaf volatile cis-3-hexenyl acetate. *New Phytol* 180:722–34.

Fujita, M., Y. Fujita, Y. Noutoshi, F. Takahashi, Y. Narusaka, K. Yamaguchi-Shinozaki, and K. Shinozaki. 2006. Crosstalk between abiotic and biotic stress responses: a current view from the points of convergence in the stress signaling networks. *Curr Opin Plant Biol* 9:436–42.

Gadd, G. M. 1988. Carbon nutrition and metabolism. In *Physiology of Industrial Fungi*, ed. D. R. Berry, 21–57. Blackwell Scientific Publications, Oxford.

Garbeva, P., W. H. G. Hol, A. J. Termorshuizen, G. A. Kowalchuk, and W. de Boer. 2011. Fungistasis and general soil biostasis – a new synthesis. *Soil Biol Biochem* 43:469–77.

Garbeva, P., C. Hordijk, S. Gerards, and W. de Boer. 2014. Volatile-mediated interactions between phylogenetically different soil bacteria. *Front Microbiol* 5:289.

Gomes, A. A. M., M. V. Queiroz, and O. L. Pereira. 2015. Mycofumigation for the biological control of post-harvest diseases in fruits and vegetables: a review. *Austin J Biotechnol Bioeng* 2(4):1051.

Gommers, C. M. M., E. J. W. Visser, K. R. St Onge, L. A. C. J. Voesenek, and R. Pierik. 2013. Shade tolerance: when growing tall is not an option. *Trends Plant Sci* 18:65–71.

Grace, J. B., and D. Tilman. 1990. *Perspectives on Plant Competition*. Academic Press, New York.

Groenhagen, U., R. Baumgartner, A. Bailly, A. Gardiner, L. Eberl, S. Schulz, and L. Weisskopf. 2013. Production of bioactive volatiles by different *Burkholderia ambifaria* strains. *J Chem Ecol* 39:892–906.

Gruber, M. Y., N. Xu, L. Grenkow, X. Li, J. Onyilagha, J. J. Soroka, N. D. Westcott, and D. D. Hegedus. 2009. Responses of the crucifer flea beetle to *Brassica* volatiles in an olfactometer. *Environ Entomol* 38:1467–79.

Gürtler, H., R. Pedersen, U. Anthoni, C. Christophersen, P. H. Nielsen, E. M. Wellington, C. Pedersen, and K. Bock. 1994. Albaflavenone, a sesquiterpene ketone with a zizaene skeleton produced by a streptomycete with a new rope morphology. *J Antibiot* 47:434–39.

Han, S. H., S. J. Lee, J. H. Moon, K. H. Park, K. Y. Yang, B. H. Cho, K. Y. Kim, Y. W. Kim, M. C. Lee, A. J. Anderson, and Y. C. Kim. 2006. GacS-dependent production of 2R,3R-butanediol by *Pseudomonas chlororaphis* O6 is a major determinant for eliciting systemic resistance against *Erwinia carotovora* but not against *Pseudomonas syringae* pv. *tabaci* in tobacco. *Mol Plant Microbe Interact* 19:924–30.

Hasanuzzaman, M., K. Nahar, M. M. Alam, R. Roychowdhury, and M. Fujita. 2013. Physiological, biochemical, and molecular mechanisms of heat stress tolerance in plants. *Int J Mol Sci* 14(5):9643–84.

Heil, M., and C. Kost. 2006. Priming of indirect defences. *Ecol Lett* 9:813–17.

Heil, M., and J. C. Silva Bueno. 2007. Within-plant signaling by volatiles leads to induction and priming of an indirect plant defense in nature. *Proc Natl Acad Sci USA* 104:5467–72

Hilker, M., J. Schwachtje, M. Baier, S. Balazadeh, I. Bäurle, S. Geiselhardt, et al. 2016. Priming and memory of stress responses in organisms lacking a nervous system. *Biol Rev Cambridge* 91:1118–33.

Hinton, A. Jr., and M. E. Hume. 1995. Antibacterial activity of the metabolic by-products of a *Veillonella* species and *Bacteroides fragilis*. *Anaerobe* 1:121–27.

Horie, T., F. Hauser, and J. I. Schroeder. 2009. HKT transporter-mediated salinity resistance mechanisms in *Arabidopsis* and monocot crop plants. *Trends Plant Sci* 14:660–68.

Hung, R., S. Lee, and J. W. Bennett. 2015. Fungal volatile organic compounds and their role in ecosystems. *Appl Microbiol Biotechnol* 99:3395–405.

Hüve, K., M. M. Christ, E. Kleist, R. Uerlings, Ü. Niinemets, A. Walter, and J. Wildt. 2007. Simultaneous growth and emission measurements demonstrate an interactive control of methanol release by leaf expansion and stomata. *J Exp Bot* 58:1783–93.

Insam, H., and M. S. Seewald. 2010. Volatile organic compounds (VOCs) in soils. *Biol Fert Soils* 46(3):199–213.

Ionenko, I. F., and A. V. Anisimov. 2001. Effect of water deficit and membrane destruction on water diffusion in the tissues of maize seedlings. *Biol Plantarum* 44:247–52.

Irmisch, S., A. Clavijo McCormick, J. Günther, A. Schmidt, G. A. Boeckler, J. Gershenzon, S. B. Unsicker, and T. G. Köllner. 2014. Herbivore-induced poplar cytochrome P450 enzymes of the CYP71 family convert aldoximes to nitriles which repel a generalist caterpillar. *Plant J* 80:1095–107.

Izaguirre, M. M., C. A. Mazza, M. S. Astigueta, A. M. Ciarla, and C. L. Ballaré. 2013. No time for candy: passionfruit (*Passiflora edulis*) plants downregulate damage-induced extra floral nectar production in response to light signals of competition. *Oecologia* 173:213–21.

Jones, S. E., L. Ho, C. A. Rees, J. E. Hill, J. R. Nodwell, and M. A. Elliot. 2017. *Streptomyces* exploration is triggered by fungal interactions and volatile signals. *Elife* 6:e21738.

Joshi, J. R., N. Khazanov, H. Senderowitz, S. Burdman, A. Lipsky, and I. Yedidia. 2016. Plant phenolic volatiles inhibit quorum sensing in pectobacteria and reduce their virulence by potential binding to ExpI and ExpR proteins. *Sci Rep* 6:38126.

Kaddes, A., M. L. Fauconnier, K. Sassi, B. Nasraoui, and M. H. Jijakli. 2019. Endophytic Fungal volatile compounds as solution for sustainable agriculture. *Molecules* 24(6):1065.

Kai, M., E. Crespo, S. M. Cristescu, F. J. M. Harren, W. Francke, and B. Piechulla. 2010. *Serratia odorifera*: analysis of volatile emission and biological impact of volatile compounds on *Arabidopsis thaliana*. *Appl Microbiol Biotechnol* 88:965–76.

Kai, M., M. Haustein, F. Molina, A. Petri, B. Scholz, and B. Piechulla. 2009. Bacterial volatiles and their action potential. *Appl Microbiol Biotechnol* 81:1001–12.

Kai, M., A. Vespermann, and B. Piechulla. 2008. The growth of fungi and *Arabidopsis thaliana* is influenced by bacterial volatiles. *Plant Signal Behav* 3:482–84.

Kanchiswamy, C. N., M. Mainoy, and M. E. Maffei. 2015a. Chemical diversity of microbial volatiles and their potential for plant growth and productivity. *Front Plant Sci* 6:151.

Kanchiswamy, C. N., M. Malnoy, and M. E. Maffei. 2015b. Bioprospecting bacterial and fungal volatiles for sustainable agriculture. *Trends Plant Sci* 20(4):206–11.

Kang, S. M., A. L. Khana, M. Waqasa, Y. H. You, J. H. Kim, J. G. Kim, M. Hamayun, and I. J. Lee. 2014. Plant growth-promoting rhizobacteria reduce adverse effects of salinity and osmotic stress by regulating phytohormones and antioxidants in *Cucumis sativus*. *J Plant Interact* 9:673–82.

Karban, R., J. Maron, G. W. Felton, G. Ervin, and H. Eichenseer. 2003. Herbivore damage to sagebrush induces resistance in wild tobacco: evidence for eavesdropping between plants. *Oikos* 100:325–32.

Karban, R., K. Shiojiri, S. Ishizaki, W. C. Wetzel, and R. Y. Evans. 2013. Kin recognition affects plant communication and defence. *P Roy Soc Lond B Bio* 280:20123062.

Karban, R., W. C. Wetzel, K. Shiojiri, S. Ishizaki, S. R. Ramirez, and J. D. Blande. 2014. Deciphering the language of plant communication: volatile chemotypes of sagebrush. *New Phytol* 204:380–85.

Karban, R. 2007. Experimental clipping of sagebrush inhibits seed germination of neighbours. *Ecol Lett* 10:791–97.

Kegge, W., and R. Pierik. 2010. Biogenic volatile organic compounds and plant competition. *Trends Plant Sci* 15:126–32.

Kegge, W., B. T. Weldegergis, R. Soler, M. V. Eijk, M. Dicke, L. A. C. J. Voesenek, and R. Pierik. 2013. Canopy light cues affect emission of constitutive and methyl jasmonate-induced volatile organic compounds in *Arabidopsis thaliana*. *New Phytol* 200:861–74.

Kigathi, R. N., W. W. Weisser, M. Reichelt, J. Gershenzon, and S. B. Unsicker. 2019. Plant volatile emission depends on the species composition of the neighboring plant community. *BMC Plant Biol* 19:58.

Kishimoto, K., K. Matsui, R. Ozawa, and J. Takabayashi. 2007. Volatile 1-octen-3-ol induces a defensive response in *Arabidopsis thaliana*. *J Gen Plant Pathol* 73:35–37.

Kong, H. G., T. S. Shin, T. H. Kim, and C. M. Ryu. 2018. Stereoisomers of the bacterial volatile compound 2,3-Butanediol differently elicit systemic defense responses of pepper against multiple viruses in the field. *Front Plant Sci* 9:90.

Koornneef, A., and C. M. J. Pieterse. 2008. Cross talk in defense signaling. *Plant Physiol* 146:839–44.

Korpi, A., J. Järnberg, and A. L. Pasanen. 2009. Microbial volatile organic compounds. *Crit Rev Toxicol* 39(2):139–93.

Korpi, A. 2001. *Faculty of Natural and Environmental Sciences C129. Fungal Volatile Metabolites and Biological Responses to Fungal Exposure* [thesis]. Kuopio University, Finland, pp. 1–77.

Kost, C., and M. Heil. 2008. The defensive role of volatile emission and extrafloral nectar secretion for lima bean in nature. *J Chem Ecol* 34:2–13.

Kottb, M., T. Gigolashvili, D. K. Großkinsky, and B. Piechull. 2015. *Trichoderma* volatiles effecting *Arabidopsis*: from inhibition to protection against phytopathogenic fungi. *Front Microbiol* 6:1–14.

Lambert, R. J. W., P. N. Skandamis, P. J. Coote, and G. J. Nychas. 2001. A study of the minimum inhibitory concentration and mode of action of oregano essential oil, thymol and carvacrol. *J Appl Microbiol* 91(3):453–62.

Lee, B., M. A. Farag, H. B. Park, J. W. Kloepper, S. H. Lee, and C. M. Ryu. 2012. Induced resistance by a long-chain bacterial volatile: elicitation of plant systemic defense by a C13 volatile produced by *Paenibacillus polymyxa*. *PLOS ONE* 7: e48744.

Lee, S., M. Yap, G. Behringer, R. Hung, and J. W. Bennett. 2016. Volatile organic compounds emitted by *Trichoderma* species mediate plant growth. *Fungal Biol Biotechnol* 3(1):7.

Lee, S. O., H. Y. Kim, G. J. Choi, H. B. Lee, K. S. Jang, Y. H. Choi, and J. C. Kim. 2009. Mycofumigation with *Oxyporus latemarginatus* EF069 for control of postharvest apple decay and rhizoctonia root rot on moth orchid. *J Appl Microbiol* 106:1213–19.

Li, N., A. Alfiky, W. Wang, M. Islam, K. Nourollahi, X. Liu, and S. Kang. 2018. Volatile Compound-mediated recognition and inhibition between *Trichoderma* biocontrol agents and *Fusarium oxysporum*. *Front Microbiol* 9:2614.

Li, T., J. K. Holopainen, H. Kokko, A. I. Tervahauta, and J. D. Blande. 2012. Herbivore-induced aspen volatiles temporally regulate two different indirect defences in neighbouring plants. *Funct Ecol* 26(5):1176–85.

Liu, X.-M., and H. Zhang. 2015. The effects of bacterial volatile emissions on plant abiotic stress tolerance. *Front Plant Sci* 6:774.

Liu, Y., L. Li, J. An, L. Huang, R. Yan, C. Huang, H. Wang, Q. Wang, M. Wang, and W. Zhang. 2018. Estimation of biogenic VOC emissions and its impact on ozone formation over the Yangtze River Delta region, China. *Atmos Environ* 186:113–28.

Loreto, F., M. Mannozzi, C. Maris, P. Nascetti, F. Ferranti, and S. Pasqualini. 2001. Ozone quenching properties of isoprene and its antioxidant role in leaves. *Plant Physiol* 126:993–1000.

Luna-Diez, E. 2016. Using green vaccination to brighten the agronomic future. *Outlook Pest Manage* 27:136–40.

Lundborg, L., L. Sampedro, A. K. Borg-Karlson, and R. Zas. 2019. Effects of methyl jasmo-
    nate on the concentration of volatile terpenes in tissues of Maritime pine and Monterey
    pine and its relation to pine weevil feeding. *Trees* 33(1):53–62.
Lutz, M. P., S. Wenger, M. Maurhofer, G. Défago, and B. Duffy. 2004. Signaling between
    bacterial and fungal biocontrol agents in a strain mixture. *FEMS Microbiol Ecol*
    48:447–55.
Macias-Rubalcava, M. L., B. E. Hernandez-Bautista, F. Oropeza, G. Duarte, M. C. Gonzalez,
    A. E. Glenn, R. T. Hanlin, and A. L. Anaya. 2010. Allelochemical effects of volatile
    compounds and organic extracts from *Muscodor yucatanensis*, a tropical endophytic
    fungus from *Bursera simaruba*. *J Chem Ecol* 36:1122–31.
Mackie, A. E., and Wheatley, R. E. 1999. Effects and incidence of volatile organic compound
    interactions between soil bacterial and fungal isolates. *Soil Biol Biochem* 31:375–85.
Mahdavi, A., P. Moradi, and A. Mastinu. 2020. Variation in terpene profiles of *Thymus vul-
    garis* in water deficit stress response. *Molecules* 25(5):1091.
Martinez-Medina, A., V. Flors, M. Heil, B. Mauch-Mani, C. M. J. Pieterse, M. J. Pozo, J.
    Ton, N. M. van Dam, and U. Conrath. 2016. Recognizing plant defense priming. *Trends
    Plant Sci* 21:818–22.
Martín-Sánchez, L., C. Ariotti, P. Garbeva, and G. Vigani. 2020. Investigating the effect of
    belowground microbial volatiles on plant nutrient status: perspective and limitations. *J
    Plant Interact* 15(1):188–95.
Mauch-Mani, B., I. Baccelli, E. Luna, and V. Flors. 2017. Defense priming: an adaptive part
    of induced resistance. *Annu Rev Plant Biol* 68:485–512.
Mei, X., Y. Liu, H. Huang, F. Du, L. Huang, J. Wu, Y. Li, S. Zhu, and M. Yang. 2019.
    Benzothiazole inhibits the growth of *Phytophthora capsici* through inducing apopto-
    sis and suppressing stress responses and metabolic detoxification. *Pesticide Biochem
    Physiol* 154:7–16.
Meldau, D. G., S. Meldau, L. H. Hoang, S. Underberg, H. Wunsche, I. T. Baldwin. 2013.
    Dimethyl disulfide produced by the naturally associated bacterium *Bacillus* sp B55
    promotes *Nicotiana attenuata* growth by enhancing sulfur nutrition. *Plant Cell*
    25:2731–47.
Mercier, J. and D.C. Manker. 2005. Biocontrol of soil-borne diseases and plant growth
    enhancement in greenhouse soilless mix by the volatile-producing fungus *Muscodor
    albus*. *Crop Prot* 24:355–62.
Mercier, J., and J. L. Smilanick. 2005. Control of green mold and sour rot of stored lemon by
    biofumigation with *Muscodor albus*. *Biol Control* 32:401–07.
Minerdi, D., S. Bossi, M. E. Maffei, M. L. Gullino, and A. Garibaldi. 2011. *Fusarium oxy-
    sporum* and its bacterial consortium promote lettuce growth and expansin A5 gene
    expression through microbial volatile organic compound (MVOC) emission. *FEMS
    Microbiol Ecol* 76:342–51.
Monson, R. 2013. Metabolic and gene expression controls on the production of biogenic vola-
    tile organic compounds. In *Biology, Controls and Models of Tree Volatile Organic
    Compound Emissions*, ed. Ü. Niinemets, and R. K. Monson, 153–79. Springer,
    Dordrecht.
Monson, R. K., and E. A. Holland. 2001. Biospheric trace gas fluxes and their control over
    tropospheric chemistry. *Annu Rev Ecol Syst* 32:547–76.
Morath, S. U., R. Hung, and J. W. Bennett. 2012. Fungal volatile organic compounds: a review
    with emphasis on their biotechnological potential. *Fungal Biol Rev* 26:73–83.
Mou, Z., X. Wang, Z. Fu, Y. Dai, C. Han, J. Ouyang, F. Bao, Y. Hu, and J. Li. 2002. Silencing
    of phosphoethanolamine N-methyltransferase results in temperature-sensitive male
    sterility and salt hypersensitivity in *Arabidopsis*. *Plant Cell* 14:2031–43.

Mumm, R., and M. Dicke. 2010. Variation in natural plant products and the attraction of bodyguards involved in indirect plant defense. *Can J Zool-Revue Canadienne De Zoologie* 88:628–67.

Munns, R. 2002. Comparative physiology of salt and water stress. *Plant Cell Environ* 25:239–50.

Naseem, H., and A. Bano. 2014. Role of plant growth-promoting rhizobacteria and their exopolysaccharide in drought tolerance of maize. *J Plant Interact* 9:689–701.

Nawrocka, J., U. Małolepsza, K. Szymczak, and M. Szczech. 2018. Involvement of metabolic components, volatile compounds, PR proteins, and mechanical strengthening in multilayer protection of cucumber plants against *Rhizoctonia solani* activated by *Trichoderma atroviride* TRS25. *Protoplasma* 255:359–73.

Naznin, H. A., D. Kiyohara, M. Kimura, M. Miyazawa, M. Shimizu, and M. Hyakumachi. 2014. Systemic resistance induced by volatile organic compounds emitted by plant growth-promoting fungi in *Arabidopsis thaliana*. *PLoS ONE* 9(1):e86882.

Nemcoviˇc, M., L. Jakubíková, I. Víden, and V. Farkaš. 2008. Induction of conidiation by endogenous volatile compounds in *Trichoderma* spp. *FEMS Microbiol Lett* 284:231–36.

Neri, F., M. Mari, S. Brigati, and P. Bertolini. 2007. Fungicidal activity of plant volatile compounds for controlling *Monilinia laxa* in stone fruit. *Plant Dis* 91:30–35.

Ngumbi, E., and J. Kloepper. 2016. Bacterial-mediated drought tolerance: current and future prospects. *Appl Soil Ecol* 105:109–25.

Niederbacher, B., J. B. Winkler, and J. P. Schnitzler. 2015. Volatile organic compounds as non-invasive markers for plant phenotyping. *J Exp Bot* 66(18):5403–16.

Niinemets, U., A. Arneth, U. Kuhn, R. K. Monson, J. Penuelas, and M. Staudt. 2010. The emission factor of volatile isoprenoids: stress, acclimation, and developmental responses. *Biogeosciences* 7:2203–23.

Nishida, S., C. Tsuzuki, A. Kato, A. Aisu, J. Yoshida, and T. Mizuno. 2011. AtIRT1, the primary iron uptake transporter in the root, mediates excess nickel accumulation in *Arabidopsis thaliana*. *Plant Cell Physiol* 52:1433–42.

Orlandini, V., I. Maida, M. Fondi, E. Perrin, M. C. Papaleo, E. Bosi, D. De Pascale, M. L. Tutino, L. Michaud, A. L. Giudice, and R. Fani. 2014. Genomic analysis of three sponge-associated *Arthrobacter* Antarctic strains, inhibiting the growth of *Burkholderia cepacia* complex bacteria by synthesizing volatile organic compounds. *Microbiol Res* 169:593–601.

Ossowicki, A., S. Jafra, and P. Garbeva. 2017. The antimicrobial volatile power of the rhizospheric isolate *Pseudomonas donghuensis* P482. *PLoS ONE* 12:e0174362.

Papaleo, M. C., M. Fondi, I. Maida, E. Perrin, A. Lo Giudice, L. Michaud, S. Mangano, G. Bartolucci, R. Romoli, and R. Fani. 2012. Sponge-associated microbial Antarctic communities exhibiting antimicrobial activity against *Burkholderia cepacia* complex bacteria. *Biotechnol Adv* 30:272–93.

Paré, P., and J. Tumlinson. 1996. Plant volatile signals in response to herbivore feeding. *Flo Entomol* 79:93–103.

Park, Y. S., S. Dutta, M. Ann, J. M. Raaijmakers, and K. Park. 2015. Promotion of plant growth by *Pseudomonas fluorescens* strain SS101 via novel volatile organic compounds. *Biochem Bioph Res Co* 461:361–65.

Peñuelas, J., D. Asensio, D. Tholl, K. Wenke, M. Rosenkranz, B. Piechulla, and J. P. Schnitzler. 2014. Biogenic volatile emissions from the soil. *Plant Cell Environ* 37:1866–91.

Picket, J. A., and Z. R. Khan. 2016. Plant volatile-mediated signalling and its application in agriculture: successes and challenges. *New Phytol* 212:856–70.

Piechulla, B., M. C. Lemfack, and M. Kai. 2017. Effects of discrete bioactive microbial volatiles on plants and fungi. *Plant Cell Environ* 40:2042–67.

Pierik, R., G. C. Whitelam, L. A. C. J. Voesenek, H. de Kroon, and E. J. M. Visser. 2004. Canopy studies on ethylene-insensitive tobacco identify ethylene as a novel element in blue light and plant–plant signalling. *Plant J* 38:310–19.

Popova, A. A., O. A. Koksharova, V. A. Lipasova, J. V. Zaitseva, O. A. Katkova-Zhukotskaya, S. Iu. Eremina, A. S. Mironov, L. S. Chernin, and I. A. Khmel. 2014. Inhibitory and toxic effects of volatiles emitted by strains of *Pseudomonas* and *Serratia* on growth and survival of selected microorganisms, *Caenorhabditis elegans*, and *Drosophila melanogaster*. *BioMed Res Int* 2014:125704.

Possell, M., and F. Loreto. 2013. The role of volatile organic compounds in plant resistance to abiotic stresses: responses and mechanisms. In *Biology, Controls and Models of Tree Volatile Organic Compound Emissions*, ed. Ü. Niinemets, and R. K. Monson, 209–35. Springer, Berlin.

Rajabi Memari, H., L. Pazouki, and Ü. Niinemets. 2013. The biochemistry and molecular biology of volatile messengers in trees. In *Biology, Controls and Models of Tree Volatile Organic Compound Emissions*, ed. Ü. Niinemets, and R. K. Monson, 47–93. Springer, Berlin.

Rani, K., S. S. Arya, S. Devi, and V. Kaur. 2017. Plant volatiles and defense. In *Volatiles and Food Security*, ed. Choudhary et al., 113–34. Springer Nature, Singapore Pte Ltd, Switzerland.

Raza, W., N. Ling, D. Liu, Z. Wei, Q. Huang, and Q. Shen. 2016b. Volatile organic compounds produced by *Pseudomonas fluorescens* WR-1 restrict the growth and virulence traits of *Ralstonia solanacearum*. *Microbiol Res* 192:103–13.

Raza, W., N. Ling, L. Yang, Q. Huang, and Q. Shen. 2016a. Response of tomato wilt pathogen *Ralstonia solanacearum* to the volatile organic compounds produced by a biocontrol strain *Bacillus amyloliquefaciens* SQR-9. *Sci Rep* 6:24856.

Rosenkranz, M., and J.-P. Schnitzler. 2013. Genetic engineering of BVOC emissions from trees. In *Biology, Controls and Models of Tree Volatile organic Compound Emissions*, ed. Ü. Niinemets, and R. K. Monson, 95–118. Springer, Berlin.

Rudrappa, T., M. L. Biedrzycki, S. G. Kunjeti, N. M. Donofrio, K. J. Czymmek, and P. W. Pare. 2010. The rhizobacterial elicitor acetoin induces systemic resistance in *Arabidopsis thaliana*. *Commun Integr Biol* 3:130–38.

Ryan, A. C., C. N. Hewitt, M. Possell, C. E. Vickers, A. Purnell, P. M. Mullineaux, W. J. Davies, and I. C. Dodd. 2014. Isoprene emission protects photosynthesis but reduces plant productivity during drought in transgenic tobacco (*Nicotiana tabacum*) plants. *New Phytol* 201:205–16.

Ryu, C.-M., M. A. Farag, C.-H. Hu, M. S. Reddy, J. W. Kloepper, and P. W. Pare. 2004. Bacterial volatiles induce systemic resistance in *Arabidopsis*. *Plant Physiol* 134:1017–26.

Ryu, C.-M., M. A. Farag, C.-H. Hu, M. S. Reddy, H.-X. Wei, P. W. Pare, and J. W. Kloepper. 2003. Bacterial volatiles promote growth in *Arabidopsis*. *Proc Natl Acad Sci USA* 100:4927–32.

Sanchez-Lopez, A. M., M. Baslam, N. De Diego, F. J. Munoz, A. Bahaji, G. Almagro, A. Ricarte-Bermejo, P. García-Gómez, J. Li, J. F. Humplik, O. Novák, L. Spíchal, K. Doležal, E. Baroja-Fernández, and J. Pozueta-Romero. 2016. Volatile compounds emitted by diverse phytopathogenic microorganisms promote plant growth and flowering through cytokinin action. *Plant Cell Environ* 39:2592–608.

Schalchli, H., G. R. Tortella, O. Rubilar, L. Parra, E. Hormazabal, and A. Quiroz. 2014. Fungal volatiles: an environmentally friendly tool to control pathogenic microorganisms in plants. *Crit Rev Biotechnol* 36(1):144–52.

Schmid, C., S. Bauer, and M. Bartelheimer. 2015. Should I stay or should I go? Roots segregate in response to competition intensity. *Plant Soil* 391:283–91.

Schmidt, R., V. Cordovez, W. de Boer, J. Raaijmakers, and P. Garbeva. 2015. Volatile affairs in microbial interactions. *ISME J* 9:2329–35.

Schmidt-Busser, D., M. Von Arx, and P. M. Guerin. 2009. Host plant volatiles serve to increase the response of male European grape berry moths, *Eupoecilia ambiguella*, to their sex pheromone. *J Comp Physiol A* 195:853–64.

Schnürer, J., J. Olsson, and T. Börjesson. 1999. Fungal volatiles as indicators of food and feeds spoilage. *Fungal Genet Biol* 27:209–17.

Schöller, C., S. Molin, and S. Wilkins. 1997. Volatile metabolites from some Gram-negative bacteria. *Chemosphere* 35:1487–95.

Schulz, S., and J. S. Dickschat. 2007. Bacterial volatiles: the smell of small organisms. *Nat Prod Rep* 24:814–42.

Schulz-Bohm, K., L. Martín-Sánchez, and P. Garbeva. 2017. Microbial volatiles: small molecules with an important role in intra and inter-kingdom interactions. *Front Microbiol* 8:2484.

Semighini, C. P., J. M. Hornby, R. Dumitru, K. W. Nickerson, and S. D. Harris. 2006. Farnesol-induced apoptosis in *Aspergillus nidulans* reveals a possible mechanism for antagonistic interactions between fungi. *Mol Microbiol* 59(3):753–64.

Semighini, C. P., N. Murray, and S. D. Harris. 2008. Inhibition of *Fusarium graminearum* growth and development by farnesol. *FEMS Microbiol Lett* 279(2):259–64.

Sharifi, R., S.-M. Lee, and C. M. Ryu. 2018. Microbe-induced plant volatiles. *New Phytol* 220:655–58.

Sharifi, R., and C. M. Ryu. 2016. Are bacterial volatile compounds poisonous odors to a fungal pathogen *Botrytis cinerea*, alarm signals to *Arabidopsis* seedlings for eliciting induced resistance, or both? *Front Microbiol* 7:196.

Sharifi, R., and C. M. Ryu. 2017. Chatting with a tiny belowground member of the holobiome: communication between plants and growth promoting rhizobacteria. *Adv Bot Res* 82:135–60.

Sharifi, R., and C. M. Ryu. 2018. Revisiting bacterial volatile-mediated plant growth promotion: lessons from the past and objectives for the future. *Ann Bot* 122(3):349–58.

Shiojiri, K., R. Ozawa, K. Matsui, M. W. Sabelis, and J. Takabayashi. 2012. Intermittent exposure to traces of green leaf volatiles triggers a plant response. *Sci Rep* 2:1–5.

Shulaev, V., P. Silverman, and I. Raskin. 1997 Airborne signalling by methyl salicylate in plant pathogen resistance. *Nature* 385(6618):718–21.

Sikkema, J., J. A. de Bont, and B. Poolman. 1995. Mechanisms of membrane toxicity of hydrocarbons. *Microbiol Mol Biol Rev* 59(2):201–22.

Simas, D. L. R., S. H. B. M. de Amorim, F. R. V. Goulart, C. S. Alviano, D. S. Alviano, and A. J. R. da Silva. 2017. Citrus species essential oils and their components can inhibit or stimulate fungal growth in fruit. *Ind Crop Prod* 98:108–15.

Simon, A. G., D. K. Mills, and K. G. Furton. 2017. Chemotyping the temporal volatile organic compounds of an invasive fungus to the United States, *Raffaelea lauricola*. *J Chromatogr A* 1487:72–6.

Smith, H. 2000. Phytochromes and light signal perception by plants – an emerging synthesis. *Nature* 407:585–91.

Smolander, A., R. A. Ketola, T. Kotiaho, S. Kanerva, K. Suominen, and V. Kitunen. 2006. Volatile monoterpenes in soil atmosphere under birch and conifers: effects on soil N transformations. *Soil Biol Biochem* 38(12):3436–42.

Song, C., R. Schmidt, V. de Jager, D. Krzyzanowska, E. Jongedijk, K. Cankar, J. Beekwilder, A. van Veen, W. De Boer, J. A. van Veen, and P. Garbeva. 2015. Exploring the genomic traits of fungus-feeding bacterial genus *Collimonas*. *BMC Genomics* 16:1103.

Song, G. C., and C. M. Ryu. 2013. Two volatile organic compounds trigger plant self-defense against a bacterial pathogen and a sucking insect in cucumber under open field conditions. *Int J Mol Sci* 14:9803–19.

Song, G. C., and C. M. Ryu. 2018. Evidence for volatile memory in plants: boosting defense priming through recurrent application of plant volatiles. *Mol Cells* 41:724–32.

Staudt, M., and L. Lhoutellier. 2011. Monoterpene and sesquiterpene emissions from *Quercus coccifera* exhibit interacting responses to light and temperature. *Biogeosciences* 8:2757–71.

Steinebrunner, F., F. P. Schiestl, and A. Leuchtmann. 2008. Ecological role of volatiles produced by *Epichloë*: differences in antifungal toxicity. *FEMS Microbiol Ecol* 64(2):307–16.

Stensmyr, M. C., H. K. M. Dweck, A. Farhan, I. Ibba, A. Strutz, L. Mukunda, J. Linz, V. Grabe, K. Steck, S. Lavista-Llanos, D. Wicher, S. Sachse, M. Knaden, P. G. Becher, Y. Seki, and B. S. Hansson. 2012. A conserved dedicated olfactory circuit for detecting harmful microbes in *Drosophila*. *Cell* 151(6):1345–57.

Stoppacher, N., B. Kluger, S. Zeilinger, R. Krska, and R. Schuhmacher. 2010. Identification and profiling of volatile metabolites of the biocontrol fungus *Trichoderma atroviride* by HS-SPME-GC-MS. *J Microbiol Meth* 81:187– 193.

Sugimoto, K., K. Matsui, Y. Iijima, Y. Akakabe, S. Muramoto, R. Ozawa, M. Uefune, R. Sasaki, K. Md. Alamgir, S. Akitake, T. Nobuke, I. Galis, K. Aoki, D. Shibata, and J. Takabayashi. 2014. Intake and transformation to a glycoside of (Z)-3-hexenol from infested neighbors reveals a mode of plant odor reception and defense. *Proc Natl Acad Sci USA* 111:7144–49.

Sunarpi, T. Horie, J. Motoda, M. Kubo, H. Yang, K. Yoda, R. Horie, W. Y. Chan, H. Y. Leung, K. Hattori, and M. Konomi. 2005. Enhanced salt tolerance mediated by AtHKT1 transporter-induced Na unloading from xylem vessels to xylem parenchyma cells. *Plant J* 44:928–38.

Sunesson, A. L. 1995. *Volatile Metabolites from Microorganisms in Indoor Environments-Sampling, Analysis and Identification* [thesis]. Ume°a University and National Institute for Working Life, Sweden, pp. 1–88.

Tahir, H. A. S., Q. Gu, H. Wu, W. Raza, A. Safdar, Z. Huang, F. U. Rajer, and X. Gaol. 2017. Effect of volatile compounds produced by *Ralstonia solanacearum* on plant growth promoting and systemic resistance inducing potential of *Bacillus volatiles*. *BMC Plant Biol* 17:133.

Takahashi, H., S. Kopriva, M. Giordano, K. Saito, and R. Hell. 2011. Sulfur assimilation in photosynthetic organisms: molecular functions and regulations of transporters and assimilatory enzymes. *Ann Rev Plant Biol* 62:157–84.

Tilocca, B., A. Cao, and Q. Migheli. 2020. Scent of a Killer: microbial volatilome and its role in the biological control of plant pathogens. *Front Microbiol* 11:41.

Timmusk, S., I. A. Abd El Daim, L. Copolovici, T. Tanilas, A. Kannaste, L. Behers, E. Nevo, G. Seisenbaeva, E. Stenstrom, and U. Niinemets. 2014. Drought tolerance of wheat improved by rhizosphere bacteria from harsh environments: enhanced biomass production and reduced emission of stress volatiles. *PLoS One* 9(5):e96086.

Tirranen, L. S., and I. I. Gitelson. 2006. The role of volatile metabolites in microbial communities of the LSS higher plant link. *Adv Space Res* 38:1227–32.

Tomsheck, A. R., G. A. Strobel, E. Booth, B. Geary, D. Spakowicz, B. Knighton, C. Floerchinger, J. Sears, O. Liarzi, and D. Ezra. 2010. Hypoxylon sp., an endophyte of *Persea indica*, producing 1,8- cineole and other bioactive volatiles with fuel potential. *Microbial Ecol* 60:903e914.

Tyagi, S., K. J. Lee, P. Shukla, and J. C. Chae. (2020). Dimethyl disulfide exerts antifungal activity against Sclerotinia minor by damaging its membrane and induces systemic resistance in host plants. *Sci Rep* 10:6547.

Tyc, O., V. C. L. de Jager, M. van den Berg, S. Gerards, T. K. S. Janssens, N. Zaagman, M. Kai, A. Svatos, H. Zweers, C. Hordijk, H. Besselink, and W. De Boer. 2017a. Exploring bacterial interspecific interactions for discovery of novel antimicrobial compounds. *Microbial Biotechnol* 10:910–25.

U.S. EPA (United States Environmental Protection Agency). 2018. 2014 National Emissions Inventory, Version 2, technical support document. https://www.epa.gov/sites/productio n/files/2018-07/documents/nei2014v2_tsd_05 jul 2018.pdf (PDF) (414 pp, 9.7MB).

Vaishnav, A., S. Kumari, S. Jain, A. Varma, and D. Choudhary. 2015. Putative bacterial volatile-mediated growth in soybean (*Glycine max* L. *merrill*) and expression of induced proteins under salt stress. *J Appl Microbiol* 119:539–51.

Vallat, A., H. N. Gu, and S. Dorn. 2005. How rainfall, relative humidity and temperature influence volatile emissions from apple trees in situ. *Phytochemistry* 66:1540–50.

van Hulten, M., M. Pelser, L. C. van Loon, C. M. Pieterse, and J. Ton. 2006. Costs and benefits of priming for defense in *Arabidopsis*. *Proc Natl Acad Sci USA* 103:5602–07.

Vespermann, A., M. Kai, and B. Piechulla. 2007. Rhizobacterial volatiles affect the growth of fungi and *Arabidopsis thaliana*. *Appl Environ Microbiol* 73:5639–41.

Vickers, C. E., J. Gershenzon, M. T. Lerdau, and F. Loreto. 2009. A unified mechanism of action for volatile isoprenoids in plant abiotic stress. *Nat Chem Biol* 5:283–91.

Vivaldo, G., E. Masi, C. Taiti, G. Caldarelli, and S. Mancuso. 2017. The network of plants volatile organic compounds. *Sci Rep* 7(1):11050.

Vuorinen, T., A. M. Nerg, and J. K. Holopainen. 2004. Ozone exposure triggers the emission of herbivore-induced plant volatiles, but does not disturb tritrophic signalling. *Environ Pollut* 131:305–11.

Wagner, P., and W. Kuttler. 2014. Biogenic and anthropogenic isoprene in the near-surface urban atmosphere — A case study in Essen, Germany. *Sci Total Environ* 475:104–15.

Wahid, A., S. Gelani, M. Ashraf, and M. Foolad. 2007. Heat tolerance in plants: an overview. *Environ Exp Bot* 61:199–223.

Webster, B., T. Bruce, S. Dufour, C. Birkemeyer, M. Birkett, J. Hardie, and J. Pickett. 2008. Identification of volatile compounds used in host location by the black bean aphid, *Aphis fabae*. *J Chem Ecol* 34:1153–61.

Wenke, K., D. Wanke, J. Kilian, K. Berendzen, K. Harter, and B. Piechulla. 2012. Volatiles of two growth-inhibiting rhizobacteria commonly engage AtWRKY18 function. *Plant J* 70:445–59.

Wilkins, K., E. M. Nielsen, and P. Wolkoff. 1997. Patterns in volatile organic compounds in dust from moldy buildings. *Indoor Air* 7(2):128–34.

Wrigley, D. M. 2004. Inhibition of *Clostridium perfringens* sporulation by *Bacteroides fragilis* and short-chain fatty acids. *Anaerobe* 10:295–300.

Xie, S., H. Wu, H. Zang, L. Wu, Q. Zhu, and X. Gao. 2014. Plant growth promotion by spermidine-producing *Bacillus subtilis* OKB105. *MPMI* 27:655–63.

Xie, S., J. Liu, S. Gu, X. Chen, H. Jiang, and T. Ding. 2020. Antifungal activity of volatile compounds produced by endophytic *Bacillus subtilis* DZSY21 against *Curvularia lunata*. *Ann Microbiol* 70:2.

Yancey, P. H. 1994. Compatible and counteracting solutes. In *Cellular and Molecular Physiology of Cell Volume Regulation*, ed. K. Strange, 81–109. CRC Press, Boca Raton, FL.

Yang, X., L. Xue, T. Wang, X. Wang, J. Gao, S. Lee, D. R. Blake, F. Chai, and W. Wang. 2018. Observations and explicit modeling of summertime carbonyl formation in Beijing: identification of key precursor species and their impact on atmospheric oxidation chemistry. *J Geophys Res Atmos* 123(2):1426–40.

Yi, H. S., M. Heil, R. M. Adame-Alvarez, D. J. Ballhorn, and C. M. Ryu. 2009. Airborne induction and priming of plant defenses against a bacterial pathogen. *Plant Physiol* 151:2152–61.

Yunus, F. N., M. Iqbal, K. Jabeen, Z. Kanwal, and F. Rashid. 2016. Antagonistic activity of *Pseudomonas fluorescens* against fungal plant pathogen *Aspergillus niger. Sci Lett* 4:66–70.

Zhang, H., M. S. Kim, V. Krishnamachari, P. Payton, Y. Sun, M. Grimson, M. A. Farag, C. M. Ryu, R. Allen, I. S. Melo, and P. W. Paré. 2007. Rhizobacterial volatile emissions regulate auxin homeostasis and cell expansion in *Arabidopsis. Planta* 226:839–51.

Zhang, H., M. S. Kim, Y. Sun, S. E. Dowd, H. Shi, and P. W. Paré. 2008. Soil bacteria confer plant salt tolerance by tissue-specific regulation of the sodium transporter HKT1. *Mol Plant Microbe Interact* 21:737–44.

Zhang, H., C. Murzello, Y. Sun, M. S. Kim, X. Xie, R. M. Jeter, J. C. Zak, S. E. Dowd, and P. W. Paré. 2010. Choline and osmotic-stress tolerance induced in *Arabidopsis* by the soil microbe *Bacillus subtilis* (GB03). *Mol Plant Microbe Interact* 23:1097–104.

Zhu, B. C. R., G. Henderson, F. Chen, H. X. Fei, and R. A. Laine. 2001. Evaluation of vetiver oil and seven insect-active essential oils against the Formosan subterranean termite. *J Chem Ecol* 27:1617–25.

Zuo, Z., S. M. Weraduwage, A. T. Lantz, L. M. Sanchez, S. E. Weise, J. Wang, K. L. Childs, and T. D. Sharkey. 2019. Isoprene acts as a signaling molecule in gene networks important for stress responses and plant growth. *Plant Physiol* 180:01391.

5 Role of Microorganisms in Mitigating Plant Biotic and Abiotic Stresses

Syeda Ulfath Tazeen Kadri,
Adinath N. Tavanappanavar,
Mohammed Azharuddin Savanur,
Nagesh Babu R, Sanjay Kumar Gupta,
Ram Naresh Bharagava, Anyi Hu,
Paul Olusegun Bankole, and Sikandar I. Mulla*

CONTENTS

---

* Corresponding author.

DOI: 10.1201/9781003213864-5

5.1   INTRODUCTION

Abiotic and biotic stresses are major constraints for loss of crop yield worldwide. Biotic stress refers to the attack by bacterial, fungal, and viral phytopathogens as well as nematode infection and insect predation. Abiotic stress, on the other hand, is due to conditions of drought, salinity, flooding, unfavorable temperatures, high heavy metal concentration, and organic contaminants. These stresses lead to drastic crop loss. Plant diseases cause about 20%–40% loss in crop yields (Savary et al., 2012), while drought conditions lead to an estimated 9%–10% drop in production (Lesk, 2016). Drought is usually dealt with by intensive irrigation, and, as most water resources are highly saline, this results in soil salinization. Presently, about 20% of the global irrigated soils have increased salinity (Negrao et al., 2017). To avoid and overcome the deleterious effects of these stresses and increase the crop yield, fertilizers, chemical pesticides and fungicides, breeding, and genetic engineering have been used. These approaches either have undesirable side effects, such as environmental contamination, or are expensive, slow, highly specific, and less practical (Reid, 2011).

Thus, exploration of alternative approaches to combat these stresses is needed. Among the alternatives is the utilization of plant-associated microorganisms that aid the plant in fighting infections and resisting abiotic stresses. Three such plant-microbe interactions are association with plant growth-promoting rhizobacteria (PGPR), arbuscular mycorrhizal fungi (AMF), and endophytes. These interactions augment the ability of the plants to overcome biotic and abiotic stresses by evoking diverse metabolic and defense responses (Nguyen et al., 2016). This chapter reviews the plant-microbe interactions depicted in Figure 5.1, emphasizing their role in plant stress alleviation and summarizes the recent research contributions made in this direction.

5.2   APPLICATION OF MICROBIAL BIOCONTROL AGENTS IN PREVENTING PLANT PATHOGENICITY

Biological control of plant pathogens refers to a reduction of inoculum or disease producing activity of a pathogen accomplished by the action of other microorganisms, often resulting in multiple interactions, such as suppressing the pest organism using other organisms or the application of antagonistic microorganisms to suppress diseases. The application of natural products and chemical compounds extracted from different sources, such as plant extracts, natural or modified organisms, or gene products, are other examples of biological control. Using beneficial microorganisms for this purpose is called as microbial biocontrol, and it appears to be the best option for the development of low cost, eco-friendly, and sustainable management approaches for protecting plants and crops against pathogen infection.

Based on their origin, microbial biocontrol agents (MBCAs) have been broadly categorized as bacterial, fungal, viral, and nematodes biocontrol agents and are highly diverse in terms of their targets as well. For instance, the MBCA could be a fungus but show biocontrol activity against other pathogenic fungi or bacteria or nematodes.

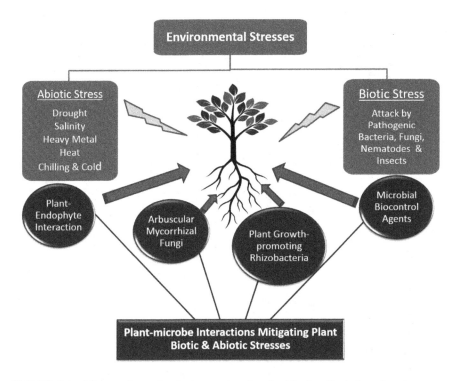

FIGURE 5.1   Diverse plant-microbe interactions involved in the alleviation of environmental stresses in plants.

This is further demonstrated in Table 5.1 by listing different types of MBCAs and their diverse target pathogens. The microbes that are considered ideal for use as biological control agents are the ones that can grow in rhizospheres, where the soil is described to be microbiologically suppressive to pathogens, as this area provides a frontline defense for the roots against various pathogenic attacks. Root colonization by beneficial microbes delivers their pathogen-antagonizing metabolites into the root system, where they directly suppress pathogenic bacterial growth. For example, rhizobacteria include antibiotic-producing strains, such as *Bacillus* sp. producing iturin A and surfactin, *Agrobacterium* sp. producing agrocin 84, *Pseudomonas* spp. producing phenazine derivatives, pyoleutorin and pyrrolnitrin, and *Erwinia* sp. producing herbicolin A (Rahman et al., 2017). Also, these MBCAs are able to activate plant defenses in plants to reduce the activity of deleterious microorganisms and then initiate induced systemic resistance (ISR) that is mediated by jasmonic acid and ethylene signaling (Pieterse et al., 2014).

The development of an MBCA is an extremely complex process that includes many phases, such as discovery, production, product development, efficacy testing, registration, and finally commercialization. Developmental time can be between 3 to 5 years, and likewise for registration. It is, therefore, imperative that enough resources are available to explore such an expensive and time-consuming method.

TABLE 5.1

Case Studies of Diverse Microbial Biocontrol Agents and Their Reported Biocontrol Activity

	Biocontrol Agent	Active against	Host Plant	Reference
Bacteria	*Paenibacillus polymyxa* AC-1	*Pseudomonas syringae* (Bacteria)	*Arabidopsis thaliana.*	Hong, Kwon, & Park, 2016
	Rhodococcus sp. KB6	*Ceratocystis fimbriata* (Fungi causing black rot disease)	*Ipomoea batatas* (Sweet Potato)	Hong et al., 2016
	Bacillus velezensis and *B. amyloliquefaciens*	*Plasmodiophora brassicae* (Protist causing Clubroot disease in crucifers)	*Brassica napus*	Zhu et al., 2020
	Bacillus thuringiensis	Insects belonging to the orders Lepidoptera, Diptera, and Coleoptera	Crucifers, cucurbits, corn, legumes, cotton, and solanaceous vegetables.	Singh et al., 2019
	Bacillus, Burkholderia Hydrogenophaga, Pseudomonas and *Streptomyces* species	*Meloidogyn* species, *Heterodera glycines*, and *Globodera pallida* (nematodes)	Variety of cultivated plants	Huang, Zhang, Yu, and Li, 2015
Fungi	*Trichoderma* species	*Macrophomina phaseolina* (fungus causing damping off, seedling blight, and charcoal rot in plants)	*Zea mays* and *Sorghum bicolor*	Mendoza et al., 2015
	Metarhizium brunneum and *Beauveria bassiana*	*Solenopsis invicta* (Fire ants)	Foraging fire ants found on numerous plants and trees worldwide	Rojas et al., 2018
	Paecilomyces lilacinus and *Trichoderma viride*	*Meloidogyne* sp. (nematodes causing root-knot)	*Cucumis sativus* (Cucumber)	Yankova et al., 2014

(Continued)

TABLE 5.1 (CONTINUED)
Case Studies of Diverse Microbial Biocontrol Agents and Their Reported Biocontrol Activity

Biocontrol Agent	Active against	Host Plant	Reference
Entomopathogenic Nematodes (EPNs)	*Steinernema glaseri* (nematode) and *Xenorhabdus Poinarii* (symbiont bacteria)	Lepidoptera, cutworms, corn rootworms, turf and Japanese beetles, flea beetles, soil insects, white grubs, black vine weevils, & citrus root weevils	Commercialized formulations of these EPNs are applied to diverse plants infected with the host insects.	Puza, 2015
	Heterorhabditis bacteriophora (nematode) and *Photorhabdus luminescens* (symbiont bacteria)	Root weevils, cutworms, fleas, banana root borers, and fungal gnats, white grubs (Popillia sp.)		Singh et al., 2019

Further examples and details of the mechanism of action are given by Lugtenberg, B. (2014) and Giri, B. (2019).

Ravensberg (2011) has described the detailed procedure, from ideation to commercialization, of using microbes as biocontrol agents against insect pathogens.

5.3   PLANT GROWTH-PROMOTING RHIZOBACTERIA (PGPR)

Soil is home to innumerable bacteria that take part in plant-microbe interactions that can be either advantageous, disadvantageous, or may have no effect on plants (Shahzad et al., 2014). However, in the case of PGPR, which are a group of free-living bacteria found to be associated with plant roots, the association is found to be beneficial, promoting plant growth (Kloepper et al., 1986). In addition to this, PGPR are also involved in mitigating the effect of biotic (de Souza et al., 2003; Zolla et al., 2013; Badri et al., 2013) and abiotic stress in plants (Sharma et al., 2003; Chang et al., 2007; Arshad et al., 2008). In view of the environmental concerns raised because of the ecological damage caused by chemical supplements and fertilizers, various alternatives are being considered. PGPR, being eco-friendly, are a promising alternative in modern-day agriculture. Based on their association with plants, PGPR can be classified as ePGPR (rhizospheric) and iPGPR (endophytic) and are briefly described in Figure 5.2.

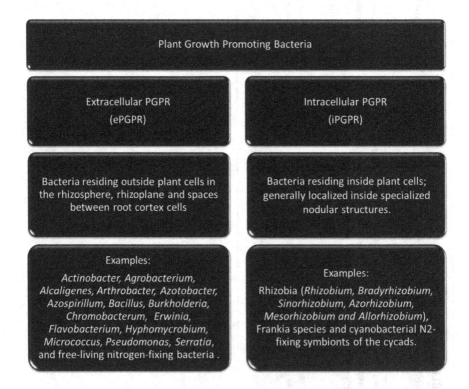

FIGURE 5.2   Extracellular and intracellular PGPR (Gray and Smith, 2005).

PGPR promote growth in plants either directly or by indirect means (Glick et al., 1995). The main processes influenced directly by PGPR are nitrogen fixation, phosphate solubilization, and production of various components, such as siderophores, ammonia, and vitamins. On the other hand, 1-aminocyclopropane-1-carboxylic acid (ACC) deaminase activity, production of cell wall degrading enzymes, antibiotic production, and induced systemic resistance are some of the processes wherein the PGPR has an indirect influence leading to growth promotion in plants.

5.3.1 PGPR AND BIOTIC STRESS

PGPR have been successfully used as biocontrol agents and are found to elicit the disease-resisting capacity of plants against a wide variety of disease-causing organisms. In addition to this, PGPR also act as growth inhibitors and reduce the virulence potential of pathogens. Single strains, as well as bacterial consortia, have been proven to be effective in providing protection to plants from several diseases. One of the many examples of PGPR as effective biocontrol agents include a combination of *Bacillus subtilis* and *Bacillus liquifaciens* used as a biocontrol agent in crops against the soil contaminant *Aspergillus parasiticus* in which the PGPR inhibit fungal growth by the synthesis of aflatoxin B1 and G1 (Siahmoshteh et al., 2018). Furthermore, B. *liquifaciens* has shown biocontrol potential, even when used singly against *Rhizoctonia solani* (Srivastava et al., 2016). Inoculation of *Colletrichum orbiculare* provided immunity against anthracnose disease in cucumber caused by *Serratia marcescens* through induced systemic resistance and reduction of internal root population (Press et al., 2001). Yet another interesting example is the inoculation of *Acienatobacter* spp. as a biocontrol agent in treating the wilt disease of tomatoes caused by *Ralstonia solanaceanum* in which, in addition to a greater biocontrol efficacy, an overall increase in yield was also observed (Xue et al., 2009). Furthermore, *Enterobacter asburiae* has shown to upregulate the expression of defense-related genes and antioxidant enzymes, thereby inducing resistance against tomato yellow leaf curl Sardina virus (TYLCSV) in tomato plants (Li et al., 2016).

5.3.2 PGPR AND ABIOTIC STRESS

PGPR have proven their effectiveness in mitigating abiotic stress in plants in various ways (Sharma et al. 2017). PGPR are used as bioinoculants, as they possess the intricate machinery to help the plant withstand the adverse effects of stress. For instance, drought is one of the pertinent issues related to crop loss and failure all around the world. PGPR have proven their ability in increasing the adaptability and conferring tolerance in plants facing drought stress (Enebe & Babalola, 2018). PGPR, in the above situation, through auxiliary production of phytohormones, plays an exemplary role in the induction of root growth, which in turn improves the water-absorbing capacity of roots. PGPR, such as *Enterobacter, Pseudomonas,* and *Stenotrophomonas,* when used as biofertilizers, have shown to modify the morphology of roots in plants exposed to abiotic stress by phytohormone secretion. A couple of examples showing the influences of PGPR in mitigating water stress include (1) the adaptation of poplar to water stress produced an overall 28% increase in biomass in poplar plants with PGPR compared to control plants (Khan et al., 2016); and (2) basil plants suffering from water stress showed an increased chlorophyll content and higher antioxidant activity in the presence of microbial consortium consisting of *Azospirillum brasilens, Bacillus lentus,* and *Pseudomonas* sp. (Heidari and Golpayegani, 2012). The use of PGPR in alleviating salt stress in rice is also noteworthy; inoculation of *Enterobacter* sp. was seen to promote rice seedling growth, reduce the production of ethylene, and boost activity of antioxidant enzymes (Sarkar et al., 2018).

On the other hand, when the soil was inoculated with *Bacillus* sp., increased production of indole acetic acid and deaminase was observed, which increased the biomass and diminished the effect of salinity in rice (Misra et al., 2017). PGPR's role in relieving stress related to elevated temperatures has also been reported. For instance, sorghum crops were found to withstand heat stress better in the presence of *Pseudomonas* sp. AMK-P6 (Ali et al., 2009). PGPR are also recognized in reducing cold stress as well. For example, canola plants showed better growth under low temperature in the presence of *P. putida* (Chang et al., 2007). There are reports of PGPR mitigating some other abiotic stresses as well, such as the use of *Pseudomonas putida* in withstanding stress due to floods, *Bacillus polymyxa,* and *Pseudomonas alcaligenes* in overcoming nutrient deficiency (Grichko and Glick, 2001).

5.4   ROLE OF ARBUSCULAR MYCORRHIZAL FUNGI (AMF) IN PLANT STRESS AMELIORATION

Arbuscular mycorrhizal fungi (AMF) are soil-borne fungi, belonging to the phylum Glomeromycota, that form multifunctional symbioses with more than 70% of vascular plants, including major crops from diverse families (Brundrett and Tedersoo, 2018). Among associations of plants with beneficial microorganisms, symbiosis with AMF is perhaps the oldest and most widespread in terms of both geographical distributions as well as phylogenetic coverage (Gutjahr and Parniske 2013). The AMF symbioses with plants are an archetypal example of a mutualistic relationship. The mycelial network of AMF outspreads under the roots of the plant, enabling uptake of water and nutrients that are otherwise unavailable to the plant. In return, the host plant provides the fungi photosynthetic products and lipids to accomplish their life cycle (Begum et al., 2019). Inoculation of AMF can, thus, significantly increase the concentration of several macro- and micro-nutrients, leading to increased photosynthesis and, consequently, increased biomass accumulation (Chen et al., 2017; Mitra et al., 2019). AMF have the ability to specifically boost phosphate uptake, as they secrete phosphatases that hydrolyze phosphate from organic phosphorus (P) compounds (Nell et al., 2010).

Besides nutritional status improvement, AMF enhance the ability of the host plant to cope with various biotic and abiotic stresses (Table 5.2). AMF promote plant growth and productivity under various environmental stresses by increased uptake of mineral nutrients; accumulation of osmo-protectants; upregulation of antioxidant enzymes; enhanced photosynthetic rate; production of metabolites such as amino

TABLE 5.2

AMF-Mediated Biotic and Abiotic Stress Mitigation in Plants

	Stress	Mycorrhizae	Host Plant	Comment	Reference
Biotic (Pathogen)	*Xanthomonas campestris* pv. *Alfalfa* (Bacteria)	*Rhizophagus irregularis* and *Gigaspora gigantea*	Alfalfa (*Medicago truncatula*)	*M. truncatula* plants showed increased resistance to "bacterial spot" through overexpression of defense genes encoding glycosyltransferase, kinase, calcium-binding protein, ubiquitin protein ligase	Liu et al. (2007)
	Sclerotinia sclerotiorum	*Rhizophagus irregularis*	French bean (*Phaseolus vulgaris*)	AMF induced protection against white mold. Disease caused by *Sclerotinia sclerotiorum* in French bean.	Mora-Romero et al. (2015)
	Begomovirus (TYLC Sardinia virus)	*Funneliformis mosseae*	Tomato (*Solanum lycopersicum*)	AMF symbiosis attenuated the symptom severity and reduced virus titre in plants infected by tomato yellow leaf curl Sardinia virus (TYLCSV)	Maffei et al. (2014)
	Meloidogyne incognita (Nematode)	*Glomus mosseae*	Tomato (*Solanum lycopersicum* cv. Marmande)	AMF significantly reduced penetration and mobility of root-knot infection causing *M. incognita*.	Vos et al. (2012)
Drought		*Glomus* spp.	Tomato (*Solanum lycopersicum*)	Application of AMF on planting media of tomato significantly improved vegetable productivity under drought conditions.	Kuswandi and Sugiyarto (2015)
		Glomus mosseae, G. fasciculatu, Gigaspora decipiens	Wheat (*Triticum aestivum* L.)	AMF singly and, in combination, improve the quality of soil and promoted plant growth in a barren land.	Pal and Pandey (2016)

(Continued)

TABLE 5.2 (CONTINUED)
AMF-Mediated Biotic and Abiotic Stress Mitigation in Plants

Stress	Mycorrhizae	Host Plant	Comment	Reference
	Funneliformis mosseae, and Paraglomus occultum	Trifoliate orange (Poncirus trifoliata)	AMF symbiosis induced greater root development and higher sucrose and proline metabolisms in order to adjust to drought stress.	Zhang et al. (2018)
Salinity	Glomus deserticola	Sweet basil (Ocimum basilicum L.).	AMF inoculation significantly improved chlorophyll content and water use efficiency in plants under salt stress	Elhindi et al. (2017)
	Rhizophagus irregularis	Black locust (R. pseudoacacia)	AMF symbiosis alleviated the salinity stress through improved photosynthesis, water status, and K+/Na+ homeostasis.	Chen et al. (2017)
	Glomus etunicatum, G. mosseae, and G. intraradices	Cucumber (Cucumis sativus L.)	AMF triggered salt stress tolerance in cucumber by regulating the oxidative system, hormones, and ionic equilibrium.	Hashem et al. (2018)
Heavy Metal	Funneliformis mosseae and Rhizophagus intraradices	Black locust (R. pseudoacacia)	The identified AMF species from heavy lead–zinc contaminated soil have potential for use in phytoremediation of heavy metals.	Yang et al. 2015
	Rhizophagus irregularis	Pigeon pea (Cajanus cajan L.)	AMF inoculation, along with silicon, improved growth, nutrient, water status, and yield under cadmium and zinc stress.	Garg and Singh (2017)
	Glomus monosporum, G. clarum, Gigaspora nigra, and Acaulospora Laevis	Fenugreek (Trigonella foenum-graecum L.)	AMF inoculation imparted cadmium tolerance in trigonella plants through increased antioxidant enzymes activity.	Abdelhameed and Rabab (2019)
Temperature	Rhizophagus irregularis, Funneliformis mosseae, and F. geosporum, Claroideoglomus claroideum	Wheat (Triticum aestivum L.)	Increased grain number and nutrient composition was observed under heat stressed (35°C) AMF–wheat plants.	Cabral et al. (2016)

(Continued)

TABLE 5.2 (CONTINUED)
AMF-Mediated Biotic and Abiotic Stress Mitigation in Plants

Stress	Mycorrhizae	Host Plant	Comment	Reference
	Glomus tortuosum	Corn (*Zea mays*)	AMF inoculation improved the nutritional status and enhanced the performance of maize plants under low temperature condition of 15°C	Liu and Chen (2016)
Soil Compaction	*Glomus mosseae*	Wheat (*Triticum aestivum* L.)	AMF inoculation reduced the stressful effects of soil compaction on wheat growth.	Miransari and Bahrami (2008)
	Glomus etunicatum, *G. mosseae*, and *G. intraradices*	Corn (*Zea mays*)	Highly compacted soil decreased corn growth, but AMF inoculation significantly enhanced plant growth under compaction	Miransari (2009); Miransari (2013)
Organic pollutants	*Glomus intraradices*	Ryegrass (*Lolium multiflorum*)	*G. intraradices* along with *Sphingomonas paucimobilis* enhanced phytoremediation of petroleum-contaminated soil.	Alarcón et al. (2008)
	Glomus hoi	Barley (*Hordeum vulgare*)	AMF association enabled the plant to cope with combined drought and xenobiotic (acetami nophen) stress	Khalvati et al. (2010)
Diesel stress	*Glomus constrictum* Trappe	Maize (*Zea mays*)	AMF inoculated seedlings significantly exhibited lower malondialdehyde and free proline content and higher superoxide dismutase and catalase activities compared to control plants under stress.	Tang et al. (2009)

acids, vitamins, phytohormones; and/or solubilization and mineralization processes (Begum et al., 2019). The multiple mechanisms used by AMF to mitigate stress-induced adverse effects are comprehensively reviewed by Nadeem et al. (2014). In addition to the above individual stresses, AMF association is also found to alleviate a combination of stresses. For example, Bauddh and Singh (2012) showed that AMF inoculation enabled castor and mustard plants to thrive under combined drought, salt, and cadmium stress. Similarly, tomato plants inoculated with *Scolecobasidium constrictum* showed higher biomass, leaf–water relations, and stomatal conductance compared to non-inoculated plants under combined drought and heat stress (Duc et al., 2018). A detailed assessment of the roles of AMF in alleviating drought, temperature, waterlogging, salt, and heavy metal stress is given by Wu (2017). However, to date, far less information is available on the performance of AMF under concurrent stresses, and further investigations are needed.

In terms of biotic stress management, the establishment of AMF symbiosis enhances host plants' resistance to phytopathogens and pests (Table 5.2). Several studies have reported the bioprotection potential of AMF against nematodes (Schouteden et al., 2015), soil-borne fungal and bacterial pathogens (Wehner et al., 2010; Cameron et al., 2013; Baum et al., 2015; Olowe et al., 2018), and aerial phytopathogens (Comby et al., 2017). The proposed mechanisms of action are direct competition between AMF and pathogens, modified production of root exudates, production of antimicrobial compounds, slowing disease progression by root structure transformation, and induced systemic response (Wehner et al., 2010). But in the case of bioprotection against viral diseases, AMF exhibit contrasting effects. For example, Maffei et al. (2014) reported that AMF symbiosis attenuated the symptom severity and reduced virus titre in tomato plants infected by tomato yellow leaf curl Sardinia virus (TYLCSV) while Miozzi et al. (2011) observed increased tomato spotted wilt virus (TSWV) titre in infected mycorrhizal tomato plants compared to non-mycorrhizal controls. This diverse behavior of AMF is probably due to the complex interaction between the virus, AMF, and the host plant, where factors such as viral makeup, plant nutritional status, and timing of interaction determine whether the final association is positive or negative. Recently Miozzi et al. (2019) have reviewed and listed the studies reporting the protective effect of AMF against viral infection and those reporting a detrimental effect.

While it is true that AMF inoculants can be active stress relievers for the host plant, in many cases, crop yield improvements following inoculation remain unreliable, with yield reductions reported in 14.6% of trials (Hijri, 2016). In other words, AMF inoculation is not necessarily beneficial in all scenarios, as plant responses to AMF colonization vary significantly depending on the host/fungal genotype (Hoeksema et al., 2010; Watts-Williams et al., 2019). AMF can fail to establish symbiosis if the genotype of the fungus is not compatible with the host genome or if they cannot compete with the soil's native microflora (Dabrowska et al., 2014; Berruti et al., 2016). In spite of all the reports on the potential of AMF in combating biotic and abiotic stresses, their actual use as biological control agents and biostimulants is still not a routine agricultural practice. This is perhaps because of the aforementioned variability in their performance. This makes careful initial selection and optimization of AMF species a necessity before use in agricultural practice (Elliott et al., 2020). Thus, further research into their modes of action will help to better establish their application.

5.5 IMPACT OF PLANT–ENDOPHYTE INTERACTIONS ON PLANT BIOTIC AND ABIOTIC STRESS

Endophytes are endosymbiont bacteria or fungi that reside in different parts of the plant without causing any infection in the host tissues (Bulgarelli et al., 2013). The mutualistic behavior of the endophytes and plants is demonstrated from observations that the host (plant) gives shelter and nutrients, and the endophytes increase the chance of survival of the host plant by alleviating plant biotic and abiotic stresses. Microbial endophytes have been investigated recently for plant growth-enhancing properties by producing secondary active compounds such as growth regulators and hormones. Endophytes manage plant growth under adverse conditions, such as salinity, drought, temperature, heavy metal stress, and nutrient stress (Anand et al., 2006; Aly et al., 2010; Kaul et al., 2012). A multitude of mechanisms are proposed to explain the beneficial effects of endophytes, such as production of phytohormones, mobilization of phosphorus; induced resistance/tolerance against abiotic stresses; biological nitrogen fixation; plant defense mechanisms by antagonistic substances (siderophores, HCN); suppression of stress-related plant ethylene synthesis by 1-aminocyclopropane-1-carboxylate (ACC) deaminase activity; or through competition for colonization sites and nutrients (Zhao et al., 2011; Munters, 2014). Endophytes isolated from *Capsicum annuum* L. showed enhanced root development due to IAA production (Sziderics et al., 2007). Endophytes from soybean exhibited phosphate assimilation, increasing the ability of plants to absorb phosphorus from the soil (Rosenblueth et al., 2006). Endophytes associated with *Saccharum officinarum* (sugarcane) have nitrogen-fixation genes such as nifH, which helped *S. officinarum* obtain the majority of its nitrogen from biological nitrogen fixation (BNF) without nodulation (Anand et al., 2013). Grasses growing in nutrient-poor sand dunes were contributed with nitrogen by endophytes *Pseudomonas*, *Burkholderia*, and *Stenotrophomonas*, as evident by the detection of nitrogenase with antibodies in roots within cell walls of stems and rhizomes (Dalton et al., 2004). Other beneficial effects on plant growth attributed to endophytes include osmotic adjustment, stomatal regulation, and modification of root morphology (Compant et al., 2005; Ryan et al., 2008). Continuous interaction between endophytes and plants could also result in an exchange of genetic material leading to sustained synthesis of microbial-derived bioactive compounds through generations in the plant system (Puri et al., 2006; Wang and Dai, 2011).

5.5.1 ENDOPHYTE-MEDIATED BIOTIC STRESS TOLERANCE

In many instances, the growth and development of plants is compromised while overcoming environmental stresses. Considerable effort has been made to understand the array of endophytic species in plants and their roles in defense mechanisms against abiotic and biotic stress via production of different secondary metabolites.

As a secondary mechanism of defense, plants can perceive different stress stimuli and induce local and systemic defense responses (Hilleary et al., 2018). Even with the intrinsic and well-evolved defense mechanisms, plants might require additional sophisticated defense strategies supported by their microbial partners to defend and rescue themselves from various biotic stresses. Initially, when the plant and endo-phytes interact, the infection of endophytes provokes the plant's defense similar to pathogen infection, but eventually, the endophytes escape the defense and colonize in the host cell (Zamioudis et al., 2012). However, the defense responses triggered by the establishment of endophytes in plants provide a priming effect and enhance resistance against other phytopathogens. This phenomenon in plants is known as induced systemic resistance (ISR) and usually is seen in bacteria–plant endophytic symbiosis (Robert-Seilaniantz et al., 2011; Zamioudis et al., 2012). In potato plants, treatment of endophytes from genera *Pseudomonas* and *Methylobacterium* resulted in enhanced resistance against the necrotrophic pathogen *Pectobacterium atrosep-ticum* via ISR (Ardanov et al., 2012). *Bacillus amyloliquefaciens*, a bacterial endo-phyte segregated from corn plants, exhibited in vitro antifungal activities against multiple phytopathogens, including *Aspergillus flavus*, *Fusarium moniliforme*, and *Colletotrichum gloeosporioides*. Further, the pretreatment of *B. amyloliquefaciens* to corn seedlings induced the expression of defense-related genes against pathogen infection in comparison to the non-treated controls (Gond et al., 2015).

While bacterial endophytes are the masters of manipulating the plant defense and imitating a priming defense effect against phytopathogens via ISR, the fungal endophytes usually do not showcase ISR-mediated defense responses in their hosts (Blodgett et al., 2007; Bae et al., 2011). On the other hand, the fungal endophytes produce growth-inhibiting chemical compounds against invading pathogens to protect their host plant. These compounds include phenols, alkaloids, flavonoids, terpenoids, quinols, peptides, steroids, and polyketones (Hardoim et al., 2015). Endophytic actinomycetes are characterized for producing antimicrobial compounds including kakadumycinx, munumbicins, and coronamycin (Hardoim et al., 2015). The endophytic *Streptomyces* sp. HKI0595 produces multicyclic indolosesquiterpene, which has antibacterial activity in *Kandelia candel* (Ding et al., 2011). In orchid plants, the endophyte *Streptosporangium oxazolinicum* K07-0450T produces spoxazomicins A–C having antitrypanosomal activity (Inahashi et al., 2011). These compounds, with various bioactive properties, are utilized for clinical or agricultural purposes, whereas their definitive action in plant-microbe interactions requires further investigation (Brader et al., 2014).

5.5.2 ENDOPHYTE-MEDIATED ABIOTIC STRESS TOLERANCE

Microorganisms are known to enhance the tolerance of plants to abiotic stresses such as drought, salinity, and metal toxicity. The probable explanation to enhance plant growth during environmental stresses is the biosynthesis of antistress biochemicals by endophytes. Plant growth-promoting endophytes produce osmolytes in response to drought stress, which act concomitantly to plant-based osmolytes and enhance plant growth (Paul et al., 2008). Potential protectants and osmoregulators involved in alleviating abiotic stresses are soluble sugars, sugar alcohols, and alkaloids secreted by endophyte. They protect macromolecules from denaturation and scavenge reactive oxygen species (ROS) associated with abiotic stress (Choudhury et al., 2017; Singh et al., 2011) Proline, which acts as an osmoprotectant, is a key metabolite that is synthesized during abiotic stresses and regulates osmotic adjustment, stabilizes subcellular structures, and scavenges free radicals (Naveed et al., 2014). Plant growth-promoting endophyte, *B. phytofirmans* PsJN increased the levels of free proline, starch, and phenolics in grapevine plantlets and enhanced cold tolerance (Sziderics et al., 2007). In endophyte colonized plants, drought tolerance is attributed to the expression of drought-responsive enzymes, such as catalase, peroxidase, and superoxide dismutase (Naveed et al., 2014). Exposure of plants to drought stress leads to the generation of ROS, such as hydroxyl radicals (OH). Plants develop antioxidant defense systems comprising both non-enzymatic and enzymatic components that help to prevent ROS accumulation and negate the oxidative damage produced due to abiotic stress conditions (Miller et al., 2010). Enzymatic components include catalase (CAT), glutathione reductase (GR), superoxide dismutase (SOD) and ascorbate peroxidase (APX). Non-enzymatic components contain glutathione, cysteine, and ascorbic acid (Kaushal and Wani, 2015). Endophyte *Sebacina vermifera*, when in symbiosis with barley exposed to heat stress and salinity, activates antioxidant enzymes and ethylene biosynthesis to counter stress tolerance (Barazani et al., 2007). Deeper understanding of endophyte plant interactions would help achieve the plant growth potential of efficient plant endophyte partnerships for developing and modifying endophytes to promote the sustainable yield production during stressed environmental conditions.

5.6 SYNERGISTIC ACTION OF DIFFERENT PLANT-MICROBE INTERACTIONS IN MITIGATING PLANT STRESS

5.6.1 SYNERGISTIC EFFECT OF PGPR AND AMF

In AMF symbiosis, the plant–fungus interactions occur in the root zone, and, in this zone, fungus also interacts with PGPR. The synergistic interaction between them, in addition to promoting plant growth, also promotes the population of each other (Artursson, 2005; Yusran et al., 2009). For example, inoculation of *Paenibacillus brasilensis* (a PGPR) increased the extent of root colonization by the *G. mosseae* (AMF) on clover (Artursson, 2005). Egberongbe et al., (2010) showed that dual inoculation of *G. mosseae* and *Trichoderma* spp. increased soybean yield and seed quality. However, antagonistic interactions between AMF and PGPR may also occur because of nutrient competition and certain secondary metabolites (Antoun and Prevost, 2005; Trivedi et al., 2012). Apart from normal conditions, combined inoculation is reported to be very effective under stress conditions as well. Perez-de-Luque et al. (2017) investigated the interactive effects of mycorrhizal fungus *Rhizophagus irregularis* and the rhizobacterial strain *Pseudomonas putida* KT2440 in enhancing defenses against pathogens in wheat and concluded that both AMF and PGPR showed an additive effect in priming host immunity. In another study, under drought stress conditions, dual inoculation of *Glomus* spp. and *Pseudomonas mendocina Palleroni* significantly enhanced root phosphatase activity, proline accumulation, and antioxidant enzyme activities in lettuce leaves (Kohler et al., 2008), but, under salinity stress, although it enhanced plant biomass, the co-inoculation failed to combat the stress and greatly decreased aggregate soil stability (Kohler et al., 2010). This inconsistency in action dictates further research in this direction. Despite better performance of co-inoculation of AMF and PGPR, there are, as yet, certain aspects – such as performance in actual field conditions as opposed to controlled laboratory conditions and performance under stresses other than drought and salt – that need further investigation. Several more case studies on the combined action of AMF and PGPR are discussed by Nadeem et al. (2014) and Nanjundappa et al. (2019) in their review.

5.6.2 SYNERGISTIC EFFECT OF AMF AND ENDOPHYTES

To date, very few studies have investigated the interactive effect of endophytes and AMF and their bio-ameliorative potential against plant stress tolerance (Hashem et al., 2016; Zhou et al., 2018). Hashem et al., (2016) studied the synergistic interactions of endophytic *B. subtilis* and AMF in *Acacia gerrardii* under salt stress and reported that co-inoculated *A. gerrardii* showed significantly higher biomass, leghemoglobin content, and nodulation when compared to those inoculated singly with AMF or *B. subtilis*. They also reported that *B. subtilis* promoted AMF colonization in the roots of *A. gerrardii* and concluded that the endophytic bacteria and AMF are coordinately involved in the adaptation of *A. gerrardii* to stress.

5.7 CONCLUSIONS

Abiotic and biotic stresses are rampant worldwide, hampering agricultural growth and productivity. Extended use of inorganic fertilizers, pesticides, and fungicides to cope with these stresses leads to various problems related to soil, plants, and human health. Sustainable techniques, enhancing crop productivity without increasing pressure on the environment, are needed. In this chapter, current information related to three different types of plant-microbe interactions that aid in the alleviation of plant biotic and abiotic stresses has been combined in a coherent way. The plant-microbe interactions discussed in this chapter can be utilized singly or in different combinations as alternatives to synthetic fertilizers and other chemicals. These microbial solutions can protect crops from pests and diseases and enhance plant productivity under various environmental stresses and help to meet the demand for more sustainable agriculture.

ACKNOWLEDGMENTS

Nagesh Babu R acknowledges support from DST-FIST and DBT-India. Syeda Ulfath Tazeen Kadri is supported by University Grants Commission (UGC) Junior Research Fellowship. Sikandar I. Mulla would like to thank all colleagues from the Department of Biochemistry, School of Applied Science, REVA University, Bangalore.

REFERENCES

Abdelhameed, R. E., and Rabab, A. M. 2019. Alleviation of Cadmium Stress by Arbuscular Mycorrhizal Symbiosis. *International Journal of Phytoremediation* 21: 663–71.

Alarcón, A., Autenrieth, R. L., and Zuberer, D. A. 2008. Arbuscular Mycorrhiza and Petroleum-Degrading Microorganisms Enhance Phytoremediation of Petroleum-Contaminated Soil. *International Journal of Phytoremediation* 10: 251–63.

Ali, S. Z., Sandhya, V., Grover, M., Kishore, N., Rao, L. V., and Venkateswarlu, B. 2009. Pseudomonas Sp. Strain AKM-P6 Enhances Tolerance of Sorghum Seedlings to Elevated Temperatures. *Biology and Fertility of Soils* 46 (1): 45–55.

Aly, A. H., Debbab, A., Kjer, J., and Proksch, P. 2010. Fungal Endophytes from Higher Plants: A Prolific Source of Phytochemicals and Other Bioactive Natural Products. *Fungal Diversity* 4: 1–16.

Anand, P., Isar, J., Saran, S., and Saxena, R. K. 2006. Bioaccumulation of Copper by *Trichoderma viride*. *Bioresource Technology* 97: 1018–25.

Anand, R., Grayston, S., and Chanway, C. 2013. N2-Fixation and Seedling Growth Promotion of Lodgepole Pine by Endophytic *Paenibacillus polymyxa*. *Microbial Ecology* 66: 369–74.

Antoun, A., and Prevost, D. 2005. Ecology of Plant Growth Promoting Rhizobacteria. In Z. A. Siddique (ed.) *PGPR: Biocontrol and Biofertilization*. Springer, Dordrecht, the Netherlands.

Ardanov, P., Sessitsch, A., Haggman, H., Kozyrovska, N., and Pirttila, A. M. 2012. Methylobacterium Induced Endophyte Community Changes Correspond with Protection of Plants against Pathogen Attack. *PLoS One* 7 (10): 46802.

Arshad, M., Shaharoona, B., and Mahmood, T. 2008. Inoculation with Pseudomonas Spp. Containing ACC-Deaminase Partially Eliminates the Effects of Drought Stress on Growth, Yield, and Ripening of Pea (*Pisum Sativum* L.). *Pedosphere* 18 (5): 611–20.

Artursson, V. 2005. *Bacterial–Fungal Interactions Highlighted Using Microbiomics: Potential Application for Plant Growth Enhancement* (Dissertation). Swedish University of Agricultural Sciences.

Badri, D.V., Zolla, G., Bakker, M. G., Manter, D. K., and Vivanco, J. M. 2013. Potential Impact of Soil Microbiomes on the Leaf Metabolome and on Herbivore Feeding Behavior. *New Phytologist* 198 (1): 264–73.

Bae, H., Roberts, D. P., Lim, H. S., Strem, M. D., Park, S. C., Ryu, C. M., Melnick, R. L., and Bailey, B. A. 2011. Endophytic Trichoderma Isolates from Tropical Environments Delay Disease Onset and Induce Resistance against Phytophthora Capsici in Hot Pepper Using Multiple Mechanisms. *Molecular Plant-Microbe Interactions* 24: 336–51.

Barazani, O., Von Dahl, C. C., and Baldwin, I. T. 2007. *Sebacina vermifera* Promotes the Growth and Fitness of *Nicotiana attenuata* by Inhibiting Ethylene Signaling. *Plant Physiology* 144: 1223–32.

Bauddh, K., and Singh, R. P. 2012. Growth: Tolerance Efficiency and Phytoremediation Potential of *Ricinus communis* (L.) and *Brassica juncea* (L.) in Salinity and Drought Affected Cadmium Contaminated Soil. *Ecotoxicology and Environmental Safety* 85: 13–22.

Baum, C., El-Tohamy, W., and Gruda, N. 2015. Increasing the Productivity and Product Quality of Vegetable Crops Using Arbuscular Mycorrhizal Fungi: A Review. *Scientia Horticulturae* 187: 131–41.

Begum, N., Qin, C., Ahanger, M. A., Raza, S., Khan, M. I., Ashraf, M., and Zhang, L. 2019. Role of Arbuscular Mycorrhizal Fungi in Plant Growth Regulation: Implications in Abiotic Stress Tolerance. *Frontiers in Plant Science* 10: 1068.

Berruti, A., Lumini, E., Balestrini, R., and Bianciotto, V. 2016. Arbuscular Mycorrhizal Fungi as Natural Biofertilizers: Let's Benefit from Past Successes. *Frontiers in Microbiology* 6 (1): 1–13.

Blodgett, J.T., Eyles, A., and Bonello, P. 2007. Organ-Dependent Induction of Systemic Resistance and Systemic Susceptibility in *Pinus nigra* Inoculated with *Sphaeropsis sapinea* and *Diplodia scrobiculata*. *Tree Physiology* 27: 511–17.

Brader, G., Compant, S., Mitter, B., Trognitz, F., and Sessitsch, A. 2014. Metabolic Potential of Endophytic Bacteria. *Current Opinion in Biotechnology* 27: 30–37.

Brundrett, M.C., and Tedersoo, L. 2018. Evolutionary History of Mycorrhizal Symbioses and Global Host Plant Diversity. *New Phytologist* 220 (4): 1108–15.

Bulgarelli, D., Chlaeppi, K. S., Spaepen, S., Ver Loren Themaat, E., and Schulze-Lefert, P. 2013. Structure and Functions of the Bacterial Microbiota of Plants. *Annual Review of Plant* 64: 807–38.

Cabral, C., Sabine, R., Ivanka, T., and Bernd, W. 2016. Arbuscular Mycorrhizal Fungi Modify Nutrient Allocation and Composition in Wheat (*Triticum Aestivum* L.) subjected to Heat-Stress. *Plant Soil* 408 (1–2): 385–99.

Cameron, D. D., Neal, A. L. and vanWees S. C. 2013. Mycorrhiza-induced resistance: more than the sum of its parts? *Trends in Plant Science* 18: 539–45.

Chang, W. S., Van De Mortel, M., Nielsen, L., De Guzman, G. N., Li, X., and Halverson, L. J. 2007. Alginate Production by *Pseudomonas putida* creates a Hydrated Microenvironment and Contributes to Biofilm Architecture and Stress Tolerance under Water-Limiting Conditions. *Journal of Bacteriology* 189 (22): 8290–99.

Chen, J., Zhang, H., Zhang, X., and Tang, M. 2017. Arbuscular Mycorrhizal Symbiosis Alleviates Salt Stress in Black Locust through Improved Photosynthesis, Water Status, and K+/Na+ Homeostasis. *Frontiers in Plant Science* 8: 1739.

Choudhury, F. K., Rivero, R. M., Blumwald, E., and Mittler, R. 2017. Reactive Oxygen Species, Abiotic Stress and Stress Combination. *The Plant Journal* 90 (5): 856–67.

Comby, M., Mustafa, G., Magnin-Robert, M., Randoux, B., Fontaine, J., Reignault, P., and Lounès-Hadj Sahraoui, A. 2017. Arbuscular Mycorrhizal Fungi as Potential Bioprotectants Against Aerial Phytopathogens and Pests. In Wu, Q.–S. *Arbuscular Mycorrhizas and Stress Tolerance of Plants*. Springer Nature, Singapore.

Compant, S., Reiter, B., and Sessitsch, A. 2005. Endophytic Colonization of *Vitis vinifera* L. by a Plant Growth-Promoting Bacterium, *Burkholderia sp. Strain PsJN*. *Applied and Environmental Microbiology* 71: 1685–89.

Dabrowska, G., Baum, C., Trejgell, A., and Hrynkiewicz, K. 2014. Impact of Arbuscular mycorrhizal Fungi on the Growth and Expression of Gene Encoding Stress Protein metallothionein BnMT2 in the Non-Host Crop *Brassica napus* L. *Journal of Plant Nutrition and Soil Science* 177: 459–67.

Dalton, D. A., Kramer, S., and Azios, N. 2004. Endophytic Nitrogen Fixation in Dune Grasses (*Ammophila arenaria* and *Elymus mollis*) from Oregon. *FEMS Microbiology Ecology* 49: 469–79.

DeSouza, J. T., Weller, D. M., and Raaijmakers, J. M. 2003. Frequency, diversity, and activity of 2, 4-diacetylphloroglucinol-producing fluorescent *Pseudomonas* spp. in Dutch take-all decline soils. *Phytopathology* 93 (1): 54–63.

Ding, L., Maier, A., Fiebig, H. H., Lin, W. H., Peschel, G., and Hertweck, C. 2011. Kandenols A-E, Eudesmenes from an Endophytic Streptomyces Sp. of the Mangrove Tree *Kandelia candel*. *Journal of Natural Products* 75: 2223–27.

Duc, N. H., Csintalan, Z., and Posta, K. 2018. Arbuscular Mycorrhizal Fungi Mitigate Negative Effects of Combined Drought and Heat Stress on Tomato Plants. *Plant Physiology and Biochemistry* 132: 297–307.

Egberongbe, H. O., Akintokun, A. K., Babalola, O. O., and Bankole, M. O. 2010. The Effect of *Glomus mosseae* and *Trichoderma harzianum* on Proximate Analysis of Soybean seed Grown in Sterilized and Unsterilised Soil. *Journal of Agricultural Extension and Rural Development* 2: 54–58.

Elhindi, K.M., El-Din, A. S., and Elgorban, A. M. 2017. The Impact of Arbuscular Mycorrhizal Fungi in Mitigating Salt-Induced Adverse Effects in Sweet Basil (*Ocimum basilicum* L). *Saudi Journal of Biological Sciences* 24: 170–179.

Elliott, A. J., Daniell, T. J., Cameron, D. D., and Field, K. J. 2020. A Commercial Arbuscular Mycorrhizal Inoculum Increases Root Colonization across Wheat Cultivars but Does Not Increase Assimilation of Mycorrhiza-Acquired Nutrients. *Plants, People, Planet* 1: 37–49

Enebe, M. C., and Babalola, O. O. 2018. The Influence of Plant Growth-Promoting Rhizobacteria in Plant Tolerance to Abiotic Stress: A Survival Strategy. *Applied Microbiology and Biotechnology* 102 (18): 7821–35.

Garg, N., and Singh, S. 2017. Arbuscular Mycorrhiza *Rhizophagus irregularis*, and Silicon Modulate Growth, Proline Biosynthesis and Yield in *Cajanus cajan*, L. (Pigeon Pea) Genotypes under Cadmium and Zinc Stress. *Journal of Plant Growth Regulation* 37: 1–18.

Giri, B., Prasad, R., Wu, Q.-S., and Varma, A. 2019. *Biofertilizers for Sustainable Agriculture and Environment*. Springer Nature, Switzerland AG.

Glick, B. R., Karaturovic, D. M., and Newell, P. C. 1995. A Novel Procedure for Rapid Isolation of Plant Growth Promoting Pseudomonads. *Canadian Journal of Microbiology* 41 (6): 533–36.

Gond, S. K., Bergen, M. S., Torres, M. S., White, J. F., and Kharwar, R. N. 2015. Effect of Bacterial Endophyte on Expression of Defense Genes in Indian Popcorn against *Fusarium moniliforme*. *Symbiosis* 66: 133–40.

Gray, E. J., and Smith, D. L. 2005. Intracellular and Extracellular PGPR: Commonalities and Distinctions in the Plant–Bacterium Signaling Processes. *Soil Biology and Biochemistry* 37 (3): 395–412.

Grichko, V. P., and Glick, B. R. 2001. Amelioration of Flooding Stress by ACC Deaminase-Containing Plant Growth-Promoting Bacteria. *Plant Physiology and Biochemistry* 39 (1): 11–17.

Gutjahr, C., and Parniske, M. 2013. Cell and Developmental Biology of Arbuscular Mycorrhiza Symbiosis. *Annual Review of Cell and Developmental Biology* 29: 593–617.

Hardoim, P. R., van Overbeek, L. S., Berg, G., Pirttilä, A. M., Compant, S., Campisano, A., Döring, M., and Angela Sessitsch, A. 2015. The Hidden World within Plants: Ecological and Evolutionary Considerations for Defining Functioning of Microbial Endophytes. *Microbiology and Molecular Biology Reviews* 79: 293–320.

Hashem, A., Abdallah, A. A. Alqarawi, A. A., Huqail, A. A. F., and Egamberdieva, D. 2016. The Interaction between Arbuscular Mycorrhizal Fungi and Endophytic Bacteria Enhances Plant Growth of *Acacia gerrardii* under Salt Stress. *Frontiers in Plant Science* 7: 1089.

Hashem, A., 2018. Alqarawi, A. A., Radhakrishnan, R., Al-Arjani, A. F., Aldehaish, H. A., and Egamberdieva, D. Arbuscular Mycorrhizal Fungi Regulate the Oxidative System, Hormones and Ionic Equilibrium to Trigger Salt Stress Tolerance in *Cucumis sativus* L. *Saudi Journal of Biological Sciences* 25 (6): 1102–14.

Heidari, M., and Golpayegani, A. 2012. Effects of Water Stress and Inoculation with Plant Growth Promoting Rhizobacteria (PGPR) on Antioxidant Status and Photosynthetic Pigments in Basil (*Ocimum basilicum* L.). *Journal of the Saudi Society of Agricultural Sciences* 11 (1): 57–61.

Hijri, M. 2016. Analysis of a Large Dataset of Mycorrhiza Inoculation Field Trials on Potato Shows Highly Significant Increases in Yield. *Mycorrhiza* 26 (3): 209–14.

Hilleary, R., and Gilroy, S. 2018. Systemic Signaling in Response to Wounding and Pathogens. *Current Opinion in Plant Biology* 43: 57–62.

Hoeksema, J. D., Chaudhary, V. B., Gehring, C. A., Johnson, N. C., Karst, J., Koide, R. T., and Umbanhowar, J. 2010. A Meta-Analysis of Context-Dependency in Plant Response to Inoculation with Mycorrhizal Fungi. *Ecology Letters* 13 (3): 394–407.

Hong, C. E., Jeong, H., Jo, S. H., Jeong, J. C., Kwon, S. Y., An, D., and Park, J. M. 2016. A Leaf-Inhabiting Endophytic Bacterium, Rhodococcus Sp. KB6, Enhances Sweet Potato Resistance to Black Rot Disease Caused by *Ceratocystis fimbriata*. *Journal of Microbiology and Biotechnology* 26 (3): 488–92.

Hong, C. E., Kwon, S. Y., and Park, J. M.2016. Biocontrol Activity of *Paenibacillus polymyxa* AC-1 against *Pseudomonas syringae* and Its Interaction with *Arabidopsis thaliana*. *Microbiological Research* 185: 13–21.

Huang, X., Zhang, K., Yu, Z., and Li, G. 2015. Microbial Control of Phytopathogenic Nematodes. In Lugtenberg, Ben, ed., *Principles of Plant-Microbe Interactions*. Springer International, Switzerland, pp. 155–164.

Inahashi, Y., Matsumoto, A., Omura, S., and Takahashi, Y. 2011. Streptosporangium Oxazolinicum Sp. Nov., a Novel Endophytic Ectinomycete Producing New Antitrypanosomal Antibiotics, Spoxazomicins. *Journal of Antibiotics* 64: 297–302.

Kaul, S., Gupta, S., Ahmed, M., and Dhar, M. K. 2012. Endophytic Fungi from Medicinal Plants: A Treasure Hunt for Bioactive Metabolites. *Phytochemistry Reviews* 11: 487–505.

Kaushal, M., and Wani, S. P. 2016. Plant-Growth-Promoting Rhizobacteria: Drought Stress Alleviators to Ameliorate Crop Production in Drylands. *Annals of Microbiology*, 66: 35–42.

Khalvati, M., Bartha, B. and Dupigny, A. 2010. Arbuscular Mycorrhizal Association Is Beneficial for Growth and Detoxification of Xenobiotics of Barley under Drought Stress. *Journal of Soils and Sediments* 10: 54–64.

Khan, Z., Rho, H., Firrincieli, A., Hung, S. H., Luna, V., Masciarelli, O., and Doty, S. L. 2016. Growth Enhancement and Drought Tolerance of Hybrid Poplar upon Inoculation with Endophyte Consortia. *Current Plant Biology* 6: 38–47.

Kloepper, J. W., Scher, F. M., Laliberte, M., and Tipping, B. 1986. Emergence-Promoting Rhizobacteria: Description and Implications for Agriculture. In *Iron, Siderophores, and Plant Diseases*, 155–64. Springer, Boston, MA.

Kohler, J., Caravaca, F., and Roldan, A. 2010. An AM Fungus and a PGPR Intensify the Adverse Effects of Salinity on the Stability of Rhizosphere Soil Aggregates of *Lactuca sativa*. *Soil Biology and Biochemistry* 42: 429–34.

Kohler, J., Hernandez, J. A., Caravaca, F., and Roldan, A. 2008. Plant-Growth Promoting Rhizobacteria and Arbuscular Mycorrhizal Fungi Modify Alleviation Biochemical Mechanisms in Water Stressed Plants. *Functional Plant Biology* 35: 141–51.

Kuswandi, P.C., and Sugiyarto, L. 2015. Application of Mycorrhiza on Planting Media of Two Tomato Varieties to Increase Vegetable Productivity in Drought Condition. *Jurnal Sains Dasar* 4: 17–22.

Lesk, C. 2016. Influence of Extreme Weather Disasters on Global Crop Production. *Nature* 529 (7584): 84–87.

Li, Y., Wang, Q., Wang, L., He, L. Y., and Sheng, X. F. 2016. Increased Growth and Root Cu Accumulation of *Sorghum sudanense* by Endophytic Enterobacter Sp. K3- 2: Implications for *Sorghum sudanense* Biomass Production and Phytostabilization. *Ecotoxicology and Environmental Safety* 124: 163–68.

Liu, J., Maldonado-Mendoza, I., and Lopez-Meyer, M. 2007. Arbuscular Mycorrhizal Symbiosis Is Accompanied by Local and Systemic Alterations in Gene Expression and an Increase in Disease Resistance in the Shoots. *Plant Journal* 50: 529–44.

Liu, N., and Chen, X. 2016. Effects of Arbuscular Mycorrhiza on Growth and Nutrition of Maize Plants under Low Temperature Stress. *Philippine Agricultural Scientist* 99 (3): 246–52.

Lugtenberg, B. 2014. *Principles of Plant-Microbe Interactions*. Springer International, Switzerland.

Maffei, G., Miozzi, L., Fiorilli, V., Novero, M., Lanfranco, L., and Accotto, G. P. 2014. The Arbuscular Mycorrhizal Symbiosis Attenuates Symptom Severity and Reduces Virus Concentration in Tomato Infected by Tomato Yellow Leaf Curl Sardinia Virus (TYLCSV). *Mycorrhiza* 24: 179–86.

Mendoza, J. L. H., Pérez, M. I. S., Prieto, J. M. G., Velásquez, J. D. C. Q., Olivares, J. G. G., and Langarica, H. R. G. 2015. Antibiosis of *Trichoderma spp* Strains Native to Northeastern Mexico against the Pathogenic Fungus *Macrophomina phaseolina*. *Brazilian Journal of Microbiology* 46 (4): 1093–1101.

Miller, G., Susuki, N., Ciftci-Yilmaz, S., and Mittler, R. 2010. Reactive Oxygen Species Homeostasis and Signalling during Drought and Salinity Stresses. *Plant, Cell & Environment* 33: 453–67.

Miozzi, L., Catoni, M., Fiorilli, V., Mullineaux, P. M., Accotto, G. P., and Lanfranco, L. 2011. Arbuscular Mycorrhizal Symbiosis Limits Foliar Transcriptional Responses to Viral Infection and Favors Long-Term Virus Accumulation. *Molecular Plant-Microbe Interactions* 24: 1562–72.

Miozzi, L., Vaira, A. M., Catoni, M., Fiorilli, V., Accotto, G. P., and Lanfranco, L. 2019. Arbuscular Mycorrhizal Symbiosis: Plant Friend or Foe in the Fight Against Viruses? *Frontiers in Microbiology* 10.

Miransari, M. 2013. Corn (Zea Mays L.) Growth as Affected by Soil Compaction and Arbuscular Mycorrhizal Fungi. *Journal of Plant Nutrition* 36: 1853–67.

Miransari, M., and Bahrami, H. A. 2008. Using Arbuscular Mycorrhiza to Alleviate the Stress of Soil Compaction on Wheat (*Triticum aestivum* L.) Growth. *Soil Biology and Biochemistry* 40: 1197–206.

Miransari, M., 2009. Effects of Soil Compaction and Arbuscular Mycorrhiza on Corn (*Zea mays* L.) Nutrient Uptake. *Soil and Tillage Research* 103: 282–90.

Misra, S., Dixit, V. K., Khan, M. H., Mishra, S. K., Dviwedi, G., Yadav, S., and Chauhan, P. S. 2017. Exploitation of Agro-Climatic Environment for Selection of 1-Aminocyclopropane-1-Carboxylic Acid (ACC) Deaminase Producing Salt Tolerant Indigenous Plant Growth Promoting Rhizobacteria. *Microbiological Research* 205: 25–34.

Mitra, D., Navendra, U., Panneerselvam, U., Ansuman, S., Ganeshamurthy, A. N., and Divya, J. 2019. Role of Mycorrhiza and Its Associated Bacteria on Plant Growth Promotion and Nutrient Management in Sustainable Agriculture. *International Journal of Life Sciences and Applied Science* 1: 1–10.

Mora-Romero, G. A., and Cervantes-Gámez, R. G. 2015. Mycorrhiza-Induced Protection against Pathogens Is Both Genotype-Specific and Graft-Transmissible. *Symbiosis* 66: 55–64.

Munters, A. R. 2014. *The Foliar Bacterial Endophyte Community in Native Pinus radiata: A Role for Protection against Fungal Disease?* PhD Dissertation. Uppasala University.

Nadeem, S. M., Ahmad, M., Zahir, Z. A., Javaid, A., and Ashraf, M 2014. The Role of Mycorrhizae and Plant Growth Promoting Rhizobacteria (PGPR) in Improving Crop Productivity under Stressful Environments. *Biotechnology Advances* 32 (2): 429–48.

Nanjundappa, A., Bagyaraj, D. J., and Saxena, A. K. 2019. Interaction between Arbuscular Mycorrhizal Fungi and Bacillus Spp. in Soil Enhancing Growth of Crop Plants. *Fungal Biology and Biotechnology* 6 (23).

Naveed, M., Mitter, B., Reichenauer, T. G., Wieczorek, K., and Sessitsch, A. 2014. Increased Drought Stress Resilience of Maize through Endophytic Colonization *by Burkholderia phytofirmans* PsJN and *Enterobacter* sp. FD17. *Environmental and Experimental Botany* 97: 30–9.

Negrao, S., Schmockel, S. M., and Tester, M. 2017. Evaluating Physiological Responses of Plants to Salinity Stress. *Annals of Botany* 119: 1–11.

Nell, M., Wawrosch, C., Steinkellner, S., Vierheilig, H., Kopp, B., and Lössl, A. 2010. Root Colonization by Symbiotic Arbuscular Mycorrhizal Fungi Increases Sesquiterpenic Acid Concentrations in *Valeriana officinalis* L. *Planta Medica* 76: 393–98.

Nguyen, D., Rieu, I., Mariani, C., and van Dam, N. M. 2016. How Plants Handle Multiple Stresses: Hormonal Interactions Underlying Responses to Abiotic Stress and Insect Herbivory. *Plant Molecular Biology* 91: 727–40.

Olowe, O. M., Olawuyi, O. J., Sobowale, A. A., and Odebode, A. C. 2018. Role of Arbuscular Mycorrhizal Fungi as Biocontrol Agents against *Fusarium verticillioides* Causing Ear Rot *of Zea mays* L. (Maize). *Current Plant Biology* 15: 30–37.

Pal, A., and Pandey, S. 2016. Role of Arbuscular Mycorrhizal Fungi on Plant Growth and Reclamation of Barren Soil with Wheat (*Triticum aestivum* L.) Crop. *International Journal of Soil Science* 12: 25–31.

Paul, M. J., Primavesi, L. F., Jhurreea, D., and Zhang, Y. 2008. Trehalose Metabolism and Signaling. *Annual Review of Plant Biology* 59: 417–41.

Perez-de-Luque, A., Tille, S., and Johnson, I. 2017. The Interactive Effects of Arbuscular Mycorrhiza and Plant Growth-Promoting Rhizobacteria Synergistically Enhance Host Plant Defences against Pathogens. *Scientific Reports* 7: 16409.

Pieterse, C. M., Zamioudis, C., Berendsen, R. L., Weller, D. M., Van Wees, S. C., and Bakker, P. A. 2014. Induced Systemic Resistance by Beneficial Microbes. *Annual Review of Phytopathology* 52: 347–75.

Press, C. M., Loper, J. E., and Kloepper, J. W. 2001. Role of Iron in Rhizobacteria-Mediated Induced Systemic Resistance of Cucumber. *Phytopathology* 91 (6): 593–98.

Puri, S. C., Nazir, A., Chawla, R., Arora, R., Riyaz-ul-Hasan, S., Amna, T., Ahmed, B., Verma, V., Singh, S., Sagar, R., and Sharma, A. 2006. The Endophytic Fungus *Trametes hirsuta* as a Novel Alternative Source of Podophyllotoxin and Related Aryl Tetralinlignans. *Journal of Biotechnology* 122: 494–510.

Puza, V. 2015. Control of Insect Pests by Entomopathogenic Nematodes. In Lugtenberg, Ben, ed., *Principles of Plant-Microbe Interactions*. Springer International, Switzerland.

Rahman, S. F. S. A., Singh, E., Pieterse, C. M. J., and Schenk, P. M. 2017. Emerging Microbial Biocontrol Strategies for Plant Pathogens. *Plant Science* 267: 102–111.

Ravensberg, W. J. 2011. *A Roadmap to the Successful Development and Commercialization of Microbial Pest Control Products for Control of Arthropods*. Springer, Dordrecht.

Reid, A. 2011. Microbes Helping to Improve Crop Productivity. *Microbe* 6: 435–39.

Robert-Seilaniantz, A., Grant, M., and Jones, J. D. G. 2011. Hormone Crosstalk in Plant Disease and Defense: More than Just Jasmonate-Salicylate Antagonism. *Annual Review of Phytopathology* 49: 317–43.

Rojas, M. G., Elliott, R. B., and Morales-Ramos, J. A. 2018. Mortality of *Solenopsis invicta* Workers (Hymenoptera: Formicidae) after Indirect Exposure to Spores of Three Entomopathogenic Fungi. *Journal of Insect Science* 18(3): 20; 1–8.

Rosenblueth, M., and Martínez-Romero, E. 2006. Bacterial Endophytes and Their Interactions with Hosts. *Molecular Plant-Microbe Interactions* 19: 827–37.

Ryan, R. P., Germaine, K., Franks, A., Ryan, D. J., and Dowling, D. N. 2008. Bacterial Endophytes: Recent Developments and Applications. *FEMS Microbiology Letters* 278: 1–9.

Sarkar, A., Ghosh, P. K., Pramanik, K., Mitra, S., Soren, T., Pandey, S., Mondal, M. H., and Maiti, T. K. 2018. A Halotolerant Enterobacter Sp. Displaying ACC Deaminase Activity Promotes Rice Seedling Growth under Salt Stress. *Research in Microbiology* 169: 20–32.

Savary, S., Ficke, A., Aubertot, J. N., and Hollier, C. 2012. Crop Losses Due to Diseases and Their Implications for Global Food Production Losses and Food Security. *Food Security* 4: 519–537.

Schouteden, N., De Waele, D., and Panis, B. 2015. Arbuscular Mycorrhizal Fungi for the Biocontrol of Plant-Parasitic Nematodes: A Review of the Mechanisms Involved. *Frontiers in Microbiology* 6: 1280.

Shahzad, S. M., Khalid, A., Arif, M. S., Riaz, M., Ashraf, M., Iqbal, Z., and Yasmeen, T. 2014. Co-Inoculation Integrated with P-Enriched Compost Improved Nodulation and Growth of Chickpea (*Cicer arietinum* L.) Under Irrigated and Rainfed Farming Systems. *Biology and Fertility of Soils* 50 (1): 1–12.

Sharma, A., Johri, B. N., Sharma, A. K., and Glick, B. R. 2003. Plant Growth-Promoting Bacterium Pseudomonas Sp. Strain GRP3 Influences Iron Acquisition in Mung Bean (*Vigna radiata* L.). *Soil Biology and Biochemistry* 35 (7): 887–94.

Sharma, I. P., Chandra, S., Kumar, N., and Chandra, D. 2017. PGPR: Heart of Soil and Their Role in Soil Fertility. In Meena, V., Mishra, P., Bisht, J., Pattanayak, A. (eds) *Agriculturally Important Microbes for Sustainable Agriculture*. Springer, Singapore.

Siahmoshteh, F., Hamidi-Esfahani, Z., Spadaro, D., Shams-Ghahfarokhi, M., and Razzaghi-Abyaneh, M. 2018. Unraveling the Mode of Antifungal Action of *Bacillus subtilis* and *Bacillus amyloliquefaciens* as Potential Biocontrol Agents against Aflatoxigenic *Aspergillus parasiticus*. *Food Control* 89: 300–307.

Singh, A., Bhardwaj, R., and Singh, I. K. 2019. Biocontrol Agents: Potential of Biopesticides for Integrated Pest Management. In Bhoopander Giri (ed.) *Biofertilizers for Sustainable Agriculture and Environment*. Springer, Cham. 413–433.

Singh, L. P., Gill, S. S., and Tuteja, N. 2011. Unravelling the Role of Fungal Symbionts in Plant Abiotic Stress Tolerance. *Plant Signaling & Behavior* 6: 175–91.

Srivastava, S., Bist, V., Srivastava, S., Singh, P. C., Trivedi, P. K., Asif, M. H., and Nautiyal, C. S. 2016. Unraveling Aspects of *Bacillus amyloliquefaciens* Mediated Enhanced Production of Rice under Biotic Stress of *Rhizoctonia solani*. *Frontiers in Plant Science* 7: 587.

Sziderics, A. H., Rasche, F., Trognitz, F., Sessitsch, A., and Wilhelm, E. 2007. Bacterial Endophytes Contribute to Abiotic Stress Adaptation in Pepper Plants (*Capsicum annuum* L.). *Canadian Journal of Microbiology* 53 (1195): 202.

Tang, M., Chen, H., Huang, J. C., and Tian, Z. Q. 2009. AM Fungi Effects on the Growth and Physiology of *Zea mays* Seedlings under Diesel Stress. *Soil Biology and Biochemistry* 41: 936–40.

Trivedi, P., Pandey, A., and Palni, L. M. S. 2012. Bacterial Inoculants for Field Applications under Mountain Ecosystem: Present Initiatives and Future Prospects. In Maheshwari D. (ed.) *Bacteria in Agrobiology: Plant Probiotics*. Springer, Berlin, Heidelberg.

Vos, C., Claerhout, S., Mkandawire, R., Panis, B., Waele, D. D., and Elsen, A. 2012. Arbuscular Mycorrhizal Fungi Reduce Root-Knot Nematode Penetration through Altered root Exudation of Their Host. *Plant Soil* 354: 335–45.

Wang, Y., and Dai, C. C. 2011. Endophytes: A Potential Resource for Biosynthesis, Biotransformation, and Biodegradation. *Annals of Microbiology* 61: 207–15.

Watts-Williams, S. J., Cavagnaro, T. R., and Tyerman, S. D. 2019. Variable Effects of Arbuscular Mycorrhizal Fungal Inoculation on Physiological and Molecular Measures of Root and Stomatal Conductance of Diverse *Medicago truncatula* Accessions. *Plant Cell and Environment* 1: 285–294.

Wehner, J., Antunes, P. M., and Powell, J. 2010. Plant Pathogen Protection by Arbuscular Mycorrhizas: A Role for Fungal Diversity? *Pedobiologia* 53: 197–201.

Wu, Q. S. 2017. *Arbuscular Mycorrhizas and Stress Tolerance of Plants.* Springer, Singapore. https://doi.org/10.1007/978-981-10-4115-0

Xue, Q. Y., Chen, Y., Li, S. M., Chen, L. F., Ding, G. C., Guo, D. W., and Guo, J. H. 2009. Evaluation of the Strains of Acinetobacter and Enterobacter as Potential Biocontrol Agents against Ralstonia Wilt of Tomato. *Biological Control* 48 (3): 252–58.

Yang, Y., Song, Y., and Scheller, H. V. 2015. Community Structure of Arbuscular Mycorrhizal Fungi Associated with *Robinia pseudoacacia* in Uncontaminated and Heavy Metal Contaminated Soils. *Soil Biology and Biochemistry* 86: 146–58.

Yankova, V., Markova, D., Naidenov, M., and Arnaoudov, B. 2014. Management of Root-Knot Nematodes (*Meloidogyne* Spp.) in Greenhouse Cucumbers Using Microbial Products. *Türk Tarım ve Doğa Bilimleri Dergisi* 1(2): 1569–1573.

Yusran, Y., Roemheld, V., and Mueller, T. 2009. Efects of *Pseudomonas* sp."Proradix" and *Bacillus amyloliquefaciens* FZB42 on the Establishment of amf Infection, Nutrient Acquisition and Growth of Tomato Affected by *Fusarium oxysporum Schlecht* f. sp. Radicis-Lycopersici Jarvis and Shoemaker. UC Davis: Department of Plant Sciences, UC Davis. https://escholarship.org/uc/item/8g70p0zt

Zamioudis, C., and Pieterse, C. M. J. 2012. Modulation of Host Immunity by Beneficial Microbes. *Molecular Plant-Microbe Interactions* 25: 139–50.

Zhang, F., Jia-Dong, H. E., Qiu-Dan, N. I., Qiang-Sheng, W. U., and Zou, Y. N. 2018. Enhancement of Drought Tolerance in Trifoliate Orange by Mycorrhiza: Changes in Root Sucrose and Proline Metabolisms. *Notulae Botanicae Horti Agrobotanici Cluj-Napoca* 46: 270.

Zhao, L., Xu, Y., Sun, R., Deng, Z., and Yang, W. 2011. Identification and Characterization of the Endophytic Plant Growth Prompter Bacillus Cereus Strain MQ23 Isolated from *Sophora alopecuroides* Root Nodules. *Brazilian Journal of Microbiology* 42: 567–75.

Zhou, Y., Li, X., Gao, Y., Liu, H., Gao, Y. B., van der Heijden, M. G. A., and Ren, A. Z. 2018. Plant Endophytes and Arbuscular Mycorrhizal Fungi Alter Plant Competition. *Functional Ecology* 32 (5): 1168–79.

Zhu, M., He, Y., Li, Y., Ren, T., Liu, H., Huang, J., and Zheng, L.. 2020. Two New Biocontrol Agents against Clubroot Caused by *Plasmodiophora brassicae. Frontiers in Microbiology* 10: 3099.

Zolla, G., Badri, D. V., Bakker, M. G., Manter, D. K., and Vivanco, J. M. 2013. Soil Microbiomes Vary in Their Ability to Confer Drought Tolerance to Arabidopsis. *Applied Soil Ecology* 68: 1–9.

6 Plant-Microbial Interactions in Natural/ Organic Cultivation of Horticultural Plants

*Baljeet Singh Saharan, Jagdish Parshad,
Dinesh Kumar, Kanika, and Nidhi Sharma*

CONTENTS

6.1   INTRODUCTION

A major challenge of the twenty-first century is "sustainable agricultural practice," which includes the increased efficient usage of resources and regulation of biological processes under integrated agricultural management with the use of green technologies (Weekley et al., 2012). India is second in horticultural production in the world (Pathak et al., 2017). The demand for horticultural products is increasing rapidly,

DOI: 10.1201/9781003213864-6

which is also important for the economic growth of our country. There is a need to adopt green technologies with a sustainable approach and longevity that ensure an increase in production without affecting the environment and the nutritional quality of the product. India is one of the countries where biofertilizers are available commercially and supported by government policies regarding the sustainable development of biofertilizers such as *Azotobacter, Rhizobium,* and *Pseudomonas,* etc. For exports of horticultural products, the Ministry of Commerce and Industry has implemented the National Programme for Organic Production (NPOP) under the Foreign Trade Development Regulations Act. It includes policies for development, accreditation of certification agencies, and certification of organic products with compliance to the national standards and encourages organic cultivation in the country (October, 2001; APEDA stat 2017–2018). In India, the implementation of the NPOP has led to the growth of organic farming at a steady pace. Indian organic cultivation and its products have marked their presence in the global market with their increasing demand. According to the Agriculture and Processed Food Production Export Development Authority (APEDA) India (2017–2018), the amount of land under organic cultivation is around 3.56 million hectares (ha), and the total organic production is 1.70 million tons, which is increasing day by day because of increasing awareness. The plant rhizosphere is a region involving plant-microbe interaction in close association with the roots (Kennedy and Luna, 2005; Kusum and Saharan, 2014; Meena et al., 2017; Verma and Saharan, 2019). The interaction may be mutual, symbiotic, or proto-cooperative. The soil micro-flora plays an important role in the development and growth of the fruit plant. Microbes (bacteria, viruses, fungi, protozoa, helminths, and nematodes) interact directly and indirectly with plants. They facilitate plants in the uptake of nutrients (as biofertilizers) and can also be referred to as bioinoculants or biostimulators. These may be defined as a formulated cocktail of specific living microbes, which, when applied to seeds and crop surfaces, colonize in the rhizosphere and help recycle nutrients, increase availability of nutrients, release plant growth hormones, and protect from insect. Biofertilizers increase the fertility and moisture-holding capacity of soil, improve soil texture, and decrease the number of phytopathogens, which, ultimately, increase the biological yield of the plant (Narula et al., 2005; Rana et al., 2012). Microbes act on the dead and decaying organic matter, converting them into simpler forms. They also reduce environmental pollution by detoxification of hazardous compounds (Santal et al., 2016). The bioinoculants may act as bioindicators because of their relationship to agricultural microbiology, being realistic, unifying, empathetic, and easy to detect, and their biocatalysts may be considered as "biological fingerprints" of earlier agricultural practices and may relate to soil cultivation and its status (Utobo and Tewari, 2014). Various beneficial microbes and their groups have been shown in Figure 6.1.

6.2 POSSIBLE ROLE OF MICROBES IN HORTICULTURE

The increasing trend of organic farming has increased the use of biofertilizer and biocontrol agents, which are cost effective for farmers. Microbes play a vital role in biofortification and controlling diseases and produce antibiotics and enzymes that

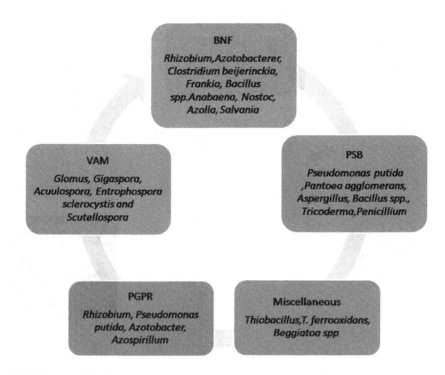

FIGURE 6.1   Microbes contributing to horticultural crops and their groups.

inhibit pests and pathogens. Their use results in the regeneration of nutrients in the soil that are easily assimilated by the plant. This is referred to as "green technology." The production of biofertilizer is cost effective; the substrates, such as whey, by-products, or waste products of sugar or beverage industries, can be used as raw material. Fruit plant-microbe interaction is also affected by abiotic sources, such as light response, that result in the altered yield and physiological changes in the plant, as the phyllospheric environment is variable (Figure 6.2) and different microbes respond to different intensity of light. For example, fungal plant interaction is considered more effective under low light than bacterial plant interactions (Alsanius et al., 2019).

Apart from being nutritional supplements for human and animals, seaweeds *Laminaria digitata* show efficiency as a biostimulator and are applied as foliar spray on plants. *Laminaria digitata*, *Ascophyllum nodosum*, *Plasmopora viticola* show a wide range of applications, such as resistance against phytopathogens (bacteria, virus, and fungi), increase in biometric parameters, and resistance to abiotic stress. Biofertilizers are more efficient when inoculated in combination or as a cocktail of microorganisms. According to Sarma et al. (2015), the application of biofertilizers, with a combination of *Azotobacter*, farmyard manure (FYM), and phosphate solubilizing bacteria (PSB) on carrots increased root girth, functional leaf count, stalk weight, fresh weight of root, root length, and cost efficiency. *Glomus deserticola* and *Rhizobium trifoli*, when used together, increase the number of nodules in

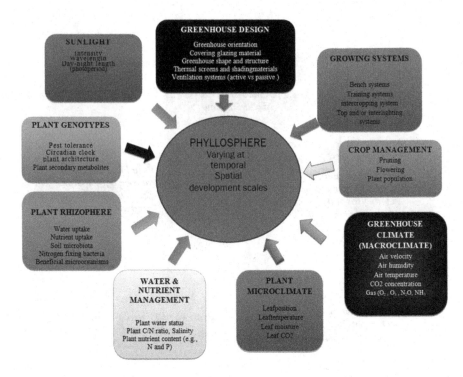

FIGURE 6.2   Growth parameters affecting the phyllosphere of horticultural crops.

plants by four times and increase the nitrogen fixing ability (Slangen and Kerkhoff, 1984; Malusa et al., 2011). Microbes, namely *Serratia, Rhizobium, Rhizobacter, Pseudomonas, Phosphobacter, Paenibacillus, Lysobacter, Klebsiella, Gordonia, Enterobacter, Bacillus, Azotobacter*, and *Azospirillum* show the potential for pesticide biodegradation. On the other hand, *Xanthomonas, Variovox, Sinorhizobium, Rhizobium meliloti, Ralstonia* sp., *Psycrobacter, Pseudomonas* sp., *Pseudomonas fluorescens, P. aeruginosa, Ochrobactrum, Mesorhizobium* sp., *Kluyvera* sp., *Brevibacillus* sp., *Bradyrhizobium, Bacillus subtilis, Bacillus megaterium, Azotobacter chroococcum*, and *Achromobacter xylosoxidans* are among some of the important rhizospheric bacteria that play a pivotal role in bioremediation (Saharan and Nehra, 2011; Mahanty et al., 2017).

6.3   BIOLOGICAL NITROGEN FIXATION

Nitrogen is required for the biosynthesis of proteins, phytohormones, nucleic acids, enzymes, and many other biomolecules. Nitrogen (N_2) fixation is an important phenomenon that involves many microbes, e.g., blue green algae (*Nostoc, Anabaena*), *Rhizobium* (pulses), *Frankia*, symbiotic relationship between *Anabaena–Azolla*, cyanobacteria, and *Anthoceros*. Due to their N_2 fixation properties, these microbes are considered to be biological nitrogen fixation (BNF) agents. The atmospheric nitrogen is fixed by certain microbes in the presence of the biocatalyst Mo-nitrogenase, which converts elemental nitrogen into organic nitrogenous compounds. A total of 70%– 80% of elemental nitrogen is biologically fixed per year, out of which approximately 24–584 kg N/ha comes from *Rhizobium*, 2–362 kg N/ha comes from *Frankia*, and 45–450 kg N/ha comes from symbiont *Azolla–Anabaena* (Pathak et al., 2017).

Further the nitrogen is assimilated (conversion of nitrogen to nitrates and nitrites) by microbes associated with the plants. The process is carried out by *Actinomycetes* and *Bacillus* spp. (*B. ramosus*, *B. valgarius*), which act upon the dead and decaying matter and convert complex nitrogenous compound into ammonia. Bacteria such as *Nitrosomonas*, *Nitrococcus*, *Nitrosogoloea*, *Nitobacter*, and *Nitrocysts* convert ammonia into nitrites and nitrates using a process referred to as nitrification. In this way, the nitrogen pool is maintained, and nutrient recycling takes place. It is reported that *Azotobacter* is a very important organism that alone contributes up to 50% of the nitrogen requirement to the plants. It increases biometric parameters such as runners count, heavier fruit, fruit length, and fruit breadth in strawberries.

6.4   PLANT GROWTH-PROMOTING RHIZOBACTERIA

Plant growth-promoting rhizobacteria are more potent than other members of the soil microflora (Ruzzi and Aroca, 2015). They provide huge benefits to the horticultural plants by producing phytohormones required for plant growth and immunity, including abscisic acid, cytokines, gibberellins, indole acetic acid (IAA), 1-amino-cyclopropane-1-carboxylic acid (ACC) deaminase, and auxin. They increase the stress tolerance of the plant during drought conditions. They also secrete volatile compounds such as butanol and pentanol-related compounds, which make the plant resistant against the phytopathogens. They produce siderophores that have a negative effect on the pathogenic microbes (Kloepper et al., 1980). The organisms may be symbiotic, free-living, and/or associative diazotrophs, i.e., the strains of *Variovorax*, *Serratia*, *Rhizobium*, *Pseudomonas*, *Pantoea*, *Comamonas*, *Burkholderia*, *Bacillus*, *Azotobacter*, *Azospirillum*, *Arthobacter*, *Alcaligens*, and *Agrobacterium*. A tremendous increase is seen in the antioxidant activity (anthocyanins, carotenoids, vitamins) and nitrogen content in tomato, basil, strawberry, and *capsicum* with the application of a cocktail of biostimulator containing *Rhizobium leguminosarum*, *Pseudomonas* sp., *P. putida*, *Bacillus licheniformis*, *B. lentus*, *Azotobacter chroococcum*, *Azospirillum lipoferum*, and *A. brasilens* (Jiménez-Gómez et al., 2017) (Table 6.1). PGPR show overall improvement in yield, length, leaf width, and yield (Saharan and Nehra, 2011; Ruzzi and Aroca, 2015).

6.5   ARBUSCULAR MYCORRHIZAE (AM)

Arbuscular-mycorrhizal symbiosis is a well-known association. About 160 taxa of *Glomeromycota* are differentiated on the basis of root infection patterns and spore morphology (Khan, 2005). Vesicular-arbuscular mycorrhiza (VAM) (association of phycomycetous fungi and angiosperm roots) and Arbuscular mycorrhizae fungi (AMF) have a mutualistic, symbiotic relationship between the fungi and the roots

TABLE 6.1

Plant Growth-Promoting Rhizobacteria and Their Advantages

Sr. No	Name(s)	Advantage(s)/traits exhibited
1.	*Azotobacter* sp.	Biological nitrogen fixation, cytokinin production
2.	*Bacillus* sp.	Auxin, cytokinin, gibberlins, K solubilization, siderophore production
3.	*Pseudomonas* sp.	Chitinase, β-glucanase, ACC deaminase production, stress resistance, siderophores production
4.	*Rhizobium* sp.	Indole acetic acid synthesis, ACC deaminase synthesis, phosphate solubilization, siderophore production, nitrogen fixation
5.	*Streptomyces* sp.	IAA (phytohormone) synthesis, production of siderophores
6.	*Phyllobacterium* sp.	Phosphate solubilization, siderophore production
7.	*Chryseobacterium* sp.	Siderophore production

of the plants. They adhere to plant rhizoid, penetrate, and form arbuscules within the root cortical and develop hyphae. Hyphae spread wide in the soil resulting in a significant increase in the rhizosphere. Mycorrhizae as a biofertilizer are very useful in organic and commercial farming. About six genera of the Endogonaceae family are involved in mycorrhizal associations: *Scutellospora*, *Sclerocystis*, *Glomus*, *Gigaspora*, *Entrophospora*, and *Acaulospora*. Sporocarps and specific spores play an important role in their characterization and identification. With the increased economic value of horticultural products, the demand for quality products of the same has also increased. When mycorhizal species such as *Glomus lipoferum*, *Glomus mossea*, and *Glomus etunicatum* are used as biofertilizer in tomatoes, it leads to an increase in lycopene. The strains of *Glomus* and *Rhizobium* help to reduce phytopathogens (Malusa et al., 2011). AMF strengthen plants' adaptability toward climate change. It provides tolerance to respective plants against stressed agro-climatic conditions, including an unfavorable temperature range, toxic/heavy metals, salinity, heat, and drought (Figure 6.3). It helps the host to upregulate tolerance mechanisms and prevent downregulation (intermediary metabolism) (Begum et al., 2019).

The horticultural plants that are grown in nurseries usually lack microflora and need to be inoculated with beneficial organisms. Plants with high economic value are inoculated with the AMF bioinoculants for better growth to meet the quantity and quality parameters (Bianciotto et al., 2018).

6.6 PHOSPHATE SOLUBILIZING MICROBES (PSM)

Phosphates are abundant in the soil, but they are present in an inorganic form that plants are unable to uptake. Phosphate compounds that are used as a chemical fertilizer stay in a soluble form for a long time and convert into an insoluble form. A wide range of bacteria and fungi, known as phosphate solubilizing microbes (PSM), convert the elemental phosphate into organic phosphate, which can be easily utilized

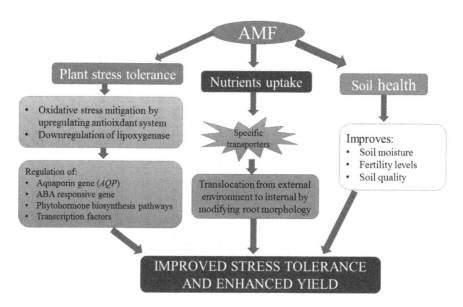

FIGURE 6.3   Abiotic stress alleviation by AMF.

by the plants (Mahanty et al., 2017). Phosphates are required for the synthesis of the nucleic acid, maintenance of the integrity of the cell membrane, and formation of a few organic acids around 0.2% of the dry weight (Muraleedharana et al., 2010). PSM include strains of *Scutellospora* sp., *Sclerocystis* sp., *Pseudomonas striata*, *Pisolithus* sp., *Penicillium* sp., *Laccaria* sp., *Glomus* sp., *Gigaspora* sp., *Boletus* sp., *Bacillus megaterium*, *Bacillus circulans*, *Aspergillus awamori*, *Amanita* sp., and *Acaulospora* sp. PSB are reported to increase crop yields by about 200–500 kg/ha. Wheat, tomatoes, tobacco, soybean, potatoes, peas, oats, and groundnuts are reported to show an average increase of 10%–15% in yields (Pathak et al., 2017).

6.7   MICROBES IN HORTICULTURE

Microbes in horticulture play a vital role and provide endless advantages. There are different modes of microbial treatments in horticultural crops (Slangen and Kerkhoff, 1984). These treatments aim to reduce the pathogens and reintroduce the growth-promoting beneficial organisms (Table 6.2). The bioinoculants have various desirable properties for which they are preferred over chemical fertilizers. The following applications are common for crops.

6.7.1   Seed Treatment

The most commonly used seed treatment method with bioinoculants on the seed coat requires the application of liquid broth or powder directly on the seed. About 20g of PSB can be used for 1kg legume seeds (Muraleedharana et al., 2010). The

TABLE 6.2
Various Plants and Their Bioinoculants

S. No.	Crop(s)	Botanical name(s)	Species/strain	Application	Reference(s)
1.	Carrot	*Daucus carota*	*Azotobacter* + farm yard manure (FYM) + RP + PSB	Seed treatment	Sarma et al., 2015
2.	Cabbage	*Brassica oleracea var. capitata*	*Pseudomonas fluorescens* and humic acid	Root treatment	Verma et al., 2017
3.	Grape vine	*Vitis vinifera*	*Ascophyllum nodosum*	Foliar spray	Saharan and Nehra, 2011
4.	Spinach	*Spinacia oleracea*	*Ascophyllum nodosum*	Root treatment	Fan et al., 2011; Saharan and Nehra, 2011
5.	Strawberry	*Fragria vesca*	*Pseudomonas* sp. strain Pf4 +AFM	Liquid inoculation	Bona et al., 2015
6.	Pomegranate	*Punica granatum*	*A. brasilense, A. chroococcum, G. fasciculatum, and G. mosseae*	Inoculation while grafting and root treatment	Aseri et al., 2008
7.	Broccoli	*Brassica oleracea*	*Brevibacillus reuszeri+ Rhizobium rubi*	Root dipping	Yildirim et al. (2011)
8.	Tomato	*Lycopersicon esculentum*	vermicompost+ PSB+ *Azotobacter* + *T. viride*	Seedling treatment	Thakur, 2017
9.	Banana	*Musa*	*Azotobacter*+ *Azospirillum* + phosphobacteria	Inoculation	Hazarika and Ansari, 2007
10.	Lettuce	*Lactuca sativa*	PGPR	Root and foliar treatment	Collaa et al., 2015
11.	Kiwi fruit	*Antinidia deliciosa*	PGPR	Folair treatment	Collaa et al., 2105
12.	Guava	*Psidium guajava*	FYM+*Azotobacter*+vermi compost	Soil treatment	Ram and Pathak, 2007
13.	Chrysanthemum	*Chrysanthemum morifolium Ramat*	*Mycorrhiza*+PSB+Phosphorous	Soil treatment	Kumari et al., 2016
14.	Onion	*Allium cepa*	*Azospirillum* + Recommended dose of NPK+FYM	Soil treatment	Talwar et al., 2017

treated seed must be dried over night before it can be used. The seed treatment shows more results in combination with microbes such as *Rhizobium*, PSB, and PGPR. This treatment is commonly used for legumes.

6.7.2   ROOT DIP (SEEDLING) METHOD

Seedling roots are treated with biofertilizers before transplantation in the field by dipping the roots in a culture of bioinoculants. Treatment shows efficiency in vegetable crops, which require transplantation of seedlings (Muraleedharana et al., 2010). This treatment is generally performed in cole crops, flowers, onions, rice, and tomatoes.

6.7.3   WHOLE FIELD

The FYM and compost are used for bioinoculants application in the field. In fruit crops, at the time of planting of saplings, the pit is treated with compost, farmyard manure, and biofertilizers. Biofertilizers may also be introduced directly in the field, but this requires 4–10 times more compost, farmyard manure, and biofertilizers.

6.7.4   SELF OR TUBER INOCULATION

Azotobacter is used for the treatment of tuber crops. In this method, tuber crops, such as sweet potato and potato (tubers), are dipped directly in bioinoculants before they are planted.

6.7.5   COMPOST

The conversion of organic solid waste, such as plant debris, leaves, peels of vegetables/fruits, municipal waste, and dung, into manure with a mixed community of microorganisms. The diversity of the microbial community varies at different stages during compositing because of change in temperature, pH, and aeration. The lignocellulolytic enzymes produced by different microorganisms – cellulase, protease, lipase, and amylase – degrade the organic material and release nitrogen (N), phosphorus (P), potassium (K), and carbon (C) as well as organic and inorganic compounds in soil (Vargas-García *et al.*, 2010). The incubation period for *Aspergillus*, *Tricoderma*, *Pencicillium*, *Azotobacter*, *Actinomycetes*, and *Bacillus* sp. is about 5–6 months. It can be spread directly onto the land and favorably influences the physiochemical properties of plants.

6.7.6   VERMICOMPOST

The organic solid waste that include leaves, peels of vegetables, dung of animals, weeds, straw of crops, and other agricultural waste are acted upon by species of earthworms, such as *Eiseniafetida* (redworm), *Eudrilus eugeniae* (African earthworm), *Perionyx excavates* (composting worm), and *Pheretima elongate*. In 5–6 weeks, compost that maintains 40% of the moisture is prepared, packed, and marketed (Pandit et al., 2012). Vermicompost and vermiwash have higher organic carbon as well as nitrogen, phosphrous, potassium, and essential macro and micro elements, resulting in higher nutritional value over other organic fertilizers. They help manage soil health and increase yield. Vermiwash consists of microbes that are beneficial for plants, whereas vermicompost is an odorless, dark brown biofertilizer that is rich in microbes and micro- and macronutrients obtained from the process of vermin culture technology. Both are an excellent soil additive and act as effective biofertilizers made up of digested compost (Soni and Sharma, 2016).

6.7.7 GENETICALLY MODIFIED (GM) CROPS

With the desired changes (traits of interests) in the genetic structure of plants by the use of biotechnology, GM crops have increased yield, salinity tolerance, and drought resistance and reduced water requirements. These crops are a gift to the world in changing environmental conditions (Mtui, 2011). Micro propagation was the first technique used for crop improvement and is still in use; it provided disease-free plants with higher yields. Transgenic crops include tomatoes (FlavrSavr), eggplants, potatoes, broccoli, cabbage, and tobacco. Edible vaccine genes are incorporated into fruits and vegetables using this technology; they are beneficial for humans and are referred to as value-added products (Gaur et al., 2018). The era of "genome editing," includes techniques such as RNAi, ZINC–Finger nucleases, and CRISPR/CAS9. Such technologies have widened the horizon for gene editing and make plants able to survive in the changing environmental conditions.

6.7.8 NANOFERTILIZER TREATMENT

As environmental conditions change drastically, there is a need for smart approaches that allow plants to survive without compromising nutritional quality. These nano-particles have several properties such as large surface area, target specificity, and excellent drug delivery, which makes them very useful in agricultural practices. Metal oxides such as zinc (Zn), silver (Ag), iron (Fe), and magnesium (Mg) are used to synthesize nanoparticles. Fertilizer is bulk packed with the nanoparticles, using the foliar spray method as the mode of application (Raliya et al., 2017). Nanofertilizer and pesticide and herbicide sensors are formulated by researchers. Carbon nanotubes are used as biosensors in agriculture (Rameshaiah et al., 2015). This protects fruits and vegetables and reduces post-harvest losses. Nanofertilizers, nanopesticides, nanoherbicides, and biosensors are used with the biofertilizers, which is a good step toward the sustainable goals of the "green era" (Kirmani et al., 2019).

6.8   ADVANTAGES OF BIOINOCULANTS ON PLANT GROWTH, DEVELOPMENT, AND YIELD

The multiple desirable properties of bioinoculants have been summarized in Figure 6.4.

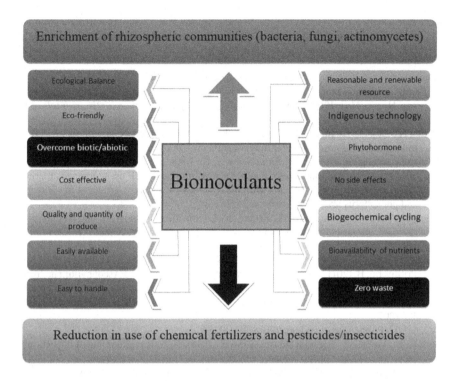

FIGURE 6.4 Desirable properties of bioinoculants for horticultural plants.

6.9 CONCLUSION AND FUTURE PERSPECTIVES

The use of biofertilizers has increased the quality of soil. It has increased the yield and nutritional level of the products, and this, ultimately, allows farmers to receive the premium price for their produce in the market. The application of biofertilizers (in different combinations) is more effective and economically viable/feasible than chemical-based fertilizers. It also protects the plants from environmental stresses. The use of biofertilizers is a solution to reduce the use of chemical fertilizers and their impact on the environment. Emerging biotechnological tools such as genome editing RNAi, CRISPR/Cas9, and nanotechnology are the gift of the intellectual scientific community to the world. The benefits of such green technologies can only be gained with the use of more green approaches, spreading awareness, and educating farmers about the techniques.

REFERENCES

Agricultural and Processed Food Products Export Development Authority (APEDA), Annual report, 2017–18.

Alsanius, B. W., Karlsson, Rosberg, A. K., Dorais, M., Naznin, M. T., Khalil, S. and Bergstrand, K. J. (2019). Light and microbial lifestyle: The impact of light quality on plant microbe interactions in horticultural production systems—A review. *Horticulturae*, 5(2): 1–24.

Aseri, G. K., Jain, N., Panwar, J., Rao, A. V. and Meghwa, P. R. (2008). Biofertilizers improve plant growth, fruit yield, nutrition, metabolism and rhizosphere enzyme activities of Pomegranate (*Punica granatum* L.) in Indian Thar Desert. *Scientia Horticulturae*, 117: 130–135.

Bianciotto, V., Victorino, I., Scariot, V. and Berruti, A. (2018). Arbuscular mycorrhizal fungi as natural biofertilizers: Current role and potential for the horticulture industry, ActaHortic. 1191. *ISHS 2018 III International Symptoms on Woody Ornamentals of the Temperate Zone*. Ed.: S. C. Hokanson.

Begum, N., Qin, C., Ahanger, M. A., Raza, S., Khan, M. I., Ahmed, N. and Zhang, L. (2019). Role of arbuscular mycorrhizal fungi in plant growth regulation: implications in abiotic stress tolerance. *Frontiers in Plant Science*, 10: 1068.

Bona, E., Lingua, G. and Manassero, P. (2015). AM fungi and PGP pseudomonads increase flowering, fruit production, and vitamin content in strawberry grown at low nitrogen and phosphorus levels. *Mycorrhiza*, 25: 181–193.

Collaa, G., Nardi, S., Cardarelli, M., Ertani, A., Lucini, L., Canaguier, R. and Rouphael, Y. (2015). Protein hydrolysates as biostimulants in horticulture. *Scientia Horticulturae*, 196: 28–38.

Fan, D., Hodges, D. M., Zhang, J., Kirby, C. W., Ji, X., Locke, S. J., Critchley, A. T. and Prithiviraj, B. (2011). Commercial extract of the brown seaweed Ascophyllum nodosum enhances phenolic antioxidant content of spinach (*Spinacia oleracea* L.) which protects *Caenorhabditis elegans* against oxidative and thermal stress. *Food Chemistry*, 124(1): 195–202.

FAO. (2018). Publications catalogue 2018, FAO USA.

Gaur, R. K., Verma, R. K. and Khurana, S. M. P. (2018). Genetic engineering of horticultural crops: present and future. *Genetic Engineering of Horticultural Crops*.

Hazarika, B. N. and Ansari, S. (2007). Biofertilizers in fruit crops - A review. *Agricultural Reviews*, 28(1): 69–74.

Jiménez-Gómez, A., Celador-Lera, L., Fradejas-Bayón, M. and Rivas, R. (2017). Plant probiotic bacteria enhance the quality of fruit and horticultural crops. *AIMS Microbiology*, 3(3): 483.

Kennedy, A. C. and Luna, L. Z. (2005). Rhizosphere, encyclopedia of soils in the environment.

Khan, A. G. (2005). Role of soil microbes in the rhizospheres of plants growing on trace metal contaminated soils in phytoremediation. *Journal of Trace Elements in Medicine and Biology*, 18: 355–364.

Kirmani, S. N., Singh, D. B., Mir, J. I., Raja, W. H. and Nabi, S. U. (2019). Nanotechnology: A novel way for enhancing horticultural crop productivity, NBL-2019.

Kloepper, J. W., Leong, J., Teintze, M. and Schroth, M. N. (1980). Enhanced plant growth by siderophores produced by plant growth-promoting rhizobacteria. *Nature*, 286: 885–886.

Kumari, A., Goyal, R. K., Choudhary, M. and Sindhu, S. S. (2016). Effects of some plant growth promoting rhizobacteria (PGPR) strains on growth and flowering of chrysanthemum. *Journal of Crop and Weed*, 12(1): 7–15.

Kusum and Saharan, B. S. (2014). Studies on plant growth promoting abilities of *Pseudomonasputida* SKR5 isolated from the rhizosphere of *Steviarebaudiana*. *AnnalsBiology*, 30(4): 613–619.

Mahanty, T., Bhattacharjee, S., Goswami, M., Bhattacharyya, P., Das, B., Ghosh, A. and Tribedi, P. (2017). Biofertilizers: A potential approach for sustainable agriculture development. *Environmental Science and Pollution Research*, 24(4): 3315–3335.

Malusa, E., Sas-Paszt, L. and Ciesielska, J. (2011). Technologies for beneficial microorganisms inocula used as biofertilizers. *The Scientific World Journal*, 2012(491206): 1–12.

Meena, Tara N. and Saharan, B. S. (2017). Plant growth promoting traits shown by bacteria Brevibacteriumfrigrotolerans SMA23 isolated from Aloe vera rhizosphere. *Agricultural Science Digest*, 37(3): 226–231.

Mtui, G. Y. (2011). Involvement of biotechnology in climate change adaptation and mitigation: Improving agricultural yield and food security. *International Journal of Biotechnology and Molecular Biology Research*, 2(13): 222–231.

Muraleedharan, H., Seshadri, S. and Perumal, K. (2010). *Booklet on biofertilizer*. Shri AMM Murugappa Chettiar Research Institute, Chennai.

Narula, N., Kumar, V., Singh, B., Bhatia, R. and Lakshminarayana, K. (2005). Impact of grain yield in spring wheat under varying fertility conditions and wheat – cotton rotation. *Archives of Agronomy and Soil Science*, 51(1):79–89.

Pandit, N. P., Ahmad, N. and Maheshwari, S. K. (2012).Vermicomposting biotechnology: An eco-loving approach for recycling of solid organic wastes into valuable biofertilizers. *Journal of Biofertilizer and Biopesticides*, 3(1): 1–8.

Parmar, N., Singh, K. H., Sharma, D., Singh, L., Kumar, P., Nanjundan, J., Khan, Y. J., Chauhan, D. K. and Thakur, A. K. (2017). Genetic engineering strategies for biotic and abiotic stress tolerance and quality enhancement in horticultural crops: A comprehensive review. 3 *Biotech.*

Pathak, D. V., Kumar, M. and Rani, K. (2017). Biofertilizer application in horticultural crops. *Microorganisms for Green Revolution*, 215–227.

Raliya, R., Saharan, V., Dimkpa, C. and Biswas, P. (2017). Nanofertilizer for precision and sustainable agriculture: Current state and future perspectives. *Journal of Agriculture and Food Chemistry*, 66: 6487–6503.

Ram, R. A. and Pathak, R. K. (2007). Integration of organic farming practices for sustainable production of guava: A case study. *Proc. Ist IS on Guava*, Eds. G. Singh et al. Acta Hort. 735, ISHS.

Rameshaiah, G. N., Pallavi, J. and Shabnam, S. (2015). Nano fertilizers and nano sensors – An attempt for developing smart agriculture. *International Journal of Engineering Research and General Science*, 3(1): 314–320.

Rana, A., Saharan, B., Nain, L., Prasanna, R. and Shivay, Y. S. (2012). Enhancing micronutrient uptake and yield of wheat through bacterial PGPR consortia. *Soil Science and Plant Nutrition*, 58(5): 573–582.

Richardson, A. E. (2001). Prospects for using soil microorganisms to improve the acquisition of phosphorus by plants. *Austrilia Journal of Plant Physiology*, 28: 897–906.

Ruzzi, M. and Aroca, R. (2015). Plant growth-promoting rhizobacteria act as biostimulants in horticulture. *Scientia Horticulturae*, 196: 124–134.

Saharan, B. S. and Nehra, V. (2011). Plant growth promoting rhizobacteria: A critical review. *Life Sciences and Medicine Research*, 21: 1–30.

Santal, A. R., Singh, N. P. and Saharan, B. S. (2016). A novel application of *Paracoccuspantotrophus* for the decolorization of melanoidins from distillery effluent under static conditions. *Journal of Environmental Management*, 169: 78–83.

Sarma, I., Phookan, D. B. and Boruah, S. (2015). Influence of manures and biofertilizers on carrot (*Daucus carota* L.) cv. Early Nantes growth, yield and quality. *Journal of Eco-friendly Agriculture*, 10(1): 25–27.

Slangen, J. and Kerkhoff, P. (1984). Nitrification inhibitors in agriculture and horticulture: A literature review. *Fertilizer Research*, 5: 1–76.

Soni, R. and Sharma, A. (2016). Vermiculture technology: A novel approach in organic farming. *Indian Horticulture Journal*, 6(1): 150–154.

Talwar, D., Singh, K. and Walia, S. S. (2017). Influence of biofertilizers on microbial count and nutrient uptake of kharif onion (*Allium cepa* L.). *International Journal of Agriculture, Environment and Biotechnology*, 10(3): 289–294.

Thakur, N. (2017). Increased soil-microbial-eco-physiological interactions and microbial food safety in tomato under organic strategies. Springer Nature Singapore Pte Ltd. V. Kumar et al. (eds.), *Probiotics and Plant Health*.

Utobo, E. B. and Tewari, L. (2014). Soil enzymes as bioindicators of soil ecosystem status. *Applied Ecology and Environmental Research*, 13(1): 147–164.

Vargas-García, M. C. Suárez-Estrella, F., López, M. J. and Moreno, J. (2010). Microbial population dynamics and enzyme activities in composting processes with different starting materials. *Waste Management*, 30: 771–778.

Verma, R., Maurya, B. R., Meena, V. S., Dotaniya, M. L. and Deewan, P. (2017). Microbial dynamics as influenced by bio-organics and mineral fertilizer in alluvium soil of Varanasi, India. *International Journal of Current Microbiology and Applied Sciences*, 6(2): 1516–1524.

Verma, S. and Saharan, B. S. (2019). Harnessing potential PGPR from rhizospheric soils of *Ocimum* sp. *Paripex – Indian Journal of Research*, 8(11): 108–110.

Weekley, V., Gabbard, J. and Nowak, J. (2012). Micro-level management of agricultural inputs. *Emerging Approaches, Agronomy*, 2: 321–357.

Yildirim, E., Karlidag, H., Turan, M., Dursun, A. and Goktepe, F. (2011). Growth, nutrient uptake, and yield promotion of broccoli by plant growth promoting rhizobacteria with manure. *Horticulture Science*, 46(6): 932–936.

7 Seed Defense Biopriming

Shivangi Negi and Narendra K. Bharat

CONTENTS

7.1 INTRODUCTION

Seeds are a fundamental and essential part of the persistent growth in agriculture production, as 90% of crops are produced from seeds that require constant treatment to prevent seed-borne and seasonal diseases and damage from insects. Seed treatment is a method of applying some definite agents to the seed before sowing to make seeds resistant to pathogens, insects, and other pests. There are multiple methods to treat seeds. Root dipping, seed coating, soil application, foliar application, and seed inoculation are the methods that are used for introducing valuable microorganisms. In seed-coating treatment, the seed is uniformly covered by suspending it in solids and liquids, and, to ensure proper coating, seed adhesives are also used (Bardin and Huang 2003). Foliar application is another method of seed treatment. In this method, essential microorganisms are not widely used for crop plants. However, the method of applying useful microorganisms to the seeds is known as seed inoculation. A carrier-based material and adhesives with beneficial microbes are also used in this method, which improves treatment application and adhesion (Elegba and Rennie 1984).

Seed priming is an advanced seed treatment technique that is used to avoid the consequences of traditional treatment procedures. In priming, seeds are subjected to effective solutions depending on the characteristics of the seeds, which improves performance under controlled environments and in field conditions (Sharma et al. 2015). Biopriming is a type of priming and an emerging technique that uses environmentally friendly beneficial biocontrol agents. Biopriming incorporates both

DOI: 10.1201/9781003213864-7

physiological and biological aspects, which show enhanced growth and disease prevention (El-Mohamedy 2004; Fath El-bab et al. 2013). The seeds are soaked in the suspension of beneficial microorganisms for a predefined period of time. This allows the imbibition of the microbes and stimulates the defense mechanism into the seed (Abuamsha et. al. 2011). After imbibition, metabolic activities take place in the seed without an emerging radicle and plumule (Anitha et al. 2013). Bioprimed seeds frequently show evidence of an enhanced total germination percentage, uniformity, and rate (Basra et al. 2005). It also improves growth attributes even in diverse environmental conditions (Lin and Sung 2001), and it has also been observed that, in the imbibition process, the rate of metabolic repair is enhanced (Bray et al. 1989). In the priming procedure, seeds should be dried after treatment to retain effectiveness; if seeds are not dried after treatment, they may not germinate or show poor results after germination (Bradford 1986; Sivasubramaniam et al. 2011).

A diverse range of fungi, which showed favorable effects on plants, have also been studied. A variety of fungal biocontrol agents, such as *Trichoderma* spp., are commonly used, and they are known as biopesticides (Singh 2006; Bisen et al. 2016). The saprophytic fungi *Trichoderma* spp. generally exists in the rhizospheric region of the plant root. The different species of *Trichoderma* showed antagonistic activity against different seeds and soil-borne pathogens. It is also known as a potential plant symbiont, and many of the commercial biopesticides include *Trichoderma* spp. as a major component (Keswani et al. 2013; Singh et al. 2014). For field application, more than 250 products of *Trichoderma* spp. are commercially available in India (Singh et al. 2012; Sharma et al. 2014; Singh 2016).

7.2 WHAT IS SEED BIOPRIMING?

A prior sowing treatment followed by seed hydration in which seeds are prepared for the first stage of germination without radicle emergence is known as seed priming. However, the process of biologically pre-sowing, which refers to the inoculation and hydration of the seed with useful microorganisms to protect the seed, is called biopriming. In seed biopriming, a number of selected fungal and bacterial combinations are used against virulent seed- and soil-borne pathogens. This is an important substitute for chemical and physical control of crop disease management (Sivasubramaniam et al. 2011). In 2013, Reddy explained biopriming as the inoculation of seeds with beneficial bacteria that protects seeds against diseases. Biopriming is also explained as applying a layer of beneficial biocontrol agents and plant growth-promoting bacteria over the surface of the seed as a protective coating.

Seed biopriming is a single, attractive, and cost-effective approach to induce disease resistance and increase germination. In a broader sense, it is a biological strategy and an effective alternative to chemical control that is unique in that it uses live beneficial microbes. Biopriming is a cheaper, safer, and easily applicable seed treatment method (El-Mougy and Abdel-Kader 2008).

During this practice, seeds are subjected to the beneficial bioagents suspension, and physical uptake of water from the suspension occurs. This is a fast process because the water potential of dry seeds is usually very low as compared to the water potential of the suspension. This process is called imbibition. The imbibition mainly depends on suspension accessibility, seed composition, and permeability of the seed coat (Eskandari 2013; Lutts et al. 2016). The process completes the first phase of seed germination and initiates the second phase of germination in which most of the metabolic processes occur, including synthesis of enzymes and proteins; synthesis and repair of DNA, RNA, and mitochondria; breakdown of storage reserves; carbohydrate metabolism; lipid metabolism; and protein metabolism (Bewley 1997; Nonogaki 2006). In this phase, the seed converts stored food reserves into compounds needed for germination. The biopriming process makes the seed ready for early emergence, which is the third phase of germination. In addition, biopriming involves bioagent inoculation and its activity on or in the seed. These bioagents promote the synthesis of growth regulatory substances, such as gibberellins, which trigger enzymes, such as proteases, α- amylase, and nucleases. These germination-related enzymes hydrolyze and assimilate the stored material, providing all the nutrients to the seedling (Bench and Sanchez 2004).

7.3   SEED BIOPRIMING WITH BACTERIAL ANTAGONISTS AND PGPRS

There are is a vast range of bacterial antagonists available to use for seed biopriming, and these agents increase yield and growth and protect the crop plants from a broad range of virulent pathogens present in the environment. After the biopriming imbibition procedure and physiological reactions take place (Anitha et al. 2013), the seedling grows with a beneficial bacterial microbe associated in the root zone, creating an ideal state to colonize and induce stimuli prior to infection (McQuilken et al, 1988), and these microbes continue to multiply around the root (Taylor and Harman 1990). There are some commonly used bacterial antagonists and PGPRs in biopriming, including *Pseudomonas fluorescens*, *Pseudomonas aureofaciens*, *Pseudomonas aeruginosa*, *Serratia polymuthica*, *Pseudomonas chlororaphis*, and *Bacillus subtilis*. Biopriming aims to enhance growth through fast and high seed germination (Moeinzadeh et al. 2010). Hence, this process is considered to be one of the most effective treatments (Harman et al. 1989). Apart from disease management and plant growth, bioprimed seeds fight with other harmful microbes present in the rhizosphere and enhance the soil nutrient level by secreting different enzymes (e.g., lytic) and metabolites (Walker et al. 2003). They also produce siderophores that are iron metabolites of beneficial microbes that reduce the growth of deleterious pathogens in the soil (Van Loon and Bakker 2005). Several species of *Rhizobium* are also used in biopriming, and these species also secrete a substance like IAA (indole acetic acid) (Ahemad and Khan 2012). Production of IAA increased the formation of nodules through the division and differentiation of the cells (Ahemad and Kibret 2014). Legume crops show a symbiotic relationship with roots and soil via the formation of nodules. Auxins play an important role in these activities, whereas, some of the PGPRs produce only nitrogen in the root zone and show a non-symbiotic relationship (Glick 2012). Other than the leguminous crops, diazotrophs, which are free-living N_2 fixing bacteria present in the soil, were observed to have a non-obligate relationship (Glick et al. 1999). The significance role of biopriming with PGPRs in various crops and their effect on PGP activities is reviewed in Table 7.1.

TABLE 7.1
Important Role of Biopriming with Plant Growth-Promoting Rhizobacteria in Different Crops and Their Effect on Plant Growth-Promotion Activities

PGPR	Crop	PGR activities	References
Azotobacter chroococcum and *Azotobacter lipoferum*	Barley	Enhanced test weight, dry matter accumulation, grain yield/plant, biological yield/plant, and harvest index	Bahram et al. 2012
Pseudomonas spp.	Safflower	Enhanced branches, number of grains/plant, number of heads/plant, number of grains/head, head diameter, test weight, oil content, and grain yield	Sharifi 2012
Azospirillum lipoferum, A. chroococcum, and *Azotobacter chroococcum*	Maize	Grain yield, crop growth rate, and dry matter accumulation	Sharifi 2011
Azotobacter and *Azospirillum* spp.	Maize	Higher plant height, grain yield, number of grains/ear and number of kernels/ear	Sharifi and Khavazi 2011
Pseudomonas fluorescens	Sunflower	Improved seed germination, promoted root length, seedling growth and seedling weight	Moeinzadeh et al. 2010
Clonostachys rosea, P. Chlororaphis, and *P. fluorescens*	Carrot and onion	Increased emergence and yield	Bennett and Whipps 2009
Pseudomonas fluorescens P-93 and *Rhizobium* strain Rb-133	Bean	Increased seed yield, number of pods per plant, weight of 100 seed, seed protein yield, and number of seeds per pod	Yadegari and Rahmani 2010
Azospirillum	Rice	Increased germination rate, seedling growth and vigor	Kokila and Bhaskaran 2016
P. fluorescens	Chilli	Increased seed germination	Chauhan and Patel 2017

Biopriming in combination with biofertilizers, such as *Azospirillum* spp., *Pseudomonas putida, P. fluorescens, Bacillus subtilis,* and *B. lentus,* has been used in wheat crops (Saber et al. 2012). Saber et al. (2012) observed that the phosphorus or nitrogen demand of primed plants is lower than non-primed plants and that other agro-morphological traits were also enhanced. Similarly, when maize seeds were bioprimed with PGPRs, the fresh weight of the seedling was enhanced under protected conditions (Gholami et al. 2009).

7.4   SEED BIOPRIMING WITH FUNGAL ANTAGONISTS

As previously discussed, a huge number of microorganisms are present in soil. These microbes interact with plants and effect their growth. Some of these microbes are useful to the plants, while some can be harmful (Jacoby et al. 2017). To improve crop production globally and provide food security to the masses, there is a dire need for research and development of ecologically safe and environmentally beneficial methods in the field of agriculture. This can be done through the integration of beneficial microbes (Abhilash et al. 2016). Many microorganisms are used as seed dressers, which colonize the rhizosphere and improve the growth of the plant after the seedling has emerged. There are other types of microorganisms that are also beneficial to the crops and can also help in eradicating harmful fungi. One major group of fungi is *Trichoderma* spp., which is generally used for seed biopriming. These species are present in the soil rhizosphere and colonize the plant root. They stimulate defense mechanisms in the root and trigger parasitism and antibiosis production, which finally controls the pathogen infection and decreases the incidence of disease. These beneficial microbes increase the systemic resistance in plants and improve growth and yield (Harman et al. 2004). Common fungal antagonists used in biopriming are *Gliocladium virens, G. catenulateum, Phlebia gigantean, Candida oleophila, Trichoderma harzianum, T. koningii, T. hamatum, T. atroviridae, Vericillium lecani, Sporidesmium scleretivorum, Penicillium* spp., *Paecilomyces lilacinus, Ampelomyces quisqualis, Coniothyrium minitans, Verticillium lecani, Sporothrix flucculosa, Chaetominum globosum, Taoromyces flavus,* etc. Among them, *Trichoderma* spp. is a widely used saprophytic fungi present in the plant rhizosphere. Now, these species commonly occupy the largest share of fungal biocontrol agents in the biopesticide industry internationally (Woo et al. 2014). *Trichoderma* spp. shows broad-spectrum antagonistic activities and promotes plant growth against different seed-borne phytopathogens (Keswani et al. 2013). For these reasons, *Trichoderma* has been commonly used for seed biopriming and because it incites stress in host plants by inducing systemic resistance and solubilization and sequestration of inorganic nutrients as well as inactivates pathogenic enzymes and improves shoot and root development (Singh et al. 2014; Bisen et al. 2016). The important role of major fungal antagonists on different crops and their impact on plant growth is reviewed in Table 7.2.

Trichoderma strains colonize the root surface and protect the plants from invading harmful microorganisms by creating a barrier and produce antimicrobial metabolites in a concentrated way in the rhizoplane (Qu et al. 2020). Similarly, root

TABLE 7.2

Important Role of Biopriming with Fungal Antagonists in Different Crops and Their Effect on Plant Growth-Promotion Activities

Fungal antagonist	Crop	PGR activities	References
T. harzianum	Maize	Improved percent germination, vigor, emergence, seed yield, and test weight	Nayaka et al. 2007
T. viride and *T. harzianum*	Radish	Increased shoot and root length, seedling fresh weight, number of leaves, area, rate of photosynthetic and percent chlorophyll content	Mukhopadhyay and Pan 2012
Trichoderma asperellum BHUT 8	Pea	Enhanced shoot length, root length, number of leaves, seedling fresh weight, seedling dry weight	Singh et al. 2016
Trichoderma viride 01PP and *Trichoderma harzianum* Th.azad	Chickpea	Increased seed germination, yield, and growth parameters	Kumar et al. 2014
T. harzianum (Th 56, 69, 75, 82, and 89)	Wheat	Increased root vigor and enhanced drought tolerance	Shukla et al. 2014
Trichoderma spp.	Pea, lentil, chickpea, and red gram	Improved seed germination, promoted seedling length, seedling fresh weight, seedling dry weight, seedling growth and seedling weight	Meshram and Sarma 2017

colonization also triggers the induced systemic resistance (ISR), as the interaction of *Trichoderma* with the host is more stable after colonization (Shores et al. 2005 and Singh 2016). *Trichoderma* species can be sold as biofertilizers and biopesticides commercially (Harman 2000; Harman et al. 2004). There are a number of benefits of using *Trichoderma* spp. for seed biopriming, such as (i) the quick germination and emergence, allowing biocontrol agents (BCAs) to colonize in the root surface and give nutrients to the seedling; (ii) a number of defense mechanisms are triggered, which prevent pathogenic microbes; (iii) disease prevention and optimizing plant health; and (iv) promotion of seedling growth (Harman et al. 2004). In a study, *T. asperellum* (BHUT8) was used for seed biopriming, which induced defense responses and plant growth. This also stimulated the fungus metabolites, such as gallic and shikimic acid, which are known as the precursor to metabolites that are used in defense reactions of bioprimed plants (Singh et al. 2013). Chickpea seeds were bioprimed with *T. viride* and *T. harzianum* (Th.Azad). The results revealed that biopriming with these fungal antagonists decreased the incidence of wilt and enhanced the germination percentage and growth as compared to chemical fungicides (Kumar et al. 2014).

7.5 SEED BIOPRIMING WITH CONSORTIUM OF BACTERIAL ANTAGONISTS AND FUNGAL ANTAGONISTS

Different beneficial bacterial and fungal species were used in the past for the biological control of pathogens (Palmieri et al. 2017). Combined applications of these strains are very effective for growth, yield, and disease suppression. A consortium of important microbes enhanced resistance against a range of phytopathogens and increased growth through the regulation of defense responses by working together (Singh et al. 2013; Singh 2016). There are three important factors for reliable biocontrol: co-inoculation and their establishment and competition in the rhizosphere. These beneficial bacterial and fungal antagonists colonize the plant root and protect it from harmful pathogens and reduce the presence of other disease-causing agents (Berg et al. 2001; Whipps 2001). However, in the bioprimed seed, the same defense mechanism takes place (Weller 1983; Mahmood et al. 2016). The consortium induces antagonistic activity against pathogenic microorganisms by synthesizing hydrolytic enzymes and other substances that killed or inhibited the growth of the contaminants on treated seeds.

The consortium of different BCAs have been used for biopriming of seeds. Among these, *Pseudomonas*, *Trichoderma*, *Bacillus*, and *Rhizobia* are generally used (Negi et al. 2019), as they have shown to enhance plant yield and growth. Soil in the rhizosphere has a vast microbial community that interacts simultaneously with the plant root (Qu et al. 2020). The consortium of compatible BCAs present in the rhizosphere work simultaneously and improve the growth and yield of the plant.

Co-inoculation of beneficial antagonists enhanced the defense responses in the seed in opposition to different phytopathogens (Singh 2016). Seed biopriming in chickpea and rajma has been done by using rhizospheric beneficial microbes such as *T. asperellum* (T42), *P. fluorescens* (OKC), and *Rhizobium* (RH1) spp. in combination and individually. The results showed that high germination and growth in bioprimed plants as compared to crops in the control. Also, the consortium shows better outcomes as compared to individuals (Yadav et al. 2013). In the case of *Phaseolus vulgaris*, the in vitro screening of PGPRs, biocontrol agents, and *Rhizobium* showed higher germination and improved seed quality and health (Negi et al. 2020). The significant role of biopriming with the consortium of microorganisms in different crops and their effect on plant growth promotion activities are reviewed in Table 7.3.

TABLE 7.3

Important Roles of Biopriming with Microbial Consortium in Different Crops and Their Effect on Plant Growth-Promotion Activities

Strains	Crop	PGR activities	References
PGPR and *Rhizobium*	Bean	Increased number of pods/plant, number of seeds/pod, 100 seed weight, weight of seeds/plant, and weight of pods/plant	Yadegari and Rahmani 2010
T. viride and *P. fluorescens*	Chickpea	Increased seedling emergence to 96%–98%	Reddy et al. 2011
T. harzainum and *P. Aeruginosa*	Chickpea	Increased germination and disease resistance	Singh et al. 2012
Azospirillum and Phosphobacteria	Maize	Increased germination speed, germination percentage, seedling length, dry matter, and vigor	Karthika and Vanangamudi 2013
P. Fluorescens (OKC), *T. asperellum* (T42), and *Rhizobium* (RH4) spp.	Chickpea French bean	Enhanced germination, seedling length, dry shoot biomass, and plant growth	Yadav et al. 2013
T. harzianum (THU0816), *P. aeruginosa* (PHU094), and *Mesorhizobium* (RL091) spp.	Chickpea	Increased the deposition of lignin, germination and yield, and activated PP pathway	Singh et al. 2013
B. cereus, B. amyloliquefaciens, Pseudomonas aeruginosa and *Trichoderma citrinoviride*	Soybean	Reduction in disease incidence, increased seed vigor, chlorophyll content and growth parameters	Thakkar and Saraf 2014
Trichoderma spp. and *B. Subtilius*	Bean	Increased seedling growth	Junges et al. 2016
Rhizobium and *Pseudomanas*	Chickpea	Increased germination, root shoot length, and vigor	Vishwas et al. 2017
T. harzainum and *B. subtilis*	Soybean	Increased plant height, pod yield, pod weight, number of seed/plant, and seed weight per plant	Mona et al. 2017
PGPR-1 and *Rhizobium* starin B_1	French bean	Improved plant growth, pod yield, seed yield, seed vigor, seed quality, and disease suppression	Negi et al. 2019

7.6   BIOPRIMING IN RELATION TO INDUCED RESISTANCE

Seed bioprimng improves crop plants' defense ability by using a number of beneficial microbes. These microbes trigger the defense mechanisms and stimulate phenolic compounds and defense enzymes, which induce resistance in seed (Chen et al.

2000; Jetiyanon 2007). When bioprimed seeds enter a harmful environment, the stimuli from microbes create warning signals to the seed to fight against the foreign destructive pathogen (Mauch-Mani et al. 2017). Seed defense enzymes showed a strong positive correlation with growth-promoting parameters and negative correlation with disease incidence. The accumulation and activation of defense-related enzymes depend on the physical, physiological, and genotypic condition of the plant and pathogen (Tuzun 2001). Many plant defense apparatus are expressed against pathogens. As previously discussed, the defense ability in bioprimed seeds is elicited by stimuli, which is a physiological state of resistance, whereas it also triggered the inherent defense potential of the seed against subsequent biotic challenges. This physiological state, also called induced resistance, is very efficient against a wide range of parasites and pathogens (Van Loon 2000). Induced resistance is mainly regulated by salicylic acid (SA), jasmonic acid (JA), ethylene (ET), and phytohormones. There are two types of defense responses: systemic acquired resistance (SAR) and induced systemic resistance (ISR) and elicitor (Van Loon et al. 1998). SAR is stimulated when bioprimed seeds are exposed to pathogenic, non-pathogenic, virulent, and virulent microorganisms. A specific time period is requisite for the establishment of SAR in which deposition of SA and pathogenesis related (PR)-proteins, viz., glucanase, chitinase occurs.

Plant growth-promoting rhizobacteria play a major role in stimulating ISR in crop plants. A wide range of PGPR are being used for biopriming that are beneficial to the root system of plants and shows a positive impact directly on the plant (Van Loon and Glick 2004). SAR shows salicylic-mediated responses, which are deleterious to biotrophic phytopathogens, whereas jasmonic-mediated ISR responses are stimulated in opposition to necrotrophic and herbivore-borne phytopathogens (Bostock 2005; Song et al. 2017). These plant defense responses are triggered by the stimuli to the plant infected by the pathogen and, ultimately, reduce disease incidence. Cucumber seeds bioprimed with BCA *T. harzianum* (T-203) showed that the BCA colonized around the roots. When the root was exposed in a detrimental rhizosphere, they triggered plant defense responses to the plant and prepared the plant for the imminent infection (Yedidia et al. 1999). SAR leads to instant enhancement in the appearance of genes related to defense systems, whereas ISR induces gene expression. According to Conrath et al. (2015), at the moment the plant senses the elicitor, stimulation of defense responses is quicker, which is called defense priming. It infers that the defense elicitors originated from the useful microorganism have a positive effect on the plant (Song et al. 2017). The induction of these defense mechanisms in plants is presented in Figure 7.1.

A number of oxidative enzymes, such as polyphenoloxidase (PPO) and peroxidase (PO), have been incorporated into defense responses in opposition to harmful pathogens. These responses catalyze the production of phenols and lignin compounds to contribute to the development of defense barriers (Avdiushko et al. 1993). Some phenylalanine ammonialyase enzymes have been used for the biosynthesis of phenolic and phytoalexin. In many crop plants, defense enzymes also have been associated with these enzymes against the disease-causing pathogens (Binutu and Cordell 2000). Some of the plant growth-promoting fungi (PGPFs) also trigger the defense biopriming. A number of studies have been done in support of induced resistance

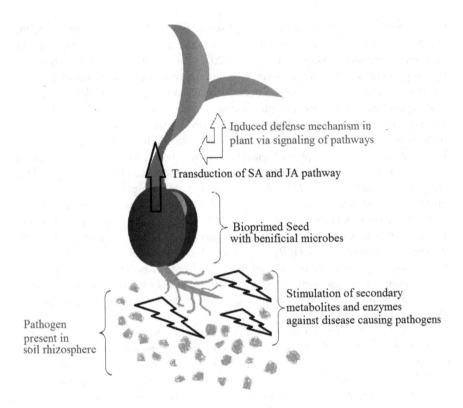

FIGURE 7.1   Induction of defense mechanism by seed biopriming with beneficial biocontrol agents.

and defense biopriming (Kloepper et al. 2004; Pieterse et al. 2014). PGPRs, PGPFs, mycorrhiza, chemical inducers, and biotic and abiotic stress showed useful interactions in the rhizosphere that enhance defense ability in plant parts (Mauch-Mani et al. 2017). Seed priming with JA and β-aminobutyric acid in tomatoes stimulates the defense mechanisms and enhances resistance to harmful pests (spider mites, caterpillars, aphids, and the necrotrophic pathogen *Botrytis cinerea*). β-aminobutyric acid seed priming used against the biotrophic pathogen reduced the incidence of powdery mildew (Worrall et al. 2012). Nakkeeran et al. (2006) used *B. subtilis* (BSCBE4) and *P. chlororaphis* (PA23) for seed biopriming against the damping off of pepper plants and found that BCAs showed effective results in suppression of *Pythium aphanidermatum* under protected vegetable production. Application of these strains at 20 g/kg of seed enhanced growth advanced the stimulation of defense responses and produced different phenylalanine ammonialyase (PAL), polyphenol oxidase (PPO) and peroxidase (PO) enzymes that ultimately suppress disease incidence. A study also reported that biopriming with the consortium of *B. amyloliquefaciens* IN937a and *B.*

pumilus IN937b has been used against diverse plant pathosystems, which triggered the defense responses in tomatoes. This also induced 25%–30% peroxidase activity as compared to the untreated control (Jetiyanon 2007).

Priming primarily includes imbibition together with a broad range of osmotic solutions, whereas priming using biologically useful microbes that increase plant immunity by secreting the defense-related enzymes is called "seed defense biopriming." A total of 1,825 species of *Bacillus* were isolated from the rhizospheric soil and heat stable metabolites were prepared from them. These metabolites are being used for biopriming in pepper and cucumber seeds. The bioprimed seeds triggered defense mechanisms and enhanced the plant immunity. Of these metabolites, seed defense biopriming with *B. gaemokensis*-PB69 significantly suppressed the diseases and enhanced the fruit yield under in vivo and in vitro conditions (Song et al. 2017). In tomatoes, the seeds were bioprimed with the *Bacillus subtilis* (BS2) strain and used against *Fusarium oxysporum* f sp. *lycopersici* in field conditions. Biopriming activates the defense mechanism in the plants and produces defense enzymes, viz., chitinase, peroxidase, ammonialyase, polyphenoloxidase, phenolics, and phenylalanine. These defense-related enzymes enhanced the plant growth, fruit set, and quality of fruit together with high amounts of lycopene at the time of harvest (Loganathan et al. 2014). Similarly, priming radish seeds with plant elicitors such as SA at 50 ppm and KNO_3 at 2% were effective for improving germination percentage, seedling growth, dry weight, vigor index, and dead seed percentage, whereas priming with SA at 75 ppm and BABA at 750 ppm was found to be efficient at reducing the seed infection percentage or seed mycoflora (Rohiwala et al. 2020).

7.7 SIGNIFICANCE OF SEED BIOPRIMING

Biological methods of pathogen suppression in crops are essential. To improve the stability of crop plants, the use of seed biopriming is one method. Biological seed priming is a model approach for increasing yield, higher growth, and disease suppression as compared to the existing methodologies. Bioprimed seeds showed harmonized and faster germination. Seedlings raised from bioprimed seeds are resistant to diseases and biotic and abiotic stresses as compared to unprimed seeds. The present chapter on seed defense biopriming emphasizes the potential role of microorganisms and their consortium in disease control and plant growth promotion.

REFERENCES

Abhilash, P. C., R. K. Dubey, V. Tripathi, V. K. Gupta, and H. B. Singh. 2016. Plant growth promoting microorganisms for environmental sustainability. *Trends Biotechnology* 10: 34–40.
Abuamsha, R., M. Salman, and R. Ehlers. 2011. Effect of seed priming with *Serratia plymuthica* and *Pseudomonas chlororaphis* to control *Leptosphaeria maculans* in different oilseed rape cultivars. *European Journal of Plant Pathology* 130: 287–295.

Ahemad, M., and M. S. Khan. 2012. Productivity of greengram in tebuconazole-stressed soil, by using a tolerant and plant growth-promoting *Bradyrhizobium* sp. MRM6 strain. *Acta Physiologiae Plantarum* 34: 245–254.

Ahemad, M., and M. Kibret. 2014. Mechanisms and applications of plant growth promoting rhizobacteria: Current perspective. *Journal King Saud University Sciences* 26: 1–20.

Anitha, D., T. Vijaya, and N. V. Reddy. 2013. Microbial endophytes and their potential for improved bioremediation and biotransformation: A review. *Indo American Journal of Pharmaceutical Research* 3: 6408–6417.

Avdiushko, S. A., X. S. Ye, and J. Kuc. 1993. Detection of several enzymatic activities in leaf prints cucumber plant. *Physiological and Molecular Plant Pathology* 42: 441–454.

Bahram, M., S. Hokmalipour, R. S. Sharifi, F. Farahvash, and A. E. K. Gadim. 2012. Effect of seed biopriming with plant growth promoting rhizobacteria (PGPR) on yield and dry matter accumulation of spring barley (*Hordium vulgare* L.) at various levels of nitrogen and phosphorus fertilizers. *Journal of Food, Agriculture and Environment* 10(3): 314–320.

Bardin, S. D., and H. C. Huang. 2003. Efficacy of stickers for seeds treatment with organic matter or microbial agents for the control of damping-off of sugar beet. *Plant Pathology Bulletin* 12: 19–26.

Basra, S. M. A., M. Farooq, and R. Tabassum. 2005. Physiological and biochemical aspects of seed vigor enhancement treatments in fine rice (*Oryza sativa* L.). *Seed Science and Technology* 33: 25–29.

Bench, A. L. R., and R. A. Sanchez. 2004. *Handbook of Seed Physiology*. Food Product Press, The Haworth Reference Press, New York/London/Oxford.

Bennett, A. J., and J. M. Whipps. 2001. Beneficial microorganism survival on seed, roots and in rhizosphere soil following application to seed during drum priming. *Biological Control* 44: 349–361.

Berg, G., A. Fritze, and N. Roskot. 2001. Evaluation of potential biocontrol rhizobacteria from different host plants of *Verticillium dahlia*. Kleb. *Journal of Applied Microbiology* 156: 75–82.

Bewley, J. D. 1997. Seed germination and dormancy. *The Plant Cell* 9: 1055–1066.

Binutu, O. A., and G. A. Cordell. 2000. Gallic acid derivative from *Mezoneuron benthamianun* leaves. *Pharmacology Biology* 38: 284–286.

Bisen, K., C. Keswani, J. S. Patel, B. K. Sarma, and H. B. Singh. 2016. *Trichoderma* spp.: Efficient inducers of systemic resistance in plants. In *Microbial-Mediated Induced Systemic Resistance in Plants*, ed. D. K. Chaudhary, and A. Verma, 185–195. Springer, Singapore.

Bostock, R. M. 2005. Signal crosstalk and induced resistance: Straddling the line between cost and benefit. *Annual Review of Phytopathology* 43: 545–580.

Bradford, K. J. 1986. Manipulation of seed water relations via osmotic priming to improve germination under stress conditions. *Horticultural Sciences* 21: 1105–1112.

Bray, C. M., P. A. Davison, M. Ashraf, and R. M. Taylor. 1989. Biochemical changes during osmopriming of leek seeds. *Annals of Botany* 63: 185–193.

Chauhan, R., and P. R. Patel. 2017. Evaluation of seed bio-priming against chilli (*Capsicum frutescence* L.) cv. GVC 111 *in vitro*. *Journal of Pharmacogynosy and Phytochemistry* 6(6): 17–19.

Chen, C., R. R. Belanger, N. Benhamou, and T. C. Paulitz. 2000. Defense enzymes induced in cucumber roots by treatment with plant growth-promoting rhizobacteria (PGPR) and *Pythium aphanidermatum*. *Physiological and Molecular Plant Pathology* 56: 13–23.

Conrath, U., G. J. Beckers, C. J. Langenbach, and M. R. Jaskiewicz. 2015. Priming for enhanced defense. *Annual Review of Phytopathology* 53: 97–119.

Elegba, M. S. and R. J. Rennie. 1984. Effect of different inoculant adhesive agents on rhizobial survival, nodulation, and nitrogenise (acetylene-reducing) activity of soybeans (*Glycine max* (L.) Merrill). *Canadian Journal of Soil Science* 64: 631–636.

El-Mohamedy, R. S. R. 2004. Bio-priming of okra seeds to control damping off and root rot diseases. *Annals of Agricultural Sciences* 49: 339–356.

El-Mougy, N. S. and M. M. Abdel-Kader. 2008. Long-term activity of bio-priming seed treatment for biological control of faba bean rot pathogens. *Australian Plant Pathology* 37: 464–471.

Eskandari, H. 2013. Effects of priming technique on seed germination properties, emergence and field performance of crops. *International Journal of Agronomy and Plant Production*, 4: 454–458.

Fath El-bab, T. S., and R. S. R. El-Mohamedy. 2013. Bio-priming seed treatment for suppressive root rot soil borne pathogens and improvement growth and yield of green bean (Phaseulas vulgaris L.,) in new cultivated lands. *Journal Applied Science Research* 9: 4378–4387.

Gholami, A., S. Shahsavani, and S. Nezarat. 2009. The effect of plant growth promoting rhizobacteria (PGPR) on germination, seedling growth and yield of maize. *Proceeding of World Academy of Science Engineering and Technology* 49: 19–24.

Glick, B. R. 2012. Plant growth promoting bacteria: Mechanism and applications. *Scientifica*: 1–15. doi:10.6064/2012/963401.

Glick, B. R., C. L. Patten, and G. Holguin. 1999. *Biochemical and Genetic Mechanisms Used by Plant Growth Promoting Bacteria*. World Scientific, Imperial College Press, London. http://doi.org/10.1142/p130.

Harman, G. E. 2000. Myths and dogmas of biocontrol: Changes in perceptions derived from research on *Trichoderma harzianum* T-22. *Plant Diseases* 84: 377–393.

Harman, G. E., C. R. Howell, A. Viterbo, I. Chet, and M. Lorito. 2004. *Trichoderma* species-opportunistic, avirulent plant symbionts. *Nature Review Microbiology* 2: 43–56.

Harman, G. E., A. G. Taylor, and T. E. Stasz. 1989. Combining effective strains of *Trichoderma harzianum* and solid matrix priming to improve biological seed treatments. *Plant Diseases* 73: 631–637.

Jacoby, R., M. Peukert, A. Succurro, A. Koprivova, and S. Kopriva. 2017. The role of soil microorganisms in plant mineral nutrition- current knowledge and future directions. *Frontiers in Plant Science* 8: 1617. doi:10.3389/fpls.2017.01617.

Jetiyanon, K. 2007. Defensive-related enzyme response in plants treated with a mixture of Bacillus strains (IN937a and IN937b) against different pathogens. *Biological Control* 42: 178–185.

Junges, E., M. F. B. Muniz, B. O. Bastos, and O. Pamela. 2016. Biopriming in bean seeds. *Acta Agriculturae Scandinavica, Section B- Soil and Plant Science* 66: 207–214.

Karthika, C., and K. Vanangamudi. 2013. Biopriming of maize hybrid COH (M) 5 seed with liquid biofertilizers for enhanced germination and vigour. *African Journal of Agricultural Research* 8: 3310–3317.

Keswani, C., S. P. Singh, and H. B. Singh. 2013. A superstar in biocontrol enterprise: *Trichoderma* spp. *Biotechnology Today* 3: 27–30.

Kloepper, J. W., C. M. Ryu, and S. Zhang. 2004. Induced systemic resistance and promotion of plant growth by *Bacillus* spp. *Phytopathology* 94: 1259–1266.

Kokila, M., and M. Bhaskaran. 2016. Standardization of *Azospirillum* concentration and duration of biopriming for rice seed vigour improvement. *International Journal of Agricultural Sciences* 12(2): 283–287.

Kumar, V., M. Shahid, A. Singh, M. Srivastava, and A. Mishra. 2014. Effect of biopriming with biocontrol agents *Trichoderma harzianum* (Th.Azad) and *Trichoderma viride* (01pp) on chickpea genotype (Radhey). *Journal of Plant Pathology and Microbiology* 5: 247–252.

Lin, J. M., and J. M. Sung. 2001. Pre-sowing treatment for improving emergence of bittergourd seedling under optimal and sub-optimal temperatures. *Seed Science and Technology* 29: 39–50.

Lutts, S., P. Benincasa, L. Wojtyla, et al. 2016. Seed priming: New comprehensive approaches for an old empirical technique. In *New Challenges in Seed Biology- Basic and Translational Research Driving Seed Technology*, ed. S. Araujo, and A. Balestrazzi, 1–46. In Tech Open, Rijeka, Croatia.

Mahmood, A., O. C. Turgay, M. Farooq, and R. Hayat. 2016. Seed biopriming with plant growth promoting rhizobacteria: A review. *FEMS Microbiology Ecology* 92: 1–13.

Mauch-Mani, B., I. Baccelli, E. Luna, and F. Victor. 2017. Defense priming: An adaptive part of induced resistance. *Annual Review of Plant Biology* 68: 485–512.

McQuilken, M.P., D. J. Rhodes, and P. Halmer. 1988. Application of microorganisms to seeds. In *Formulation of Microbial Biopesticides, Beneficial Microorganisms, Nematodes and Seed Treatments*, ed. H. D. Burges, 255–285. Kluwer Academic Publishers, Dordrecht.

Meshram, S., and B. Sarma. 2017. Comparative analysis of effects of seed biopriming on growth and development in different pulses: Pea, lentil, red gram and chickpea. *International Journal of Current Microbiology and Applied Science* 6(10): 2944–2950.

Moeinzadeh, A., F. Sharif-Zadeh, and M. Ahmadzadeh. 2010. Biopriming of sunflower (*Helianthus annuus* L.) seed with *Pseudomonas fluorescens* for improvement of seed invigoration and seedling growth. *Australian Journal of Crop Sciences* 4: 564–570.

Mona, M. M. R., A. M. A. Ashour, R. S. R. El-Mohamedy, A. A. Morsy, and E. K. Hanafy. 2017. Seed bio priming as biological approach for controlling root rot soil born fungi on soybean (*Glycine max* L.) plant. *International Journal of Agriculture Technology* 13(5): 771–788.

Mukhopadhyay, R., and S. Pan. 2012. Effect of biopriming of radish (*Raphanus sativus*) seed with some antagonistic isolates of *Trichoderma*. *Journal of Plant Protection Sciences* 4(2): 46–50.

Nakkeeran, S., K. Kavitha, G. Chandrasekar, P. Renukadevi, and W. G. D. Fernando. 2006. Induction of plant defence compounds by *Pseudomonas chlororaphis* PA23 and *Bacillus subtilis* BSCBE4 in controlling damping-off of hot pepper caused by *Pythium aphanidermatum*. *Biocontrol Science and Technology* 16: 403–416.

Nayaka, S. C., S. R. Niranjana, Uday, et al. 2007. Seed biopriming with novel strain of *Trichoderma harzianum* for the control of toxigenic *Fusarium verticillioides* and fumonisins in maize. *Archives of Phytopathology and Plant Protection* 43: 264–282.

Negi, S., N. K. Bharat, and M. Kumar. 2019. Effect of seed biopriming with indigenous PGPR, rhizobia and *Trichoderma* sp. on growth, seed yield and incidence of diseases in French bean (*Phaseolus vulgaris* L.). *Legume Research* 1–9. doi:10.18805/LR-4135.

Negi, S., N. K. Bharat, R. Kaushal, and P. Rohiwala. 2020. Screening of bioagents for seed biopriming in French bean (*Phaseolus vulgaris* L.) under laboratory condition. *International Journal of Chemical Studies* 8(3): 790–793.

Nonogaki, H. 2006. Seed germination - The biochemical and molecular mechanisms. *Breeding Science* 56: 93–105.

Palmieri, D., D. Vitullo, F. De Curtis, and G. Lima. 2017. A microbial consortium in the rhizosphere as a new biocontrol approach against fusarium decline of chickpea. *Plant and Soil* 412: 425–439.

Pieterse, C. M., C. Zamioudis, R. L. Berendsen, D. M. Weller, S. C. Van Wees, and P. A. Bakker. 2014. Induced systemic resistance by beneficial microbes. *Annual Review of Phytopathology* 52: 347–375.

Qu, Q., Z. Zhang, W. J. G. M. Peijnenburg, et al. 2020. Rhizosphere microbiome assembly and its impact on plant growth. *Journal of Agriculture and Food Chemistry* 68 (18): 5024–5038.

Reddy, P. P. 2013. *Recent Advances in Crop Protection*. Springer, New Delhi, India. doi: https://doi.org/10.1007/978-81-322-0723-8.

Reddy, A. S. R., G. B. Madhavi, K. G. Reddy, S. K. Yellareddygari, and M. S. Reddy. 2011. Effect of seed biopriming with *Trichoderma viride* and *Pseudomonas fluorescens* in chickpea (*Cicer arietinum*) in Andhra Pradesh, India. In *Plant Growth Promoting Rhizobacteria (PGPR) for Sustainable Agriculture: Proceeding of the 2nd Asian PGPR Conference*, Beijing, China. 18: 324–429.

Rohiwala, P., N. K. Bharat, S. Negi, P. Sharma, and A. Padiyal. 2020. Priming of radish seeds with plant elicitors like SA, MeJA, BABA and KNO_3 improves seed quality, seedling quality and seed health. *Journal of Pharmacognosy and Phytochemistry* 9(5): 2709–2713.

Saber, Z., H. Pirdashti, and M. Esmaeili. 2012. Response of wheat growth parameters to co-inoculation of plant growth promoting rhizobacteria (PGPR) and different levels of inorganic nitrogen and phosphorus. *World Applied Sciences Journal* 16: 213–219.

Sharifi, R. S. 2011. Study of grain yield and some of physiological growth indices in maize (*Zea mays* L.) hybrids under seed biopriming with plant growth promoting rhizobacteria (PGPR). *Journal of Food, Agriculture and Environment* 189: 3–4.

Sharifi, R. S. 2012. Study of nitrogen rates effects and seed biopriming with PGPR on quantitative and qualitative yield of Safflower (*Carthamus tinctorius* L.). *Technical Journal of Engineering and Applied Sciences* 2: 162–166.

Sharifi, R. S., and K. Khavazi. 2011. Effects of seed priming with plant growth promotion rhizobacteria (PGRP) on yield and yield attribute of maize (Zea mays L.) hybrids. *Journal of Food, Agriculture and Environment* 9: 496–500.

Sharma, K. K., U. S. Singh, P. Sharma, A. Kumar, and L. Sharma. 2015. Seed treatments for sustainable agriculture-A review. *Journal of Applied and Natural Sciences* 7: 521–539.

Sharma, P., M. Sharma, M. Raja, and V. Shanmugam. 2014. Status of *Trichoderma* research in India: A review. *Indian Phytopathology* 67: 1–9.

Shores, M., I. Yedidia, and I. Chet. 2005. Involvement of jasmonic acid/ethylene signaling pathway in the systemic resistance induced in cucumber by *Trichoderma asperellum* T203. *Phytopathology* 95:76–84.

Shukla, N., R. P. Awasthi, L. Rawat, and J. Kumar. 2014. Seed biopriming with drought tolerant isolates of *Trichoderma harzianum* promote growth and drought tolerance in *Triticum aestivum*. *Annals of Applied Biology* 166: 171–182.

Singh, A., A. Jain, B. K. Sarma, R. S. Upadhyay, and H. B. Singh. 2013. Rhizosphere microbes facilitate redox homeostasis in *Cicer arietinum* against biotic stress. *Annals of Applied Biology* 163: 33–46.

Singh, A., B. K. Sarma, H. B. Singh, and R. S. Upadhyay. 2014. *Trichoderma*: A silent worker of plant rhizosphere. In *Biotechnology and Biology of Trichoderma*, ed. V. K. Gupta, M. Schmoll, A. Herrera-Estrella, R. S. Upadhyay, I. Druzhinina, and M. G. Tuohy, 533–542. Elsevier, Amsterdam.

Singh, A., B. K. Sarma, R. S. Upadhyay, and H. B. Singh. 2013. Compatible rhizosphere microbes mediated alleviation of biotic stress in chickpea through enhanced antioxidant and phenylpropanoid activities. *Microbiology Research* 168: 33–40.

Singh, H. B. 2006. Trichoderma: A boon for biopesticides industry. *Journal of Mycology and Plant Pathology* 36: 373–384.

Singh, H. B. 2016. Seed biopriming: A comprehensive approach towards agricultural sustainability. *Indian Phytopathology* 69: 203–209.

Singh, H. B., A. Singh, B. K. Sarma, and D. N. Upadhyay. 2014. *Trichoderma viride* 2%WP (Strain No. BHU-2953) formulation suppresses tomato wilt caused by *Fusarium oxysporum* f. sp. *lycopersici* and chilli damping-off caused by *Pythium aphanidermatum* effectively under different agroclimatic conditions. *International Journal of Agriculture Environment Biotechnology* 7: 313–320.

Singh, H. B., B. N. Singh, and S. P. Singh. 2012. Exploring different avenues of *Trichoderma* as a potent biofungicidal and plant growth promoting candidate - An overview. *Review of Plant Pathology* 5: 315–426.

Singh, V., R. S. Upadhyay, and H. B. Singh. 2016. Seed bio-priming with *Trichoderma asperellum* effectively modulate plant growth promotion in pea. *International Journal of Agriculture Environment Biotechnology* 9: 361–365.

Sivasubramaniam, K., R. Geetha, K. Sujatha, K. Raja, A. Sripunitha, and R. Selvarani. 2011. Seed priming: Triumphs and tribulations. *Madras Agricultural Journal* 98: 197–209.

Song, G. C., H. K. Choi, Y. S. Kim, J. S. Choi, and C. M. Ryu. 2017. Seed defense biopriming with bacterial cyclodipeptides triggers immunity in cucumber and pepper. *Scientific Report* 7: 14209.

Taylor, A. G., and G. E. Harman. 1990. Concept and technologies of selected seed treatments. *Annual Review of Phytopathology* 28: 321–339.

Thakkar, A., and M. Saraf. 2014. Development of microbial consortia as a biocontrol agent for effective management of fungal diseases in *Glycine max* L. *Archives of Phytopathology and Plant Protection* 48(6): 1–16.

Tuzun, S. 2001. The relationship between pathogen-induced systemic resistance (ISR) and multigenic (horizontal) resistance in plants. *European Journal of Plant Pathology* 107: 85–93.

Van Loon, L., P. Bakker, and C. Pieterse. 1998. Systemic resistance induced by rhizosphere bacteria. *Annual Review of Phytopathology* 36: 453–483.

Van Loon, L. C., and P. A. H. M. Bakker. 2005. Induced systemic resistance as a mechanism of disease suppression by rhizobacteria. In *PGPR: Biocontrol and Biofertilization*, ed. Z. A. Siddiqui, 39–66. Springer, Dordrecht, The Netherlands.

Van Loon, L. C., and B. R. Glick. 2004. Increased plant fitness by rhizobacteria. In *Molecular Ecotoxicology of Plant*, ed. H. Sandermann, 177–205. Springer-Verlag, Berlin, Heidelberg.

Van Loon, L. C. 2000. Systemic induced resistance. In *Mechanisms of Resistance to Plant Diseases*, ed. A. J. Slusarenko, R. S. S. Fraser, and L. C. Van Loon, 521–574. Kluwer, Dordrechet.

Vishwas, S., A. K. Chaurasia, B. M. Bara, et al. 2017. Effect of priming on germination and seedling establishment of chickpea (*Cicer arietinum* L.) seeds. *Journal of Pharmacognosy and Phytochemistry* 6(4): 72–74.

Walker, T. S., H. P. Bais, and E. Grotewold. 2003. Root exudation and rhizosphere biology. *Plant Physiology* 132: 44–51.

Weller, D. M. 1983.Colonization of wheat roots by fluorescent pseudomonad suppressive to take-all. *Phytopathology* 73: 1548–1553.

Whipps, J. 2001. Microbial interactions and biocontrol in the rhizosphere. *Journal of Experimental Botony* 52: 487–511.

Woo, S.L., M. Ruocco, F. Vinale, et al. 2014. Trichoderma-based products and their widespread use in agriculture. *The Open Mycology Journal* 8: 71–126.

Worrall, D., G. H. Holroyd, J. P. Moore, et al. 2012. Treating seeds with activators of plant defense generates long-lasting priming of resistance to pests and pathogens. *New Phytologist* 193: 770–778.

Yadav, S. K., A. Dave, A. Sarkar, B. S. Harikesh, and B. K. Sarma. 2013. Co-inoculated biopriming with *Trichoderma, Pseudomonas* and *Rhizobium* improves crop growth in *Cicer arietinum* and *Phaseolus vulgaris*. *International Journal of Agriculture, Environments and Biotechnology* 6: 255–259.

Yadegari, M. and H. A. Rahmani. 2010. Evaluation of bean (*Phaseolus vulgaris*) seeds inoculation with *Rhizobium phaseoli* and plant growth promoting *Rhizobacteria* (PGPR) on yield and yield components. *African Journal of Agricultural Research* 5: 792–799.

Yedidia, I., N. Benhamou, and I. Chet. 1999. Induction of defense responses in cucumber plants (*Cucumis sativus* L.) by the biocontrol agent *Trichoderma harzianum*. *Applied and Environmental Microbiology* 65(3): 1061–1070.

8 Molecular Communications between Plants and Microbes

Gajendra B. Singh, Gaurav Mudgal,
Ankita Vinayak, Jaspreet Kaur, Kartar Chand,
Manisha Parashar, Amit Kumar Verma,
Ayanava Goswami, and Swati Sharma

CONTENTS

DOI: 10.1201/9781003213864-8

8.1 INTRODUCTION

Life enables us to experience the nuances on this planet and beyond. It supports our basic needs, such as nutritional and other attributes, that form a circle of life. Within the wildernesses of life one succumbs to various changes that are inherently a part of developmental phases or responses to exogenous stimuli, whether incorporated or forced. Carrying on with the natural tendencies of life, be it a unicellular or multicellular one – the smallest of which, the virus, is commonly situated on the margins of life and death, as they appear inanimate – require other lifeforms or their processes – evolutionarily conserved, well synchronized, and balanced mechanisms – to be close by. The extent to which life relies on these elements is vast and so very unimaginable and diverse that science has recently started to coin and explore the various milestones to fully unveil mother nature's magical mysteries of life and its sustenance on this planet. Progressing further into technical and scientific jargon to the understanding of life's reliance and how it shapes and balances ecosystems, which is more suitably referred to as interactions, is exemplified in various simplistic terms such as competition, parasitism, and commensalism. These interactions have become the way of life since time immemorial when life emerged on this planet and realized its need to grow, spread, and colonize, forming naturalized biocommunities.

From another view, the competition would have started amongst quorums of ancestral lives. Perhaps, DNA data speaks that, billions of years ago, eukaryotes, such as plants, evolved from prokaryotic lives, like bacteria via inherently symbiotic lifestyles and adaptations. Since then, plants have occurred in many lineages, emerging from algae, cyanobacteria, lichens, and bryophytes, and then followed the vascular homoiohydric lifestyles that included the seed-bearing gymnosperms and angiosperms. As plants diversified, so did their interactions with other lifeforms such as pollinator insects and avian species. What was common in the photosynthetic mode of life for many of these forms are the love (competition) for sunlight and the ability to minimize water loss. Later, micro- and macro-communities gradually built up, which either sequestered plants for their needful and/or stealing from them sunlight or other hard-earned resources. In this chapter, we discuss various direct or indirect associations of plants and their crosstalks within the ecological niches they inhabit.

8.2 PLANTS AFFECTED BY ASSOCIATION WITH THE MICROBIAL WORLD

All organisms living in the ecosystem have a specific method of interacting with their ambient environment. The method of this interaction depends on the nature of the organism. Multicellular organisms tend to use specially developed organs or cellular associations to sense their surroundings. In animals, interspecies interaction occurs by the means of sounds, vibrations, pheromones, specific movements, and stances to convey their thoughts, message, or greetings. However, in plants, the method of such interaction is much more subtle. The basic organization of living organisms (cell > tissue > organ > organ system > organism) suggests the existence of communication between diverse organisms. Ecological interactions, such as symbiosis, mutualism, parasitism, and pathogeny, suggest that microorganisms have the ability to form a wide range of associations with other single or multicellular organisms (Newton et al. 2010). To understand these associations, it is important to study the method of interaction between the participating organisms.

Plants communicate with microorganisms present in the soil via the root system. The micro-ecosystem produced as a result is called the Rhizosphere. The root of the plant branches out (Jones 1998). The carbohydrates in the exudates become a primary carbon source leads to microscopic cellular extensions, also known as root hairs, which become a beacon for microorganisms. In the rhizosphere, plants secrete a mixed concoction of various complex compounds produced to facilitate the growth of root hairs called exudates. The exudates released by a plant are a rich source of nutrition and, therefore, act as a chemoattractant for soil bacteria, thus accelerating the process of microbial colonization of the rhizosphere's living bacterial population to associate with the root hairs. Plants are known to avidly communicate with each other using volatile compounds released from a site of injury in the shoot of the plants. This is also true when the roots of a plant are involved (Ueda et al. 2012).

Compounds, such as flavonoids, are released through the root hair to attract microorganisms to the site where microbes use these compounds at the molecular level (Phillips and Tsai 1992). The presence of nutritive compounds in the soil increases the microbial activity, which then causes the microbes to lyse the plant cells as a source of nutrition. The growth of microbes is regulated by releasing secondary metabolites. Some plants have a specific group of microorganisms that are preferentially grown for the benefit of the plant. Molecular communication occurs as systems of certain specific compounds are synthesized by either the plant or the microbes. Membrane-bound proteins recognize the released proteins that send signals and cause the other microbes to act accordingly. Some bacterial species have developed resistance to certain secondary metabolites, which enables the plant to selectively allow the growth of specific microbial colonies to thrive while others are killed (Braga et al. 2016). Genetic analysis of microorganisms from the rhizosphere of a plant shows that gene expression of the plant, as well as the microbes, are significantly affected (Singh et al. 2004).

8.2.1 ALGAE IN ASSOCIATIONS

8.2.1.1 Communication in Lichens: Algae and Fungi

When microorganisms associate with another organism(s), both parties must approve the interaction in order to establish a symbiotic relationship. In such a condition, the two organisms have a special chemical code to communicate with each other and enable a safe and flourishing community. The communication occurs when the dominant organism produces certain biomolecules that cause the other to react in a specific way to confirm their compatibility. The most common association found naturally is that of algae and fungi, also known as lichens. Studies conducted upon the process of formation of lichens show that not all species of algae have the ability to form a symbiotic association with random fungal species. The two participating groups are termed photobiont and mycobiont. The mycobiont, as suggested by the name, is the fungal component, while the photobiont is the photosynthetic organism that may be an alga or cyanobacteria. The successful formation of lichen depends upon the compatibility of the photobiont and the mycobiont, which may occur via contact or a cascaded molecular interaction between the two organisms. Lichens produced in laboratories often die due to excessive loss of the photobiont because of predation by the fungi. However, when pre-existing lichens are cultured in vitro, they seem to propagate very well. This further suggests that the photobiont and the mycobiont interact to check their compatibility, ensuring that both organisms equally benefit from the association. Special proteins called *lectins* are used to confirm the compatibility. Lectins are a heterogeneous compound of glycol and proteins that have a non-immunogenic nature and a non-catalytically binding property to specific carbohydrates, thus aiding in molecular cellular recognition. The mycelium of the fungi releases these molecules that attach to the membrane of the algae. If the photobiont is non-compatible, predation of the algal cells causes the death of the photobiont, causing the fungal component to die out (Ahmadjian and Jacobs 1981). In a place where a lichen colony has taken root, other algae and moss are removed by the fungal mycelium. Genetic analysis of the algae and fungal components has shown that gene expression in both is different from those that are found separately. The recognition of the photobiont by the mycobiont is important in both the initial as well as the subsequent formation of new cells within the thallus of lichens. The process is the formation of a symbiotic association between a fungus and an alga is termed *lichenization*. During propagation, the two moieties undergo re-mechanization to ensure coexistence between the newly formed cells of both species. Although the exact cascade of molecular communication between fungal and algal cells is not clear, experiments conducted by Insarova and Blagoveshchenskaya (2016) show that, in a culture media containing root exudates of photobiont inoculated with mycobiont of *Fulgensia bracteate*, the type of changes in the mycelium of the fungi responsive to mechanization occurred after the addition of *ribitol* to the culture. The molecular analysis of lichens with species of *Trebouxia* indicated the presence of ribitol, but the root exudate lacked the compound because it had an aposymbiotic origin. Thus, it can be inferred that the role of ribitol is more than a source of carbohydrate, but it also acts as a signal to begin the process of lichenization.

Lectins play an important role in the final recognition of the photobiont in the process of lichenization. The mechanism of molecular signaling during lichenization is initiated after a mycobiont receives the initial signaling provided by the exudates of a potential photobiont. Lectins, due to the property of binding to specific carbohydrates, cause the agglomeration of cells but may also act as an enzyme. Lectin is produced by the mycobiont and is present on the surface of the mycelium or released into the substrate of a mucous released by the mycobiont after initial exudate signaling. The compatible photobiont has a ligand present on its surface where the lectin binds. In 1980, Bubrick and Galun isolated a protein that could bind to the photobiont of *Xanthoria pereitina,* calling it ABP (algal binding protein). However, the protein did not bind to photobionts of species belonging to a different taxonomic family. The protein showed hemagglutination, thus called a lectin, and had some catalytic activity of arginase. Glycosylated arginase released by the mycobiont attaches to glycosylated urease on the cell wall of the photobiont. Incompatibility of photobiont occurs when the photobiont lacks glycosylated urease, causing fungal arginase to enter the algal cell, leading to excessive production of putrescine. Putrescine causes the algal cell to rupture, thereby killing the photobiont (Lockhart et al. 1978).

The incorporation of cyanobacterial components to lichens occurs via chanced recognition of the cells by the hyphae of the fungus. Cellular contact occurs in a gelatinous pool where the cyanobiont can move away or toward the hyphae. The hyphae continue to lyse the cells in small numbers that do not tend to drastically affect the cyanobiont colony. The lectin mechanism of recognition initiates the association with the recognized cells closer to the hyphae and the new cells toward the outer region. Lichens often tend to involve a correlation between both a cyanobiont and a chlorobiont that lead to variation in its morphology. The addition of two different photobionts causes morphological changes in the lichen structure that are sometimes referred to as *photomorphism*. This is generally witnessed when the lectin produced by the mycobiont has a greater affinity toward the cyanobiont than the chlorobiont (Díaz et al. 2009). The differences in the affinity of compatibility of lectin toward one or the other lichen moiety cause differences in the degree of compatibility and temporary and permanent lichenization. The lichen propagule, upon reaching a new substratum to grow, may incorporate a natively growing species of algae to obtain a better nutrient source. However, the original algal population remains intact as all-new algal and fungal cells undergo re-lichenization in which lectin-based recognition occurs to ensure effective propagation and survival. Therefore, it is safe to say that active communication between all moieties of lichens is essential for the formation as well as the propagation and survival of a symbiotic relationship in lichens (Figure 8.1).

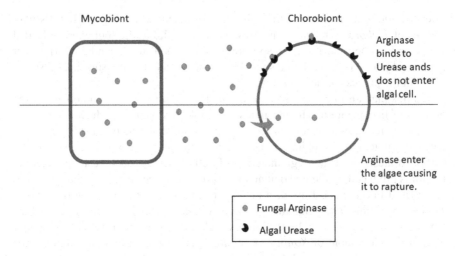

FIGURE 8.1 A schematic diagram of lichenization recognition. The upper hemisphere shows compatibility where fungal arginase binds to surface algal urease, thus, the cell survives. The lower hemisphere shows a lack of algal urease on the surface of the algal cell resulting in the entry of fungal arginase in the cell, causing rapture of the algal cell due to increased production of putrescine.

8.2.1.2 Interaction between Algae and Bacteria

Bacteria and algae interact in many ways, from parasitism to mutualism. Most of the nutrient cycle in soil depends largely upon these two forms of life. Just as is the case with the symbiosis between algae and fungi in lichens, both algae and bacteria undergo physiological changes. In some cases, the bacteria enable the flocculation of algal cells, while the algae produce and provide nutrients to the bacteria. In wastewater treatment plants, cellular aggregates of algae and bacteria are often encountered. This enables the effective breakdown of biological matter. These types of interactions are found to occupy key roles in most ecosystems. Integration of the two types of microbes causes a change in the metabolic activities, thereby enhancing the growth rate of both the species more than could be achieved individually. In the aquatic environment, phytoplankton act as the major autotrophic organism being the first tier of the food pyramid or the producers. Phytoplankton are minute organisms and constitute of mostly microalgae, cyanobacteria, and aquatic plants. These remain suspended in the water by the flow of currents in bodies of water. Phytoplankton secrete certain compounds that are produced and released into the water in the form of a dissolved organic compound. Studies conducted on dissolved organic compounds have revealed that they have the ability to enhance the growth of certain bacteria present in both the soil as well as aquatic ecosystems. Organic compounds released by microalgae often act as a source of nutrients to the bacteria because the nutrient contained in the cellular secretions include carbohydrates, proteins, and lipids. In order to increase the availability of these nutrients in the immediate surroundings, the bacteria produce growth promoters that enable the algae to grow at an increased rate, thereby proliferating both species to a healthy concentration. Phylogenetic studies of bacteria and algae have shown that species obtained from a single soil sample show mutualistic interaction resulting in genetic changes in the two species (Ramanan et al. 2015).

The symbiotic relationship between soil microbes and microalgae *Chlorella* was studied by culturing *C. sorokiniana* in media after the separation of symbionts (Watanabe et al. 2005). The studies indicated that, upon separation of the microalgae from symbiotic fungi and bacteria, the culture showed relatively slower growth. The symbionts, however, failed to grow in the absence of the extracellular organic compound released by the algae. In order to establish symbiosis, bacteria rely on specific chemical compounds that act as signaling agents to attract and form a biofilm or a consortium. The communication occurs when the algae or the bacteria secrete a mucus-like substance that contains a certain chemoattractant. Quorum sequencing is a method of molecular communication of bacteria in which specific compounds released by bacterial cells attract other bacteria, increasing their concentration in soil. Algae produce molecular equivalents of the QS compounds, such as riboflavin, enabling certain gene expression that results in the initiation of algae and bacterial interaction (Rajamani et al. 2008).

8.2.2 Fungal Association with Microbes

The microbial population often tends to form a variety of coexisting associations in order to enhance the chances of survival. Opportunistic pathogenic microbes are often seen to cause a series of simultaneous infections often causing an increase in the virulence of both individual species. The fungal pathogen *Candida albicans* is often associated to *Pseudomonas aeruginosa* although, it is found to interact with many different species of microbe based on the environment of its presence. The fungal and bacterial interactions, or BFIs, may show a symbiotic relationship, thereby altering the innate nature of both symbionts to be considerably different from their free-living population (Bjarnsholt 2013). In the case of a symbiotic relationship, the bacterial cells may be presently attached to the surface of the fungal hyphae or become incorporated within the fungal hyphae. The formation of symbiotic biofilms has been discovered where both moieties interact to produce unique physiological and structural properties. In the rhizosphere, the fungal entity protects the bacteria from the defense mechanism of the plant while the increased virulence of the pathogenic microbe allows the fungus to penetrate deeper into the plant tissues. It has also been reported that BFIs are specific to the morphology of the fungus. During the formation of the biofilm of *P. aeruginosa* and *C. albicans,* bacterial fungal associations are existent only in the hyphal form and not the yeast form of the fungus. The formation of associations between bacteria and fungi involves the chemotactic movement of both moieties toward a mutualistic relationship. Bacterial biofilms containing fungus often tend to increase the resistance of the bacteria toward antibodies. Biofilms of *Candida albicans* and *Staphylococcus aureus* tend to provide resistance to the latter against vancomycin (Harriott and Noverr 2010). The mode of communication between bacteria and fungus can be via molecular signaling or contact. A prominent example of molecular signaling between fungi and bacteria is the release of secondary metabolites. *Penicillium* sp., upon exposure to bacterial species such as *Staphylococcus*, produces antibodies against the bacteria as both a means of survival as well as cellular interaction. Although, in some cases, the interaction may also cause the release of compounds that may trigger or enhance the virulence of the bacterial species as witnessed in the case of *C. albicans* and *P. aeruginosa*. This also holds in the case of the fungal association of *Rhizopus microspores* and *Paraburkholderia rhizoxinica* in which the bacteria provide toxin to the fungi to effectively infect rice (Mondo et al. 2017). Contact communication between fungi and bacteria occurs in a process similar to that of lichenization. The fungal hyphae release exudates that chemotactically attract microbes toward the hyphae. Surface proteins and polysaccharides enable cellular recognition and attachment. The biofilm is made up mostly of a polysaccharide, although proteins and nucleic acids have also been observed (Branda et al. 2005). Fungal and bacterial interaction is evident mostly in arbuscular mycorrhizae of plants in which both fungi and bacteria interact in various ways to support the survival of the other species. The fungal association with the plant's roots increases the surface area of the root hair, thus increasing the nutrient availability of the plant. The fungus may enter the tissues of the plant by the formation of haustorium, which is used by the fungi to uptake nutrients from the plant. Bacteria act as a pathogen to the fungi and enter the fungal hyphae. This is established by quorum sequencing, which reduces the virulence of the bacteria, allowing it to remain in the filaments of the fungi. Surface polysaccharides and lectins allow the fungi and bacteria to form aggregations without causing either organism to predate the other. Bacteria in the mycorrhizal microecosystem produce growth-hormone derivates that benefit the plant.

8.2.3 Oomycetes' Association with Microbes

Oomycetes are transitional organisms that are sometimes referred to as fungi, while others are referred to as protists. Oomycetes are often found to grow in the rhizosphere along with other microorganisms. The rhizosphere is a highly active microenvironment where plants interact with a large variety of microbes often resulting in multispecies interaction with more than two participants. Bacteria possess certain genes that enable them to communicate effectively with other cells of the species and counter the defined mechanism of the host plant while the plant constantly releases a mixture of bioactive compounds that trigger the expression of genes that interfere with the signaling molecules. In the presence of a secondary pathogenic organism such as oomycetes, the bacterial cells are observed to act in a way that enhances microbial propagation.

In multispecies associations of microbes, a common pool of nutrients is formed where the efficient species show overexpression of a gene to provide the other with the necessary nutrient and benefit from the externally secreted product of the other. Pathogenic oomycetes produce externally secreted enzymes that lyse the host cells for the cells. This nutrient pool acts as a signal to other opportunistic bacteria that can cause secondary infections and aid in the suppression of the defense system of the host. Infection of plant root cells by oomycetes occurs by the formation of haustoria, a mechanism similar to that of arbuscular fungi. However, the host defense system recognizes the pathogenic oomycetes because of compounds known as *elicitors* that trigger immune responses. Elicitors are compounds that can initiate stress responses in plants, leading to increased production and release of secondary metabolites. These substances are produced as a result of metabolic processes in the various stages of the pathogen's invasion of the host. A study of the interaction between *Eurychasma dicksonii* in *Ectocarpus siliculosus* shows that the oomycetes enter the algal cell in an endophytic interaction where the sporangium is formed within the algae (Sekimoto et al. 2008). A multispecies interaction between soil bacteria and oomycetes in the mycorrhiza of a tomato plant indicates that oomycetes *Phytophthora parasitica* invade the plant root by protein digestion using pectin and the polysaccharides released to facilitate the growth of bacteria that communicate with quorum sequencing (Larousse et al. 2017).

Oomycetes and bacteria may sometimes act antagonistically in which the growth of bacteria prevents the growth of oomycetes. *Phytophthora capsici* is known to cause root rot in pepper plants, which leads to a decrease in the yield. Currently, biocontrol of pathogenic *Phytophthora* sp. is being done to effectively tackle the situation. Biocontrol is a method of disease prevention and control that involves the use of an unrelated species to affect the growth and virulence of a pathogen. An experiment conducted in Korea revealed that the inoculation of chitinolytic bacteria regulated the growth *Phytophthora capsici*, thus acting as a biocontrol agent. The growth of the plant was significantly higher than when the soil lacked such bacterial strains (Kim et al. 2008). Many living organisms use volatile compounds as a mode of communication. The compounds are released in various concentrations, which cause a different change in the behavior of the receiving organism. In molecular signaling between bacteria and plants, the volatile organic compound is a key factor. Syed-Ab-Rahman et al. (2019) conducted an experiment that studied the effect of volatile compounds released by soil bacteria that promoted the growth of pepper plants and also acted as a biocontrol for *Phytophthora capsici*. Treatment of chili pepper seeds with *Actinobacter* sp. resulted in greater root growth. Due to diffusible volatile compounds produced by *Actinobacter* sp., an inhibition test of *P. caosici* showed a significant reduction in fungal growth.

8.2.4 BRYOPHYTES' ASSOCIATION WITH MICROBES

Bryophytes are key members of the ecosystem, as they are the closest living pioneers of land plants (Read et al. 2000). The bryophytic plant body has an undeveloped root and shoots system which results into their non-vascularity, thus, they depend on moist environments for their water source. Based on fossilized records, the endophytic interaction of microorganisms mostly between fungi and bacteria found throughout the plant kingdom (Strullu-Derrien et al. 2014). Plants that are non-interactive with microbes were hardly found, as these plants are substantially prone to microbial infection and environmental stresses. The microbial–plant interaction may have beneficial (e.g., nutrients uptake, improved growth, and fitness) neutral or inimical effects on the host lifestyle. Substantial experiments of microbe–plant interaction have been performed in angiosperms, but this microbial endophytic interaction to plants shares a similarity in resource possession and administration, hormonal interaction, cell wall composition, and tissue developments.

The bryophyte plants include mainly liverworts, hornworts, and mosses. A study found that the bacterial microflora of the mosses and liverworts showed an evolutionary adaptation from aquatic to extreme drought conditions (Tang et al. 2016). To understand plant diversification from aquatic to flowering plants, bryophytes are important links for research and studies conducted to find microorganism interactions (Knack et al. 2015; Ponce de León and Montesano 2017; Carella et al. 2018). One of the key components for land colonization is nutrients absorption, and initially, it was supported by symbiosis between plants and microorganisms. A large group of microbes associated with the bryophytic plants studied are Arbuscular mycorrhiza (AM) fungi, which provide improved growth, fitness, and nutrient absorption in host plants A study found that the association of cyanobacteria (*Nostoc punctiforme*) and hornwort plants (Blasia and Caviculara) enhanced nutrient uptake and nitrogen fixing (Adams and Duggan 2008). The plant-microbial interactions also lead to the pathogenic relationship and colonization of bacteria, fungi, and viruses in bryophytic plants (Ponce de León 2011). Many necrotic-specific studies found an interaction between species of oomycetes, bacteria, and fungi with bryophytic plants (Table 8.1).

A bryophytic moss, *Physcomitrella patens*, has been studied to observe the microbial colonization and defense system of the plant (Overdijk et al. 2016; Ponce de León and Montesano 2017). A full cycle of colonization of oomycetes *Phytophthora palmivora also* studied in the liverwort plant *Marchantia polymorpha* in early land plant lineage (Carella et al. 2018)

8.2.5 PTERIDOPHYTES' ASSOCIATION WITH MICROBES

Pteridophytes are the earliest colonized vascular plants that produce spores for their reproduction, also known as a cryptogam. Pteridophytes are thought to have originated in the Devonian period (350 mya) and have a developed root and shoot system with xylem and phloem. These categories of plant vegetation were more suitably found in hilly areas such as the Himalayas in India (Chandra et al. 2008; Sukumaran et al. 2009). Many studies were also conducted to find the pteridophyte's diverse vegetation in the Western Ghats, eastern African biodiversity hotspots, and oceanic islands (Nampy 1998; Roux 2009; Kreft et al. 2010).

The Pteridophytes were commonly classified as ferns, horsetails, and lycophytes, and nearly all 48 families of pteridophytes are included in the data pole, except for Matoniaceae, Thyrsopteridaceae, and Rhachidosoraceae (Lehnert et al. 2009). Fossilized records were also found for the interactions of microbes (e.g., mycorrhiza) with pteridophytic plant samples (Dotzler et al. 2006; Remy et al. 1994). Microbial interactions with plants not only show harmful effects on plants but also some direct benefits, including resistance to harsh environments, nutrient absorption, and

TABLE 8.1

Interaction between Species of Oomycetes Bacteria and Fungi with the Bryophytic Plants

Microbes Species	Mosses species	Interactive relationship	Reference
Atradidymella muscivora (fungi)	*Aulacomnium palustre, Hylocomium splendens,* and *Polytrichum juniperinum*	Symbiosis and pathogenic	(Davey et al. 2009)
Coniochaeta velutina (Fungi)	*Funaria hygrometrica*	Symbiosis and pathogenic	(Davey et al. 2010)
Fusarium avenaceum	*Racomitrium japonicum*	Pathogenic	(Lehtonen et al. 2012)
Physcomitrella patens	*Pythium debaryanum* and *Pythium irregulare*	Pathogenic	(Mittag et al. 2015)
Phytophthora palmivora	*Marchantia polymorpha*	Symbiosis	(Carella et al. 2018)
Irpex lacteus Marchantia-infectious (MI)1, Phaeophlebiopsis peniophoroides MI2, Bjerkandera adusta MI3, and *B. adusta MI4*	*Marchantia polymorpha*	Symbiosis and pathogenic	(Matsui et al. 2020)

diversification of ecological interaction. The evidence records the direct interaction of fungi with pteridophytes, driving toward its terrestrialization about 460–480 mya (Bonfante and Genre 2008; Parniske 2008; Selosse and Le Tacon 1998). A research study found that the rhizosphere microbes help pteridophytes to improve the nutrient uptake by increasing resistance toward heavy metals and elevate growth factors (Nazir and Bareen 2011; Srivastava et al. 2012).

8.2.6 GYMNOSPERMS' ASSOCIATION WITH MICROBES

Gymnosperms are non-flowering plants that have a well-organized root and shoot. Unlike algae, bryophytes, and pteridophytes, gymnosperms are seed-producing plants and are, thus, referred to as higher plants, a name shared by angiosperms. The word gymnosperm originates from the fusion of two Greek word "*gymnos*" and "*sperma*," meaning "naked seed." Gymnosperms develop seeds on the tips of leaves or stems that are later modified into specific structures such as cones in the case of *conifers*. Gymnosperms are highly resistant plants that can grow in extreme geographical and climatic conditions. They are found to grow in cold and windy climatic belts of tundra regions of the world and snow-clad mountains as well as in dry and sandy deserts. Having been in existence long before the first flowering plant had evolved and descended from pteridophytes, the interaction of gymnosperms with soil microbiota is almost inevitable, with fungal microbial cycads and cyanobacteria being the most predominant (Zheng et al. 2018).

Cyanobacterial association allows cycads to cope with harsh climatic and soil conditions. They have two types of roots: coralloid and regular. Coralloid roots are mostly present in association with cyanobacteria than regular roots. The fungal community in all root samples is the same, but bacterial colonies change (probably due to the difference in microbial interaction with the plant with various locally present microbes as well as the condition of the soil). Coralloid roots are formed as a result of endophytic correlation of the cyanobiont with the gymnosperms. This interaction prevents other microbial interactions. However, the regular roots are free to interact with soil microbiota. Epiphytic gymnosperms found in the Panamanian region show that coralloid roots form a large variety of modified structures, which indicates the symbiotic interaction between the two moieties. Hormogonium are produced to effectively participate in the morphological changes in the structure of the plant, enabling enhanced cyanobacteria efficiency in fixation of nitrogen for the growth of the plant as it is completely arboreal, and its roots are not connected to the soil.

8.2.7 ANGIOSPERMS ASSOCIATION WITH MICROBES

Angiosperms are seed-bearing, flowering plants. They are the most terresterilized plants, appearing everywhere on the earth from forests to grasslands and sea margins to deserts, and are one of the largest primary producers with around 300,000 species. Unlike the non-vascular bryophytes, every part of the plant body is involved in development and nourishment, but angiosperms, as a vascular plant, developed a specialized structures for specific functions, such as xylem and phloem for nutrients transport and flowers for reproduction. Previously microbial interaction in lower classes of plants were discussed, and angiosperm plants are also having this kind of relationship with microbes. Most endophytic microbes are transferred vertically to the plant and ensure a long-term interaction with the plant (Hosokawa et al. 2006). An array of around 500 flowering plants that have microbial interactions, such as endophytic, symbiotic, and pathogenic relationships. Generally, the root, leaf, and stem part of the plant undergo interaction such as bacterial leaf symbiosis, formation of nodules, mycorrhizal interaction, and pathogenic invasion. Because angiosperms are a major part of our day-to-day life, a series of studies have been conducted to find the normal flora and pathogenic interaction between the angiosperm plant and microbes. A study found that the *Amblyanthus* and *Amblyanthopsis* having the nodule formation in the leaf by the bacterial endosymbionts (Miller 1990). A genome of the plant *Populus trichocarpa* submitted supposed to be the first used annual plant model system and found that the plant has a fungal endophytic mutualism (Liao et al. 2019; Singh 2020).

8.3 PLANT–INSECT INTERACTION

A dynamic and diverse system of plant–insect interaction exists in natural habitats, ranging from antagonism to reciprocity. The plant-derived chemicals play a vital role in insect–plant interactions, ranging from host acceptance to habitat selection. In a mutually beneficial relationship, insects assist plants in defense and pollination, and, in return, plants provide nutrients, shelter, and egg-laying sites for insects (Mello and Silva-Filho 2002). However, a large number of insect herbivores might be lethal for a plant. To tackle such a situation, plants have developed various defense mechanisms starting with physical and chemical barriers that act as a first line of defense, using various plant defensive proteins, enzymes, secondary metabolites, and herbivore-induced plant volatiles (HIPV). Plant-derived defensive chemicals can resist insect herbivores either by modulating feeding, molting, ovipositional behavior, or by attracting natural enemies of herbivores (War et al. 2012). Similarly, insects have also developed several counter tactics, such as detoxification of toxins, sequestration of toxic compounds, altering midgut composition etc., to break plant defense barriers.

8.3.1 Plant–Insect Interaction: Significance in Agriculture and Weed Biocontrol

Today, the managed and conserved crop production system is gaining importance to increase agricultural sustainability and address food security issues and regulatory mandates. In the past few decades, among conservation and conventional systems, the use of herbicide formed the basis of weed management (Locke et al. 2002). However, the cost, environmental problems, and health-related issues caused by the use of agrochemicals have changed people's perspectives on the use of synthetic chemicals for weed management. The use of target-specific, naturally occurring organisms, such as natural enemies, herbivorous fish, and insects for crop management is referred as the biological control of weeds (Telkar et al. 2015). The objective of the biological control method is not the complete eradication of weeds, but the reduction of weeds to a low or negligible level. The biological control of weeds using arthropods or insects has been frequently practiced throughout the world. The early success in the control of weeds through arthropods led to increasing interest in employing arthropods as biological weed control agents (Keerthi et al. 2019).

Weeds can serve as a food source for herbivorous insects and provide shelter and other ecosystem resources to the insects. They also act as an alternative host for the insects when the preferred host (crop) is absent (Norris and Kogan 2005). The alien weeds in the crop fields are readily amenable to insects because of a lack of host-specific natural enemies to keep them in check in the newly colonized area. Herbivores remove weeds' biomass, resulting in reduced fitness and growth of damaged the plant (weed). Insects damage various parts of the weed plant, which affects their photosynthetic activity, carbohydrate reserve, flowering, plant density, and reduces their growth and reproductive potential (Uludag et al. 2018). The synergistic action of two insect species on weeds may inflict more damage than either of single species could on their own. The insects that directly interact and interfere with the root system of weeds damage the weeds in a shorter period of time in comparison to flower feeders and defoliators. Both annual and perennial weeds have been effectively controlled by insects. These interactions between weeds and insects form the basis of insect-mediated weed management (Sankaran 1990).

Insects interacting with weeds make nutrients, water, and sunlight more available to field crops, thereby reducing negative impacts and competition with crops. However, there are some negative impacts linked with insect–weed interactions. Many insects act exclusively on specific weeds, but some insects, in the absence of target weeds, damage nearby crops. In other cases, if the insect is not restricted to its feeding habit, it shifts from weeds to crops, leading to crop damage (Capinera 2005). Most of the insects are themselves vulnerable to diseases, parasitism, and predation, leading to a low population of insects, which affects the weed control system. Underestimation, lack of commercialization, and practical knowledge of these natural weed-control candidates, and processes for the management of agriculture are the key constrain in the current agricultural practices. These interactions offer a sustainable, economical, and practical solution for the control of native and alien weeds in the crop fields and can serve as a "silver bullet."

8.3.2 PLANT DEFENSE AGAINST INSECT HERBIVORES

Plants growing in natural habitats always remain vulnerable to attack from insect herbivores. As an important nutrient source, insect attacks on agricultural crops are one of the major causes of reduced crop productivity. As a first line of defense, inherent plant protection systems, such as cell walls, wax, and bark, shield the plant from potential antagonists (Malinovsky et al. 2014). Like animals, plants have also developed defense mechanisms equipped with various chemicals and enzymes that are well maintained by biochemical pathways. A plant's defense system helps to detect attacking insect herbivores and neutralize them by activating appropriate defense machinery. Plant defensive enzymes and secondary metabolites are the key soldier that selectively and effectively target and counteract harmful insects. Three main groups of secondary metabolites, i.e., terpenes/terpenoids, phenolics, and alkaloids, are known to protect plants from pathogenic enemies (Kortbeek et al. 2019). Terpenoids are a large class of secondary metabolites comprising more than 22,000 individual terpenoids, classified as monoterpenes, sesquiterpenes, diterpenes, triterpenes, and polyterpenes (Tiku 2018). Terpenes have insect repellent and insecticidal properties. Various examples of such terpenes include abietic acid, limonene, limonoid, menthol, menthone, myrcene, phorbol, pyrethrins, and sterols (Tiku 2018). Saponins, a special class of glycosylated triterpenoids, found in the cell membrane of several plant cells, also exhibit insecticidal property. Phenolics, another class of secondary metabolites, are the most widely distributed defensive secondary metabolites synthesized by plants through malonate–acetate, isoprenoid, and shikimic acid pathways (Stewart and Stewart 2008). Coumarin, furanocoumarin, lignin, flavonoids, isoflavonoids, and tannins are the most common examples of phenolic compounds (Tiku 2018). Flavonoids are the secondary metabolites known for their structural diversity and ubiquitous nature. To date, more than 9,000 varieties of flavonoids are known in plants (Wang et al. 2018). Different flavonoids are known to exhibit a range of functions, including protection against biotic stresses, such as insects, pathogens, and abiotic stress including heat and UV radiation, auxin transport, signaling during nodulation, and plant interactions. They also tend to provide coloration, taste, and fragrance to fruits, seeds, and flowers that act as a signal to attract birds and insects (Kumar et al. 2018; Zunjarrao et al. 2020; Mathesius 2018). Various studies have been conducted to understand the role of flavonoids and stimulation of feeding and oviposition in insects (Zhang et al. 2017). Flavonoids protect the plants from plant-feeding insects by influencing their growth, behavior, and development. In some cases, flavonoids reduce the nutritional value of plants, alter the palatability of plants, and even can act as toxins for plant pests. Several studies on feeding behavior of insects have been reported which confer insects' sensitivity to flavonoid compound. (Aboshi et al. 2018). The next class of secondary metabolites, alkaloids, belongs to the separate class of nitrogenous secondary metabolites and are considered to be quite toxic and have insect repellent and insecticidal properties. Some of the well-studied plant defensive metabolites are mentioned in Table 8.2.

Plant-derived proteins and enzymes, such as chitinase, peroxidase, lectins, defensins, protease inhibitors, α- amylase inhibitors, β glucosidases, glucose oxidase, and MAP kinase (Table 8.2), directly or indirectly strengthen the plant defense system to fight against herbivores and other harmful enemies. Unlike secondary metabolites, the enzymatic defense system requires ample amount of energy and plant resources for their biosynthesis. Also, these responses are generally activated after the attack of insect herbivores. However, upon activation, defensive enzymes and proteins prevent herbivores from causing major damage to plants.

In order to protect themselves from insect herbivores and other invading pathogens, plants have developed an inducible defense mechanism (Figure 8.2), which can target insects at genetic, biochemical, and physiological levels (Tiku 2018).

When an insect herbivore feeds on, interacts, or attacks plants, several defense mechanisms are induced in plants. Plants respond to insect attack by either direct or indirect defense. Direct defense against herbivores is mediated by plants' characteristics, such as mechanical protective structures on the plant surface (hair, spines, thick leaves, and thorns), and through the formation of toxic secondary metabolites, such as alkaloids, terpenes, phenols, and anthocyanins. These herbivore-induced defense traits directly affect insect physiology, growth, reproductive success, and their survival (War et al., 2012). Indirect defense in plants is initiated by plant-released chemicals known as herbivore-induced plant volatiles that themselves do not have direct impact on insects, but specifically attract natural enemies (parasites or predators). Plants provide shelter and food to attract natural enemies to enhance the effectiveness and action of natural enemies against insects. Plant defenses may be expressed constitutively or after insect attack and elicitor release. Insect oral secretions and ovipositional fluid contains chemicals, such as fatty acid conjugates, which act as specific elicitors and induce plant defense. The known elicitors are enzymes, peptides, fatty acid conjugates, cell wall fragments, induced-plant volatiles, and esters (Aljbory and Chen, 2018). In some cases, elicitors formed in insects are applied or injected in plants to launch plant-induced defenses. Induced defenses in plants, known as induced resistance, are employed as a pest management strategy to minimize the use of insecticides. The chemical elicitors are used to manipulate

TABLE 8.2
Plant Defensive Machinery against Insect Herbivores

Plant defence agents	Characteristics/Function	Reference
Plant defensive proteins		
Chitinase	Attacks on exoskeleton (chitin) of insects	(Banerjee and Mandal 2019; War et al. 2012)
Peroxidase	Impairs nutrient uptake in insects, toxic for insect gut, food deterrent	(Mello and Silva-Filho 2002)
Lectins	Acts as anti-nutritive and toxin, alters hormonal and immunological status of insects	(Napoleão et al. 2019)
Protease inhibitors	Inhibition of digestive enzymes in insects, starvation of insects, impairs growth and development	(Napoleão et al. 2019)
α- amylase inhibitors	Inhibits digestive enzymes	(Mello and Silva-Filho 2002)
β glucosidases	Induces volatile terpenes	(Aljbory and Chen 2018)
Glucose oxidase	Attractant of natural enemies	(Aljbory and Chen 2018)
MAP kinase	Activates defence-related genes, accumulates defensive metabolites, regulates phytohormone level	(Hettenhausen et al. 2015; McNeece et al. 2018)
Polyphenol oxidase	Induces oxidative stress in insect gut, reduces nutritional quality in plants, increases cell wall resistance	(Mello and Silva-Filho 2002)
Lipooxygenase	Antioxidant, acts as toxin and insect repellent, activates jasmonic acid signalling pathway, reduces larval food production and growth	(War et al. 2012)
Plant defensive secondary metabolites		
Phenols	Phagostimulant, attractant for natural enemies of insects, feeding deterrents, affects growth, development and oviposition in insects	(Bennett and Wallsgrove 1994; Kumar 2019)
Tannins	Bitter and astringent in taste, deters insects, modulates digestive enzymes of insects, forms midgut lesions in insects	(Xiao et al. 2019)
Cyanogenicglucosides	Phagostimulant or feeding deterrents	(War et al. 2020)
Glucosionolates	Feeding deterrent	(Fürstenberg-Hägg et al. 2013; Kumar 2019)
Alkaloids	Reduces feeding and survival rate	(Fürstenberg-Hägg et al. 2013; Kumar 2019)
Terpenes	Modifiers of insect, acts as anti-feedant and toxin	(Aljbory and Chen 2018)
Flavonoids	Affects behaviour, growth, and development of insects	Fürstenberg-Hägg et al., 2013
Cardenolides	Alters growth, development, and oviposition in insects	(Williams et al. 2017)

(Continued)

TABLE 8.2 (CONTINUED)
Plant Defensive Machinery against Insect Herbivores

Plant defence agents	Characteristics/Function	Reference
Iridiods	Bitter in taste, reduces nutritional quality of plants, denatures protein and nucleic acid of insects, inhibits formation of leukotrienes and prostaglandins	(Williams et al. 2017)
Phytohormones		
Jasmonic acid	Induces production of alkaloids, antioxidant enzymes, volatile compounds, and proteinase inhibitors	(War et al. 2012)
Salicylic acid	Attractant of natural enemies, production of reactive oxygen species	(Aljbory and Chen 2018)
Ethylene	Induces direct and indirect plant defence response, formation of volatile compounds	(Van der Ent and Pieterse 2018)

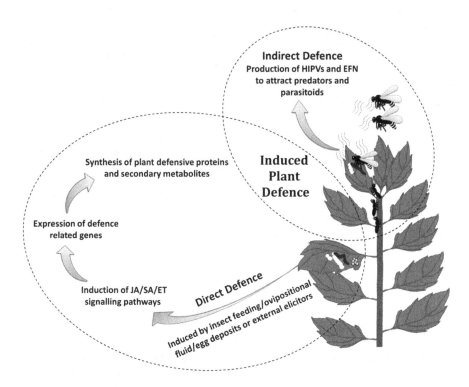

FIGURE 8.2   Direct and indirect plant defense against insect herbivores. JA: jasmonic acid; SA: salicylic acid; ET: ethylene; HIPV: herbivore-induced plant volatiles; EPF: extrafloral nectar.

induced resistance in plants to control crop loss due to insects. The elicitors are applied or sprayed on attacked plants so that the plant can build defenses against herbivores (War et al., 2012).

8.3.3 EFFECT OF ELEVATED CARBON DIOXIDE ON PLANT–INSECT INTERACTION

Carbon dioxide (CO_2) is the primary source of carbon for photosynthesis in plants, and the changes in the level of CO_2 has a remarkable effect on the physiology, plant productivity, biochemical composition, and phenotype of plants. An elevation in CO_2 concentration results in an increased rate of photosynthesis, growth, and biomass (Jiang et al. 2020). The allocation of excess carbon for the formation of structural and secondary metabolites alters the secondary chemistry of plants. Under elevated CO_2 levels, the accumulation of biomass in plants dilutes the concentration of nitrogen in plant tissues, known as nitrogen dilution effect or increased C:N ratio (Pincebourde et al. 2017). It may also increase leaf abscission, plant senescence, and decreased water content. A high C:N ratio, low levels of nitrogen, and altered secondary plant metabolism results in a reduced concentration of protein in plants and, therefore, reduced nutritive content for interacting herbivores (Kazan 2018). Under these altered conditions, an interacting pair responding for compensatory feeding, subsequently results in an increased level of defoliation and damage to plants. Insects tend to consume more leaf area of plants grown on conditions in which the CO_2 level is elevated. It allows insect larvae to maintain their growth rate (Sun et al. 2016, 2019). Long-lived leaves and plants are more vulnerable to defoliation. The plants have multiple mechanisms to compensate for defoliated leaf surface: upregulation of photosynthesis, resource allocation, production of foliage, and use of reserved storage organs (DeLucia et al. 2012). Not all insects respond in same way to the altered plant phenotype. Some insects prolong their developmental time and reduce their growth rate and food conversion efficiency. The self-induced changes in insects may reduce their richness, diversity, and abundance when compared to plants grown in optimum CO_2 conditions. Finally, altered CO_2 levels affect the behavior of insects, but, of course, that is not the case with all plants, as it varies according to plant species, interacting species, season, and age of both the plant and insects. For example, older larval instar is better adapted to compensate reduced foliar availability than younger larvae.

8.3.4 EFFECT OF ELEVATED TEMPERATURE ON PLANT–INSECT INTERACTION

In the coming years, global climate change will be one of the major challenges for the terrestrial ecosystem and its members. Plants and insects comprise 50% of the known species of the terrestrial ecosystem. Temperature is a crucial factor affecting the plant–insect interaction and, therefore, cannot be excluded when studying plant–herbivore interaction (Asfaw et al. 2019). A non-excessive level of heat is a basic requirement for ecological life. Every vital process of living beings is restricted to an optimum temperature range. Temperature affects plant growth, development, phenological events, physiological processes, reproduction, its nutritional quality, and defensive components (Pham and Hwang 2020). The major impact of elevated temperature is the alteration in the phenological events of plants and associated herbivores. Increased heat waves can cause buds to burst and flower early, disrupting interaction between plants and their visitors. The host plant phenology is a prerequisite for the successful completion of the life cycle, as asynchrony could disrupt the evolutionary and ecological relationship between them (Pureswaran et al. 2019). Because of the close relationship with their host (plants), either for nutrition or shelter, insects are also expected to suffer the consequences of elevated temperature experienced by plants (Havko et al. 2020).

Climate change alters the population size, geographical range, and feeding behavior of insects. Elevated temperature decreases the larval developmental time and increases its performance, causing asynchrony between host (plant) and insect, which may result in greater herbivore outbreak and lower plant biomass. The synchrony between plant and interacting pairs is crucial, as development of insect and plant outside their optimal time period often has fitness consequences (Akbar et al. 2016; Dyer et al. 2013). The increase in the insect metabolism driven by the increase in temperature is another major factor responsible for crop damage. The rate of food consumption and plant damage by insects not only depends on its metabolic needs but also on plant defenses and the quality of plant tissue (Havko et al. 2020). The rise in temperature also alters the formation of plant secondary metabolites, which, in turn, modifies the insect attraction and plant–insect defenses, ultimately, altering plant–insect interactions (Jamieson et al. 2017).

8.3.5 ADAPTATIONS IN PLANTS IN GAINING RESOURCES FROM INSECTS

Plant–insect interaction is generally considered as scenario in which insects feed on plants. However, there exists a situation in which plants also feed on insects. Insectivorous or carnivorous plants are known to obtain nutrients such as nitrogen from the prey (insect) they capture (Sangale 2020). These plants are usually found in nutrient poor habitats and this carnivory has evolved as an adaptation to supplement nutrients. To date, more than 800 species from 10 independent lineages of carnivorous plants are known (Cross 2019). To be called carnivorous plants, they must be able to grow in soil with little nutrient, absorb nutrients from bodies adjacent to it, and have adapted to attract, capture, and digest prey (insect), contributing to plant growth, development, and reproduction (Givnish 2015).

Carnivory involves the acquisition of several morphological and anatomical properties in plants. First, the emission of plant volatiles for communication with insects, specifically to attract pollinators and other visitors. The chemical communication between plant and prey (insect) is an integral part of plant–insect interaction (Jürgens et al. 2009). Second, is the trapping as well as the killing through a specialized form of structures. These include snap traps in Venus flytrap (*Dionaea muscipula*), pitfall traps in *Nepenthes*, flypaper traps in *Drosera*, bladder traps in *Utricularia*, and lobster pot traps in *Genlisea*. When insects are attracted to a plant through chemical attractants, such as a sweet fragrant fluid or extrafloral nectar, insects fall into the specialized evolved traps. The waxy and slippery surface of traps drowns the prey into it and prevents its escape. The bottom of trap, known as "digestive zone," is filled with multicellular gland and fluid to serve the digestive function. Finally, in the digestive zone, prey is digested by endogenous hydrolytic enzymes, and nutrients are released for plant usage (Behie and Bidochka 2013; Sangale 2020). In addition to digestive proteinase, pathogenesis-related proteins (PR) and other low molecular weight molecules are also reported in pitcher plants. It is likely that these compounds are antimicrobials that tend to reduce the growth of microorganisms, which could compete for the nutrients (Lee et al. 2016; Rischer et al. 2002).

8.3.6 Insect Adaptation against Plant Defense

Plants exhibit various defense mechanisms against herbivores, yet few insects have evolved countermeasures against all the defense mechanisms of plants. It enables insects to make their living by using plants for their benefit and as a nutrient source. Plants have a variable chemical composition, which poses a challenge for insects to adapt and elicit defense mechanisms. However, insects exhibit an array of enzymes that provide them with defenses against plants' chemical toxins. As herbivores feed on a variety of plants, they have evolved to deal with a range of chemical toxicants for the exploitation of the host. Because no single adaptation is effective against diverse toxins, insects have various detoxification strategies and have the ability to deploy countermeasures when required (Mello and Silva-Filho 2002). These permanent changes in the assemblage of the insect enzyme system permit them to effectively defeat a plant's defense system. On the feeding and plant defense strategy, insects can be classified as generalized and specialist insects. Generalist insects feed on a range of plant hosts, and, thus, their defense mechanisms are more complex. They possess general adaptive mechanism to tolerate plant defenses and possess the ability to modify plant mechanisms. Because generalized insects are polyphagous, they do not master any defense mechanism and respond to an array of plant chemicals. On the other hand, specialist insects cannot use many plants as a host, but they possess the ability to tolerate plant defenses more effectively (Ali and Agrawal 2012).

One of the important strategies employed by insects is the desensitization or detoxification of plant chemical defense molecules (Heidel-Fischer and Vogel 2015). Insects secrete various salivary effectors that reduce the risk of defense molecules. Also, the structural changes in the active sites of the molecules through oxidation, hydrolysis, reduction, and conjugation help to evade plant defenses (Mello and Silva-Filho 2002). Cytochrome P450 enzyme, found in almost all organisms, plays an important role in insect defenses, i.e., for the detoxification of host plant chemical toxins (xenobiotics) (Peng et al. 2017). Another defense adaptation to plants is the sequestration of chemical defenses through metabolic adaptations that modifies the plant chemical in the insect and renders them less toxic for the insect. The sequestered protein is then easily transported out from the insect or excreted. In some cases, after sequestration, insects employ chemicals for their own defense (Erb and Robert 2016; Heidel-Fischer and Vogel 2015). Lepidoptera sequesters terpenes, alkaloids, and other secondary metabolites of plants and then uses them as its own defense molecule against natural enemies and predators (Yoshinaga and Mori 2018). For example, flavonoids are compounds used in plant defense mechanisms, but there is evidence suggesting that flavonoids increase insect fitness during plant–insect interactions. Insects are reported to increase their fitness by sequestering flavonoids into their body cuticle or wings for protection against predators and pathogens (War et al. 2020; Zunjarrao et al. 2020). The amount of flavonoid sequestered depends on the type of flavonoid in the diet. Plant protease inhibitors are known as important components of plant defense mechanisms, as they inactivate proteases. Various insects are known to overcome the effect of protease inhibitors by altering midgut composition. In addition, insects also produce protease-inhibitor proteinase, which helps in the digestion of the inhibitor, and the product is used as an amino acid source (Zhu-Salzman and Zeng 2015).

8.4 INTERACTIONS WITHIN THE PLANT KINGDOM

It is sometimes thought that plants do not have sense organs and are sessile. Regardless, this should not deem them incapable of mediating responses in the form of positive and negative interactions, which are more pronounced within their kingdom. Of course, plants do not have sensory organs in general, with some exceptions those such as thorns, fibers, and root and shoot hairs. Perhaps, the absence of movement and sense organs in plants is more than compensated by the highly evolved inherent chemical signaling mechanisms that help them to productively showcase various responses to stimuli and move through the use of water, wind, insects, animals, and so on. In fact, they have a higher degree of repetition of some organs, such as the leaves, roots, flowers, fruits, for example. This might equate with the phenomenon of phenotypic plasticity amongst animal species that develop specialized structures, such as the emergence of a helmet in *Daphnia* while predators are close and thermally regulated sex variations in reptiles and fish. Plants could be said to have the "ability to move," as many plant seeds can cling to other plants, animals, insects, or just fly away with fibrous protrusions that very often help them glide along the wind and water turbulence. For example, some mistletoe, especially the genus *Arceuthobium*, have developed explosive seed dispersal. Phenotypic plasticity, or flexibilities, are also seen in the selective shedding of leaves at points where shade is routinely encountered on trees.

Plant interactions amongst the members of its kingdom are well understood and are exhibited in the form of parasitic, symbiotic, and commensal relations. As the ecological biomes are a sum of multitrophic communities, the existence of plants imposes variations in the structure of the physical habitat in their vicinities via direct or indirect modes. They are known to change the availability of light and water under the soil, humidity and gas buffer, and texture of the soil as well as attract or repel organisms such as insects and microbes in the close surroundings that might positively, negatively, or even neutrally affect themselves or other plants. Dying plants may even offer benefits to other plants. Such a phenomenon is seen with nurse logs in dense forests where old canopies fall to the ground and prepare an ambient condition for growth of other plant seedlings to germinate and proliferate. In these

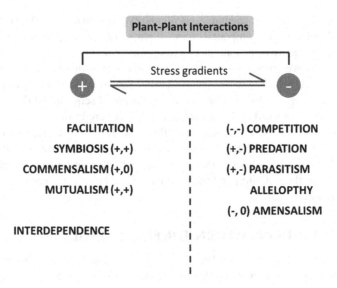

FIGURE 8.3　Plant interactions with plants.

situations, a plethora of complex physical interactions are governed and mediated by chemical crosstalk among plant types.

At the outset, it has been understood that a successful perpetuation of diverse species within plant communities in a given region is the result of summing up the effects of positive and negative interactions that each individual plant or its populations mediate and undergo. Also, positive interactions indicate interdependence among plants in plant ecosystems. Many such types of interactions have been categorized in Figure 8.3.

8.4.1　Negative Interactions among Plants

Many of plant-life's intriguing behaviors stem from competition. From the "survival of fittest" notions of Charles Darwin, it is obvious that plants sharing the same vicinity exhibit competitiveness for their various needs. Such competitions may arise for resources such as energy and molecules that drive energy metabolism, including light, water, and mineral nutrients (N, P, K, Ca, S, Mg etc.), carbon (in general CO_2, sugars, bicarbonates), radiant heat, and oxygen in soil. Many studies have predominantly analyzed the competition for photosynthetic radiation, as this is instantaneously available and presents a unidirectional commodity that diminishes with time if not tapped. Competition for minerals, such as NPK, has been widely established through pot and field experiments (Clements et al. 1929; Wilson and Newman 1987). Competition for N is more frequent than for P, whereas competition for K falls in the middle. Plants even compete with microbes in soil for minerals such as iron. Competition is realized in the form of expansion of plant appendages, such as roots, leaves, and branches, in view of gaining extended access to the resources and making it easier to tap these resources. The effect of seasonal favorability develops predominance of one plant over the other in addition to plant-led adaptations, whether it is the addition of appendages or as highly orchestrated molecular signals. Allelopathy, on the other hand, specializes the host or the partner to negatively affect the growth properties of the ally in the association by releasing certain chemical substances. This could be counteracted similarly as a chemotactic action from the target plant in view of its defense and in view of its epigenetic influence. These could also be triggered by environmental conditions. "Community-welfare" signaling mechanisms may also co-exist among plants within the same species and varieties or, perhaps, within a conserved pattern amongst various genera. This may alert plants to an insect attack. Amensalism is an example of community interactions in which one plant poses negative effect(s) on other distinct species within a region without influencing plants similar to the effector species. Commensalism occurs when one or more plants coexist in direct connection with each other in which one benefits without harming or benefitting the host. This is seen among trees allowing the orchids and bromeliads to grown on them, gaining gain water and other nutrients from the air or from the surfaces of the host without showing any attachment point on the host. Similarly, bryophytes, ferns, and lichens are also epiphytes that live upon other plants such as trees and, at times, may account for more plant material on the tree than the tree itself. Mutualism interactions may be direct via attachment points, whereby both the partners are benefitted. Other such interactions may be seen as interference, competition, and allelopathy among plants for resources. This is exemplified in weeds and parasitic plants such as dodder, mistletoes, and pinedrops. It is very rare that plants will not be affected by the presence, or even absence, of similar species, same genera, or drastically distinct orders of their classification in their near vicinity.

8.4.1.1 Examples of Negative Interaction between Plants

Competition in plants is considered a major criterion for natural selection and structuring of plant communities. Plants compete with each other for light, nutrients, water supply, etc. Plants with denser root systems have the ability to absorb more nutrients and water from the soil. Furthermore, the canopy affects the absorbance of light by plant species. So, plants that have more developed mechanisms and structure will have better chances of survival compared to less developed plants (Craine et al. 2013; Craine and Dybzinski 2013). Amensalism is seen in some plants that have significant allopathic effects on other plants. They inhibit the growth of other plants by releasing certain signaling molecules. Chon and Nelson (2013) reported that leaves of ground ivy (*Glectroma hederacea*) act as inhibitors to radish plants. Nadkarni (1981) studied predation and parasitism in some plants that produce adventitious roots beneath the epiphytes to obtain nutrition from the roots of other plants beneath the soil. Norton and Smith (1999) reported that mistletoes grow on the branches of host trees, mimic them, and obtain their nutrition from the host plants, leaving the host plant undernourished and moving toward early senescence. In fact, chemical profiles for mistletoe species may be geographically distinct at various global locations, and this is so well woven with the plant community structures (tree community context) that they have been regarded as keystones resource of biodiversity where they prevail. Host preference is another under investigated area for such plant parasitic plants (Mudgal and Mudgal 2011). Mistletoes further show strong host properties to their own types as well as many insects and microbes as well (Mudgal et al. 2011).

8.4.2 Positive Interactions among the Plants in Their Community: Facilitations

Positive interactions could be related to epigenetic effects and may be triggered from environmental disturbances as balancers. Plants cannot be considered merely an outcome of the population phenomena. Some have viewed this as one perspective. From yet another, coexistence of many different plant species at one place occurs because they have common adaptabilities to biotic and abiotic conditions (Gleason 1926). This established their presence as a typological entity, as plant communities exhibit remarkable interdependence. This is exemplified in that many species within a given zone are critical for others' growth and plentifulness in that area. This is the concept of facilitation, which goes against the primitive view that holds plant communities as an "individualistic model." Ideas on facilitation form more than three decades of empirical research and indirect interactions within plant communities (Callaway 1995, 1997). The individualistic model of plant community has led the evolution of predominating ideas on the significance of abiotic conditions and others that shape plant communities and, thus, has been pivotal. They account for founding inferences on ecology and evolution. Nonetheless, new plant ecologists realize that plants closely situated to each other can variously better the propensities of similar member. Ideally, this bifurcation of notions has led to emerging classifications, wherein negative interactions include competition, allelopathy, and predation. However, facilitation and interdependence should be viewed as positive interactions among plants in which one plant effectuates the other positively, enhancing its survival, growth, movement, and spread or may also provide protection from predation, herbivory, and pathogen attacks. It should not be confused with mutualism (+,+) in which both parties gain except under some bidirectional facilitation circumstances. The benefactor may not gain from the recipient, which might be seen as commensalism (+, 0) in some rare events; however, studies on such instances have not been conducted. Some argue that commensalism could be a commonly undetected phenomena (Futuyama 1979). On the other hand, many have investigated whether facilitation could result in benefactor's loss (+,-) or a negative interaction. This result cannot be deemed a nuance of individualists' concept of competition (+,-). The frequency of facilitation events has built up interdependence in plant communities, such that a deficiency or absence of one plant will impose negative effects on others.

Community interactions, such as facilitations, cannot be studied in conventional ways, as with plants in a glasshouse. The real stress factors that play together, as well as the dynamisms with various factors such as winds, temperature, herbivores, and humidity, cannot be replicated in glasshouse trials studying this positive interaction.

These conventional methods have overstressed individualists' ideas of competition. Facilitation can be direct as well as indirect, with the latter involving an intermediate plant.

8.4.2.1 Examples of Positive Interaction: Facilitation among Plants

Facilitation in a plant community interaction was first observed by Person in 1914, as he noticed that conifers regenerate more easily when among aspen (*Populous tremuloides*) clones. He verified this by growing Douglas fir (*Pseudotsuga menziesii*) seedlings within aspens and in open fields devoid of aspen. He also controlled the fluctuations from effects of winds, evaporation rates, and shade effects to infer that regeneration is more productive in the presence of aspens.

However, as reviewed by Callaway, positive interactions in view of facilitation were overlooked for more than a half century, even after many such works had been reported. However, it gathered pronounced acceptance in the 1980s with two works, *Positive Feedback in Natural Systems* (DeAngelis et al. 1986), which narrated on the positive effects in ecosystems and *Plants Helping Plants* (Hunter and Aarssen 1988), which vouched for considering facilitation as a vital phenomenon in plants coexisting as communities. These were also followed by many revolutionary empirical works by Gigon and Ryser (1986), Bertness (1991), Stephens and Bertness (1991), Bertness and Callaway (1994), Van Wesenbeeck et al. (2007), Callaway (1995, 1997), Stachowicz (2001), Brooker et al. (2008), and Bruno et al. (2003). Spatial relationships can consistently foster facilitative interactions. An example of this was put forth by Tirado and Pugnaire (2003). They showed that in the oceanic dunes of the south of Spain, *Asparagus* seedlings grow better within the patches of oceanic dunes that harbor *Ziziphus lotus* trees and are their survival, flowering, fruiting, and seed counts are negatively impacted upon transfer to open areas. Mature plants provide protection to the young plants and help to provide them shelter for establishing seedlings in the desert ecosystem. Facilitation can be either mutualism (both benefit) or commensalism (one benefits and the other is unaffected). Some of the classic examples of plant facilitation, such as nurse plant interactions, or facilitations during succession are commensalisms, but there are also many examples of facilitations that are mutualisms, especially in stressful habitats such as salt marshes and sand dunes. In the case of the *Asparagus– Ziziphus lotus interaction,* it is likely that this facilitation was caused by nutrient enrichment in the patches where these plants grew together. Similarly, leguminous plants are sown as a rotational crop with cash crops for the similar purpose of enriching the soil after the cash crop has been harvested. Many small plants also require conditions of shades offered by higher plants, the latter which actually are not outcompeted for growth requisites by the small plants. Large, steady plants protect the small plants living in their vicinity from wind, predators, and extreme temperatures. Young saguaro cacti (*Cereus giganteus*) require nurse plants to provide them shade until they are large enough to tolerate low moisture (Steenbergh and Lowe 1969). Commensalism between epiphytes and phorophytes was shown by Zotz (2016) in that epiphytes rest on the host tree without causing harm to the host tree. Mutualism is shown by Yong et al. (2015) in myrmecophytes in Southeast Asia that generally have a mutualistic association with their host plants. Mutualism and protocooperation are ecological interactions in which there is advantage for both partners.

8.4.3 FACTORS THAT EFFECTUATE SWITCHES IN PLANT INTERACTIONS

According to the stress gradient hypothesis (SGH), it has been reported that, under changes in environmental stress, interactions among the plants change and may switch between positive and negative events. Switching has been seen mostly in facilitative (positive) and competitive (negative) associations in various geographically distinct plant communities. Facilitations have been recorded to predominate under high abiotic stress conditions compared to those under conditions of mild abiotic stresses. An extensive study conducted on the Gongga Mountain glacier region exhibits an example of such an effect among *Populusprudomii* (Rehder) and *Salix rehderiana* (Schneider) communities growing on 20- and 40-year old soils (Song et al. 2019). This study revealed that young soils showed positive effects on plant survival while the old soil presented negative effects. Soil aging drives the shift from a positive to negative interaction event, and this further supports the SGH. Similar support is available from a study with interactions between European beech, *Fagussylvatica*, and underground soil microbiota under elevation gradients (Defossez et al. 2011). These notions on SGH are supported by the understanding of vascular plant interactions with other plants, insects, mycorrhizae, and microbes, such as fungi, and bacteria, in various environments. However, there are few studies that have been conducted otherwise.

8.4.4 PLANTS DO TALK AND LISTEN TO EACH OTHER

Individual plants are known to display directed signaling responses within their systems to light and nutritional resources. Additionally, plants also emit various signals in response to insect herbivory that may attract various carnivores (Silvertown and Gordon 1989; Farmer 2001; Karban and Baldwin 2007; Dicke and van Loon 2000). The ability to emit volatile compounds bridges their interaction with other organisms, known as "plant talking behavior" in the molecular purview. Whether plants talk among themselves in their neighborhood about their damage relevant status has been a debatable theme of research and discussion among plant biologist communities with many believing that plants usually are neutral to receiving any such responses from different neighbors or those conspecific to them in their close vicinity (Rhoades 1983; Baldwin and Schultz 1983). However, plethora of evidence has been accumulating to vouch for such a signaling phenomenon among neighboring plants and has been realized by either launching i) a direct defense, rendering them tolerant to upcoming herbivore infestation or ii) an indirect defense by recruiting arthropod bodyguards. By these mechanisms, specific plants that are capable of receiving such calls from their neighbors have an advantage over those who are incapable of tapping into these calls (Dicke and van Loon 2000). This should be interpreted as facilitation and possibly interdependence in terms of the integrity of the plant community structure and its build up and reshaping – more in the circumstances of periodic invasion from unwanted guests such as insects, weeds, and parasitic plants at such communities. Notably, these notions have built the hypothesis that plants adaptively redirect responses to chemical signals emitted from those close by (Figure 8.4). These studies have been frequently criticized, indicating that these responses have been performed vaguely and in not using scientific methods with assays that run in small vessels (Fowler and Lawton 1985), but the merits of these ideas are gaining scientific acceptance (Karban and Baldwin 1997). However, more robust field scale and genomics studies now strongly support these views.

Plants in a community structure may vary in their chemical profiles, which may be variously tittered among different genera, and many among similar species and varieties may vary in a low to high extent, depending on the abiotic and biotic factors that prevail in the biomes they inhabit. The aftermath of tree defoliation may biochemically influence neighboring flora (Tuomi et al. 1990), and this influence may be exhibited through attachment points such as mycorrhizae (Simard et al. 1997), haustorial connections of mistletoes, and other parasitic plants, such as *Orobanche* and *Striga*. Chemical signaling also may be triggered by the release and reception of plant volatiles (Dicke 1994), such as the case with many terpenoids, growth regulators, ethylene, and methyl jasmonate. Some of these induce genes responsible for

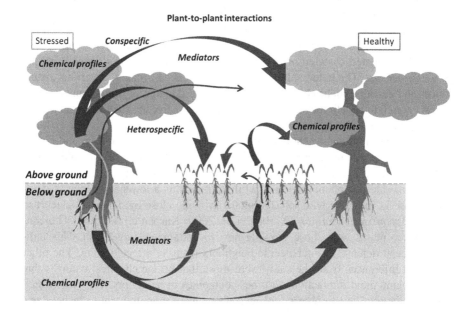

FIGURE. 8.4   Plant interactions within community settings. Here we show how plants may effectuate calls to near and far plant communities that may employ the use of hormones, intermediate mediator organisms, or other vectors. Plants can sense or send signals for help and respond to them by drastic changes in chemical profiles. This may be both amongst the similar types (conspecifics) and different plants (heterospecifics). There is more information regarding above-ground communications than below-ground crosstalk.

plant resistance to the environmental cues (Ecker and Davis 1987; Farmer and Ryan 1990) and, concomitantly, bring about variations in secondary metabolisms (Shonle and Bergelson 1995), which, in turn, exhibit as differences in chemical profiles of the plants that release them volatiles or the neighboring plants that receive them. A group in Germany (Dolch and Tscharntke 2000) studied the effects of defoliation damage in alder trees (*Alnus glutinosa*) caused by the infestation of leaf beetles (*Agelastica alni*). They witnessed that undamaged conspecifics very closely located to the defoliated tree were less infested by beetles than those at a distant location. They tested the degree of leaf damage in field trials as well as in laboratory settings and the extent of leaf herbivory and oviposition frequency, all of which correlated positively with the distance to the defoliated alder. This inferred a naturally induced resistance in defoliated trees as well as an induced defense (before the actual onset of damage) in neighboring undamaged trees. Surprisingly, this would mean the alarm was signaled by the damaged trees to their conspecifics that derive the fitness benefit from an induced defense upon such signaling. With more elaborate biochemical studies, this group of researchers also subsequently pinned down the volatiles that regulated these chemical cues (Tscharntke et al. 2001). They showed that alder trees respond to beetle herbivory by emitting ethylene and various terpenes. Chemical profiling also revealed a drastic increase in leaf polyphenols, activity of oxidative enzymes, such as polyphenoloxidase, lipoxygenase, and peroxidase as well as proteinase inhibitors. They reported a jasmonic acid signaling to work following the octadecanoid pathways (Tscharntke et al. 2001).

Signal transfers among plants may work through aerial, above-ground processes, preferably regulated by wind or diffusion as well as similar ways through below-ground processes. The latter is, however, less pronounced because of various disturbances and barriers, but has been less explored. Aerial modes of signal transfer could be fast enough to benefit the neighbors amenable to receive the chemical signal from the calling plant.

Another question one might ask would be "Do these signal transfers and receptions work only among conspecifics?" The next work shows a documented study with interspecific interactions in plants. Karban's group reported a higher level of defense enzyme polyphenol oxidase, while a there was a lower incidence of damage by insects in native wild tobacco plants growing close by damaged sagebrush species (*Artemisia tridentata*) (Karban et al. 2000; Karban et al. 2003). The control tobacco plants were those growing next to undamaged sagebrush. This indicates induced defense in interspecific neighbors from an affected plant. The other important inferences from the study were those that bridge chemical changes that mediate plant–plant interactions to fitness outcomes in neighbors. They uncovered that even though the neighboring tobacco plants were more tolerant to insect damage, they became more susceptible to frost damage with respect to controls nearing undamaged sagebrush (Karban and Maron 2002). This might also suggest to not underestimate the chances of miscommunications or interfering disturbances that nonspecifically induce the frost damage symptoms and should be treated as natural events in the community wilderness and diversity of interactions, signals, and responses that prevail. The group also showed a detrimental influence of sagebrush on the health of the tobacco plants (Karban and Maron 2002).

Gene expression studies excel at digging deeper into responses in plants to signals via volatiles among the neighboring plants. Spider-mite damaged lima beans emit odors, signaling the neighbors. Upon receipt of these signals, the neighboring non-specific plants induce the expression of a plethora of defense-responsive genes. This revelation by Arimura et al. (2000) has paved the way for enthralling investigations to be carried out as well enabling plant biologists to understand how such signaling networks work in nature within the various ecosystems. More of these and thorough works on other crops would enable efficient agriculture.

8.4.5 Underground Crosstalk among Plants

Only a few groups have emphasized investigating the underground crosstalk among plants, but surely, this area is rapidly emerging and is becoming a trend setter. In one work Dicke and Dijkman (2001) collected a systemic elicitor from spider-mite (*Tetranychusurticae*) infected lima bean leaves by placing the petioles in water. These elicitors can induce production of volatile chemicals that attract predators to spider mites. They recovered the elicitors and administered them to healthy lima bean leaves. This attracted the predatory mites (*Phytoseiulus persimilis*) against spider mites, as the plant mimicked responses of the mite-infested lima plant. The group also showed that placing the healthy lima beans in distilled water previously dipped with mite infested lima bean could still attract the predatory mites (Dicke and Dijkman 2001). Hence the plant to plant communication by means of below ground activities (chemical profile changes) cannot be underrated. In yet another work by Guerrieri et. al (2002), a similar response from faba beans that attract predators to herbivores by means of the production of volatiles, called host induced synomones, was reported. These studies indicate that plant to plant interactions may also work in rhizosperic levels and may work by systemic translocation of elicitors (Figure 8.4).

8.5 CONCLUSIONS

Understanding community interactions among plants and that of other lifeform groups is invaluable because they shape the ecological biomes and are so central to balances that maintain community diversity. Understanding the basic mechanisms behind these interactions may be the key to better produce and more eco-friendly farming without jeopardizing community settings.

ACKNOWLEDGMENTS

The authors thank the fraternity of the University Institute of Biotechnology and the University Center for Research and Development (UCRD) at Chandigarh University (CU) for support. The institute to which the authors are currently affiliated has no role in shaping the manuscript. Also, the authors declare no conflict of interest.

REFERENCES

Aboshi T, Ishiguri S, Shiono Y, Murayama T (2018) Flavonoid glycosides in Malabar spinach Basella alba inhibit the growth of Spodoptera litura larvae. *Bioscience, Biotechnology, and Biochemistry* 82 (1):9–14.

Adams DG, Duggan PS (2008) Cyanobacteria–bryophyte symbioses. *Journal of Experimental Botany* 59 (5):1047–1058.

Ahmadjian V, Jacobs JB (1981) Relationship between fungus and alga in the lichen Cladonia cristatella Tuck. *Nature* 289 (5794):169–172.

Akbar SM, Pavani T, Nagaraja T, Sharma H (2016) Influence of CO$_2$ and temperature on metabolism and development of Helicoverpa armigera (Noctuidae: Lepidoptera). *Environmental Entomology* 45 (1):229–236.

Ali JG, Agrawal AA (2012) Specialist versus generalist insect herbivores and plant defense. *Trends in Plant Science* 17 (5):293–302.

Aljbory Z, Chen MS (2018) Indirect plant defense against insect herbivores: a review. *Insect Science* 25 (1):2–23.

Arimura G-I, Ozawa R, Shimoda T, Nishioka T, Boland W, Takabayashi J (2000) Herbivory-induced volatiles elicit defence genes in lima bean leaves. *Nature* 406 (6795):512–515.

Asfaw MD, Kassa SM, Lungu EM, Bewket W (2019) Effects of temperature and rainfall in plant–herbivore interactions at different altitude. *Ecological Modelling* 406:50–59.

Baldwin IT, Schultz JC (1983) Rapid changes in tree leaf chemistry induced by damage: evidence for communication between plants. *Science* 221 (4607):277–279.

Banerjee S, Mandal NC (2019) Diversity of chitinase-producing bacteria and their possible role in plant pest control. In *Microbial Diversity in Ecosystem Sustainability and Biotechnological Applications*: Satyanarayana T, Das SK, Johri BN (Eds.), Springer, pp. 457–491.

Behie SW, Bidochka MJ (2013) Insects as a nitrogen source for plants. *Insects* 4 (3):413–424.

Bennett RN, Wallsgrove RM (1994) Secondary metabolites in plant defence mechanisms. *New Phytologist* 127 (4):617–633.

Bertness MD (1991) Interspecific interactions among high marsh perennials in a New England salt marsh. *Ecology* 72 (1):125–137.

Bertness MD, Callaway R (1994) Positive interactions in communities. *Trends in Ecology & Evolution* 9 (5):191–193.

Bjarnsholt T (2013) The role of bacterial biofilms in chronic infections. *Apmis* 121:1–58.

Bonfante P, Genre A (2008) Plants and arbuscular mycorrhizal fungi: an evolutionary-developmental perspective. *Trends in Plant Science* 13 (9):492–498.

Braga RM, Dourado MN, Araújo WL (2016) Microbial interactions: ecology in a molecular perspective. *Brazilian Journal of Microbiology* 47 Supplement 1:86–98.

Branda SS, Vik Å, Friedman L, Kolter R (2005) Biofilms: the matrix revisited. *Trends in Microbiology* 13 (1):20–26.

Brooker RW, Maestre FT, Callaway RM, Lortie CL, Cavieres LA, Kunstler G, Liancourt P, Tielbörger K, Travis JM, Anthelme F (2008) Facilitation in plant communities: the past, the present, and the future. *Journal of Ecology* 96 (1):18–34.

Bruno JF, Stachowicz JJ, Bertness MD (2003) Inclusion of facilitation into ecological theory. *Trends in Ecology & Evolution* 18 (3):119–125.

Callaway RM (1995) Positive interactions among plants. *The Botanical Review* 61 (4):306–349.

Callaway RM (1997) Positive interactions in plant communities and the individualistic-continuum concept. *Oecologia* 112 (2):143–149.

Capinera JL (2005) Relationships between insect pests and weeds: an evolutionary perspective. *Weed Science* 53 (6):892–901.

Carella P, Gogleva A, Tomaselli M, Alfs C, Schornack S (2018) Phytophthora palmivora establishes tissue-specific intracellular infection structures in the earliest divergent land plant lineage. *Proceedings of the National Academy of Sciences* 115 (16):E3846–E3855.

Chandra S, Fraser-Jenkins C, Kumari A, Srivastava A (2008) A summary of the status of threatened pteridophytes of India. *Taiwania* 53 (2):170–209.

Chon S-U, Nelson CJ (2013) Allelopathic dynamics in resource plants. In: *Allelopathy*: Cheema ZA, Farooq M, Wahid A (Eds.), Springer, pp. 81–110.

Clements FE, Weaver JE, Hanson HC (1929) Plant competition: an analysis of community functions.

Craine JM, Dybzinski R (2013) Mechanisms of plant competition for nutrients, water and light. *Functional Ecology* 27 (4):833–840.

Craine JM, Towne EG, Tolleson D, Nippert JB (2013) Precipitation timing and grazer performance in a tallgrass prairie. *Oikos* 122 (2):191–198.

Cross AT (2019) *Carnivorous Plants. A Jewel in the Crown of a Global Biodiversity Hotspot* Perth: Kwongan Foundation and the Western Australian Naturalists' Club Inc.

Davey ML, Tsuneda A, Currah RS (2009) Pathogenesis of bryophyte hosts by the ascomycete *Atradidymella muscivora*. *American Journal of Botany* 96 (7):1274–1280.

Davey ML, Tsuneda A, Currah RS (2010) Saprobic and parasitic interactions of Coniochaeta velutina with mosses. *Botany* 88 (3):258–265.

DeAngelis DL, Travis CC, Post WM (1986) Positive feedback in natural systems. (Vol. 15). Springer Science & Business Media.

Defossez E, Courbaud B, Marcais B, Thuiller W, Granda E, Kunstler G (2011) Do interactions between plant and soil biota change with elevation? A study on Fagus sylvatica. *Biology Letters* 7 (5):699–701.

DeLucia EH, Nabity PD, Zavala JA, Berenbaum MR (2012) Climate change: resetting plant-insect interactions. *Plant Physiology* 160 (4):1677–1685.

Díaz E-M, Sacristán M, Legaz M-E, Vicente C (2009) Isolation and characterization of a cyanobacterium-binding protein and its cell wall receptor in the lichen Peltigera canina. *Plant Signaling & Behavior* 4 (7):598–603.

Dicke M (1994) Local and systemic production of volatile herbivore-induced terpenoids: their role in plant-carnivore mutualism. *Journal of Plant Physiology* 143 (4–5):465–472.

Dicke M, Dijkman H (2001) Within-plant circulation of systemic elicitor of induced defence and release from roots of elicitor that affects neighbouring plants. *Biochemical Systematics and Ecology* 29 (10):1075–1087.

Dicke M, van Loon JJ (2000) Multitrophic effects of herbivore-induced plant volatiles in an evolutionary context. *Entomologia experimentalis et applicata* 97 (3):237–249.

Dolch R, Tscharntke T (2000) Defoliation of alders (Alnus glutinosa) affects herbivory by leaf beetles on undamaged neighbours. *Oecologia* 125 (4):504–511

Dotzler N, Krings M, Taylor TN, Agerer R (2006) Germination shields in Scutellospora (Glomeromycota: Diversisporales, Gigasporaceae) from the 400 million-year-old Rhynie chert. *Mycological Progress* 5 (3):178–184.

Dyer LA, Richards LA, Short SA, Dodson CD (2013) Effects of CO_2 and temperature on tritrophic interactions. *PLoS One* 8 (4):e62528.

Ecker JR, Davis RW (1987) Plant defense genes are regulated by ethylene. *Proceedings of the National Academy of Sciences* 84 (15):5202–5206.

Erb M, Robert CA (2016) Sequestration of plant secondary metabolites by insect herbivores: molecular mechanisms and ecological consequences. *Current Opinion in Insect Science* 14:8–11.

Farmer EE (2001) Surface-to-air signals. *Nature* 411 (6839):854–856.

Farmer EE, Ryan CA (1990) Interplant communication: airborne methyl jasmonate induces synthesis of proteinase inhibitors in plant leaves. *Proceedings of the National Academy of Sciences* 87 (19):7713–7716.

Fowler SV, Lawton JH (1985) Rapidly induced defenses and talking trees: the devil's advocate position. *The American Naturalist* 126 (2):181–195.

Fürstenberg-Hägg J, Zagrobelny M, Bak S (2013) Plant defense against insect herbivores. *International Journal of Molecular Sciences* 14 (5):10242–10297.

Futuyama D (1979) *Evolutionary Biology*. Sunderland, MA: Sinauer. The Routledge Encyclopedia of Philosophy.

Gigon A, Ryser P (1986) Positive Interaktionen zwischen Pflanzenarten. *Veröff Geobot Inst ETH, Stiftung Rübel* 87:372–387.

Givnish TJ (2015) New evidence on the origin of carnivorous plants. *Proceedings of the National Academy of Sciences* 112 (1):10–11.

Gleason HA (1926) The individualistic concept of the plant association. *Bulletin of the Torrey Botanical Club* 53 (1):7–26.

Guerrieri E, Poppy G, Powell W, Rao R, Pennacchio F (2002) Plant-to-plant communication mediating in-flight orientation of *Aphidius ervi*. *Journal of Chemical Ecology* 28 (9):1703–1715.

Harriott MM, Noverr MC (2010) Ability of *Candida albicans* mutants to induce Staphylococcus aureus vancomycin resistance during polymicrobial biofilm formation. *Antimicrobial Agents and Chemotherapy* 54 (9):3746–3755.

Havko NE, Kapali G, Das MR, Howe GA (2020) Stimulation of insect herbivory by elevated temperature outweighs protection by the jasmonate pathway. *Plants* 9 (2):172.

Heidel-Fischer HM, Vogel H (2015) Molecular mechanisms of insect adaptation to plant secondary compounds. *Current Opinion in Insect Science* 8:8–14.

Hettenhausen C, Schuman MC, Wu J (2015) MAPK signaling: a key element in plant defense response to insects. *Insect Science* 22 (2):157–164.

Hosokawa T, Kikuchi Y, Nikoh N, Shimada M, Fukatsu T (2006) Strict host-symbiont cospeciation and reductive genome evolution in insect gut bacteria. *PLoS Biology* 4 (10):e337.

Hunter A, Aarssen L (1988) Plants helping plants. *Bioscience* 38 (1):34–40.

Jamieson MA, Burkle LA, Manson JS, Runyon JB, Trowbridge AM, Zientek J (2017) Global change effects on plant–insect interactions: the role of phytochemistry. *Current Opinion in Insect Science* 23:70–80.

Jiang H, Huang B, Qian Z, Xu Y, Liao X, Song P, Li X (2020) Effects of increased carbon supply on the growth, nitrogen metabolism and photosynthesis of Vallisneria natans grown at different temperatures. *Wetlands* 40 (5):1–9.

Jones DL (1998) Organic acids in the rhizosphere–a critical review. *Plant and Soil* 205 (1):25–44.

Jürgens A, El-Sayed AM, Suckling DM (2009) Do carnivorous plants use volatiles for attracting prey insects? *Functional Ecology* 23 (5):875–887.

Karban R, Baldwin IT (1997) *Induced Responses to Herbivory*. University of Chicago Press.

Karban R, Baldwin IT (2007) *Induced Responses to Herbivory*. University of Chicago Press.

Karban R, Baldwin IT, Baxter KJ, Laue G, Felton GW (2000) Communication between plants: induced resistance in wild tobacco plants following clipping of neighboring sagebrush. *Oecologia* 125 (1):66–71. doi:10.1007/PL00008892

Karban R, Maron J (2002) The fitness consequences of interspecific eavesdropping between plants. *Ecology* 83 (5):1209–1213. doi:10.1890/0012-9658(2002)083[1209:tfcoie]2.0.co;2

Karban R, Maron J, Felton GW, Ervin G, Eichenseer H (2003) Herbivore damage to sagebrush induces resistance in wild tobacco: evidence for eavesdropping between plants. *Oikos* 100 (2):325–332. doi:10.1034/j.1600-0706.2003.12075.x

Kazan K (2018) Plant-biotic interactions under elevated CO_2: a molecular perspective. *Environmental and Experimental Botany* 153:249–261.

Keerthi P, Singh M, Bishnoi A (2019) Chapter-3 role of biological control of weeds and bioherbicides. *Advances in Agronomy* 4:61–79.

Kim YC, Jung H, Kim KY, Park SK (2008) An effective biocontrol bioformulation against Phytophthora blight of pepper using growth mixtures of combined chitinolytic bacteria under different field conditions. *European Journal of Plant Pathology* 120 (4):373–382.

Knack J, Wilcox L, Delaux P-M, Ané J-M, Piotrowski M, Cook M, Graham J, Graham L (2015) Microbiomes of streptophyte algae and bryophytes suggest that a functional suite of microbiota fostered plant colonization of land. *International Journal of Plant Sciences* 176 (5):405–420.

Kortbeek RW, van der Gragt M, Bleeker PM (2019) Endogenous plant metabolites against insects. *European Journal of Plant Pathology* 154 (1):67–90.

Kreft H, Jetz W, Mutke J, Barthlott W (2010) Contrasting environmental and regional effects on global pteridophyte and seed plant diversity. *Ecography* 33 (2):408–419.

Kumar D (2019) Chapter-4 defense strategies in plants against insect herbivores. *Advances in Agricultural Entomology* 7:67–119.

Kumar V, Suman U, Yadav SK (2018) Flavonoid secondary metabolite: biosynthesis and role in growth and development in plants. In: *Recent Trends and Techniques in Plant Metabolic Engineering*: Yadav SK, Kumar V, Singh SP (Eds.), Springer, pp. 19–45.

Larousse M, Rancurel C, Syska C, Palero F, Etienne C, Nesme X, Bardin M, Galiana E (2017) Tomato root microbiota and Phytophthora parasitica-associated disease. *Microbiome* 5 (1):1–11.

Lee L, Zhang Y, Ozar B, Sensen CW, Schriemer DC (2016) Carnivorous nutrition in pitcher plants (Nepenthes spp.) via an unusual complement of endogenous enzymes. *Journal of Proteome Research* 15 (9):3108–3117.

Lehnert M, Kottke I, Setaro S, Pazmiño LF, Suárez JP, Kessler M (2009) Mycorrhizal associations in ferns from Southern Ecuador. *American Fern Journal* 99 (4):292–306.

Lehtonen MT, Marttinen EM, Akita M, Valkonen JP (2012) Fungi infecting cultivated moss can also cause diseases in crop plants. *Annals of Applied Biology* 160 (3):298–307.

Liao H-L, Bonito G, Rojas JA, Hameed K, Wu S, Schadt CW, Labbé J, Tuskan GA, Martin F, Grigoriev IV (2019) Fungal endophytes of Populus trichocarpa alter host phenotype, gene expression, and rhizobiome composition. *Molecular Plant-Microbe Interactions* 32 (7):853–864.

Locke MA, Reddy KN, Zablotowicz RM (2002) Weed management in conservation crop production systems. *Weed Biology and Management* 2 (3):123–132.

Lockhart C, Rowell P, Stewart W (1978) Phytohaemagglutinins from the nitrogen-fixing lichens Peltigera canina and P. polydactyla. *FEMS Microbiology Letters* 3 (3):127–130.

Malinovsky FG, Fangel JU, Willats WG (2014) The role of the cell wall in plant immunity. *Frontiers in Plant Science* 5:178.

Mathesius U (2018) *Flavonoid Functions in Plants and Their Interactions with Other Organisms*. Multidisciplinary Digital Publishing Institute.

Matsui H, Iwakawa H, Hyon G-S, Yotsui I, Katou S, Monte I, Nishihama R, Franzen R, Solano R, Nakagami H (2020) Isolation of natural fungal pathogens from Marchantia polymorpha reveals antagonism between salicylic acid and jasmonate during liverwort–fungus interactions. *Plant and Cell Physiology* 61 (2):265–275.

McNeece BT, Sharma K, Lawrence GW, Lawrence KS, Klink VP (2018) Mitogen activated protein kinases function as a cohort during a plant defense response. *bioRxiv*:396192.

Mello MO, Silva-Filho MC (2002) Plant-insect interactions: an evolutionary arms race between two distinct defense mechanisms. *Brazilian Journal of Plant Physiology* 14 (2):71–81.

Miller IM (1990) Bacterial leaf nodule symbiosis. *Advances in Botanical Research* 17:163–234. Elsevier.

Mittag J, Šola I, Rusak G, Ludwig-Müller J (2015) Physcomitrella patens auxin conjugate synthetase (GH3) double knockout mutants are more resistant to Pythium infection than wild type. *Journal of Plant Physiology* 183:75–83.

Mondo SJ, Lastovetsky OA, Gaspar ML, Schwardt NH, Barber CC, Riley R, Sun H, Grigoriev IV, Pawlowska TE (2017) Bacterial endosymbionts influence host sexuality and reveal reproductive genes of early divergent fungi. *Nature Communications* 8 (1):1–9.

Mudgal G, Mudgal B (2011) Evidence for unusual choice of host and haustoria by Dendrophthoe falcata (Lf) Ettingsh, a leafy mistletoe. *Archives of Phytopathology and Plant Protection* 44 (2):186–190.

Mudgal G, Mudgal B, Gururani MA, Jelli V (2011) Pseudaulacaspis cockerelli (Cooley) hyperparasitizing Dendrophthoe falcata (Lf) Ettingsh. *Archives of Phytopathology and Plant Protection* 44 (3):282–286.

Nadkarni NM (1981) Canopy roots: convergent evolution in rainforest nutrient cycles. *Science* 214 (4524):1023–1024.

Nampy S (1998) *Fern Flora of South India: Taxonomic Revision of Polypodioid Ferns.* Daya Books.

Napoleão TH, Albuquerque LP, Santos ND, Nova IC, Lima TA, Paiva PM, Pontual EV (2019) Insect midgut structures and molecules as targets of plant-derived protease inhibitors and lectins. *Pest Management Science* 75 (5):1212–1222.

Nazir A, Bareen F (2011) Synergistic effect of Glomus fasciculatum and Trichoderma pseudokoningii on Heliathus annuus to decontaminate tannery sludge from toxic metals. *African Journal of Biotechnology* 10 (22):4612–4618.

Newton AC, Fitt BD, Atkins SD, Walters DR, Daniell TJ (2010) Pathogenesis, parasitism and mutualism in the trophic space of microbe–plant interactions. *Trends in Microbiology* 18 (8):365–373.

Norris RF, Kogan M (2005) Ecology of interactions between weeds and arthropods. *Annual Review of Entomology* 50:479–503.

Norton DA, Smith MS (1999) Why might roadside mulgas be better mistletoe hosts? *Australian Journal of Ecology* 24 (3):193–198.

Overdijk EJ, De Keijzer J, De Groot D, Schoina C, Bouwmeester K, Ketelaar T, Govers F (2016) Interaction between the moss Physcomitrella patens and Phytophthora: a novel pathosystem for live-cell imaging of subcellular defence. *Journal of Microscopy* 263 (2):171–180.

Parniske M (2008) Arbuscular mycorrhiza: the mother of plant root endosymbioses. *Nature Reviews Microbiology* 6 (10):763–775.

Peng L, Zhao Y, Wang H, Song C, Shangguan X, Ma Y, Zhu L, He G (2017) Functional study of cytochrome P450 enzymes from the brown planthopper (Nilaparvata lugens Stål) to analyze its adaptation to BPH-resistant rice. *Frontiers in Physiology* 8:972.

Pham TA, Hwang S-Y (2020) High temperatures reduce nutrients and defense compounds against generalist Spodoptera litura F. in Rorippa dubia. *Arthropod-Plant Interactions* 14:333–344.

Phillips D, Tsai S (1992) Flavonoids as plant signals to rhizosphere microbes. *Mycorrhiza* 1 (2):55–58.

Pincebourde S, Van Baaren J, Rasmann S, Rasmont P, Rodet G, Martinet B, Calatayud P-A (2017) Plant–insect interactions in a changing world. *Advances in Botanical Research* 81:289–332.

Ponce de León I (2011) The moss Physcomitrella patens as a model system to study interactions between plants and phytopathogenic fungi and oomycetes. *Journal of Pathogens* 2011: 719873–719873.

Ponce de León I, Montesano M (2017) Adaptation mechanisms in the evolution of moss defenses to microbes. *Frontiers in Plant Science* 8:366.

Pureswaran DS, Neau M, Marchand M, De Grandpré L, Kneeshaw D (2019) Phenological synchrony between eastern spruce budworm and its host trees increases with warmer temperatures in the boreal forest. *Ecology and Evolution* 9 (1):576–586.

Rajamani S, Bauer WD, Robinson JB, Farrow III JM, Pesci EC, Teplitski M, Gao M, Sayre RT, Phillips DA (2008) The vitamin riboflavin and its derivative lumichrome activate the LasR bacterial quorum-sensing receptor. *Molecular Plant-Microbe Interactions* 21 (9):1184–1192.

Ramanan R, Kang Z, Kim B-H, Cho D-H, Jin L, Oh H-M, Kim H-S (2015) Phycosphere bacterial diversity in green algae reveals an apparent similarity across habitats. *Algal Research* 8:140–144.

Read D, Duckett J, Francis R, Ligrone R, Russell A (2000) Symbiotic fungal associations in 'lower'land plants. *Philosophical Transactions of the Royal Society of London Series B: Biological Sciences* 355 (1398):815–831.

Remy W, Taylor TN, Hass H, Kerp H (1994) Four hundred-million-year-old vesicular arbuscular mycorrhizae. *Proceedings of the National Academy of Sciences* 91 (25):11841–11843

Rhoades DF (1983) Responses of alder and willow to attack by tent caterpillars and webworms: evidence for pheromonal sensitivity of willows. In *Plant Resistance to Insects*: Hedin PA (Ed.), ACS Publications, pp. 55–68.

Rischer H, Hamm A, Bringmann G (2002) Nepenthes insignis uses a C2-portion of the carbon skeleton of L-alanine acquired via its carnivorous organs, to build up the allelochemical plumbagin. *Phytochemistry* 59 (6):603–609.

Roux JP (2009) *Synopsis of the Lycopodiophyta and Pteridophyta of Africa, Madagascar and Neighbouring Islands*. South African National Biodiversity Institute.

Sangale P (2020) Carnivorous plants and its mechanisms, BO. 4.2.

Sankaran T (1990) Biological control of weeds with insects: a dynamic phenomenon of insect-plant interaction. *Proceedings: Animal Sciences* 99 (3):225–232.

Sekimoto S, Beakes GW, Gachon CM, Müller DG, Küpper FC, Honda D (2008) The development, ultrastructural cytology, and molecular phylogeny of the basal oomycete Eurychasma dicksonii, infecting the filamentous phaeophyte algae Ectocarpus siliculosus and Pylaiella littoralis. *Protist* 159 (2):299–318.

Selosse M-A, Le Tacon F (1998) The land flora: a phototroph-fungus partnership? *Trends in Ecology & Evolution* 13 (1):15–20.

Shonle I, Bergelson J (1995) Interplant communication revisited. *Ecology* 76 (8):2660–2663.

Silvertown J, Gordon DM (1989) A framework for plant behavior. *Annual Review of Ecology and Systematics* 20 (1):349–366.

Simard SW, Perry DA, Jones MD, Myrold DD, Durall DM, Molina R (1997) Net transfer of carbon between ectomycorrhizal tree species in the field. *Nature* 388 (6642):579–582.

Singh BK, Millard P, Whiteley AS, Murrell JC (2004) Unravelling rhizosphere–microbial interactions: opportunities and limitations. *Trends in Microbiology* 12 (8):386–393.

Singh GB (2020) Potential of endophytes at treating tuberculosis. *Plant Cell Biotechnology and Molecular Biology* 21 (17-18):218–224.

Song M, Yu L, Jiang Y, Korpelainen H, Li C (2019) Increasing soil age drives shifts in plant-plant interactions from positive to negative and affects primary succession dynamics in a subalpine glacier forefield. *Geoderma* 353: 435–448. doi:10.1016/j.geoderma.2019.07.029

Srivastava S, Verma PC, Singh A, Mishra M, Singh N, Sharma N, Singh N (2012) Isolation and characterization of Staphylococcus sp. strain NBRIEAG-8 from arsenic contaminated site of West Bengal. *Applied Microbiology and Biotechnology* 95 (5):1275–1291.

Stachowicz JJ (2001) Mutualism, facilitation, and the structure of ecological communities: positive interactions play a critical, but underappreciated, role in ecological communities by reducing physical or biotic stresses in existing habitats and by creating new habitats on which many species depend. *Bioscience* 51 (3):235–246.

Steenbergh WF, Lowe CH (1969) Critical factors during the first years of life of the saguaro (Cereus giganteus) at Saguaro National Monument, Arizona. *Ecology* 50 (5):825–834.

Stephens EG, Bertness MD (1991) Mussel facilitation of barnacle survival in a sheltered bay habitat. *Journal of Experimental Marine Biology and Ecology* 145 (1):33–48.

Stewart AJ, Stewart RF (2008) Phenols. In *Encyclopedia of Ecology*: Jorgensen SE, Fath BD (Eds.), Academic Press, pp. 2682–2689.

Strullu-Derrien C, Kenrick P, Pressel S, Duckett JG, Rioult JP, Strullu DG (2014) Fungal associations in H orneophyton ligneri from the R hynie C hert (c. 407 million year old) closely resemble those in extant lower land plants: novel insights into ancestral plant–fungus symbioses. *New Phytologist* 203 (3):964–979.

Sukumaran S, Jeeva S, Raj A (2009) Diversity of Pteridophytes in miniature sacred forests of Kanyakumari district, southern Western Ghats. *Indian Journal of Forestry* 32 (3):285–290.

Sun Y, Guo H, Ge F (2016) Plant–aphid interactions under elevated CO_2: some cues from aphid feeding behavior. *Frontiers in Plant Science* 7:502.

Sun Y, Guo H, Ge F (2019) Medicago truncatula–pea aphid interaction in the context of global climate change. In *The Model Legume Medicago truncatula*: F. de Bruijn (Ed.), John Wiley & Sons, pp. 369–376.

Syed-Ab-Rahman SF, Carvalhais LC, Chua ET, Chung FY, Moyle PM, Eltanahy EG, Schenk PM (2019) Soil bacterial diffusible and volatile organic compounds inhibit Phytophthora capsici and promote plant growth. *Science of the Total Environment* 692:267–280.

Tang JY, Ma J, Li XD, Li YH (2016) Illumina sequencing-based community analysis of bacteria associated with different bryophytes collected from Tibet, China. *BMC Microbiology* 16 (1):276.

Telkar S, Gurjar G, Dey JK, Kant K, Solanki SPS (2015) Biological weed control for sustainable agriculture. *International Journal of Economic Plants* 2 (4):181–183.

Tiku AR (2018) Antimicrobial compounds and their role in plant defense. In *Molecular Aspects of Plant-Pathogen Interaction*: Singh A, Singh IK (Eds.), Springer, pp. 283–307.

Tirado R, Pugnaire FI (2003) Shrub spatial aggregation and consequences for reproductive success. *Oecologia* 136 (2):296–301.

Tscharntke T, Thiessen S, Dolch R, Boland W (2001) Herbivory, induced resistance, and interplant signal transfer in Alnus glutinosa. *Biochemical Systematics and Ecology* 29 (10):1025–1047.

Tuomi J, Niemelä P, Siren S (1990) The Panglossian paradigm and delayed inducible accumulation of foliar phenolics in mountain birch. *Oikos* 59:399–410.

Ueda H, Kikuta Y, Matsuda K (2012) Plant communication: mediated by individual or blended VOCs? *Plant Signaling & Behavior* 7 (2):222–226.

Uludag A, Uremis I, Arslan M (2018) Biological weed control. In *Non-Chemical Weed Control*: Jabran K, Chauhan BS (Eds.), Elsevier, pp. 115–132.

Van der Ent S, Pieterse CM (2018) Ethylene: multi-tasker in plant–attacker interactions. *Annual Plant Reviews Online*:343–377.

Van Wesenbeeck BK, Crain CM, Altieri AH, Bertness MD (2007) Distinct habitat types arise along a continuous hydrodynamic stress gradient due to interplay of competition and facilitation. *Marine Ecology Progress Series* 349:63–71.

Wang T-y, Li Q, Bi K-s (2018) Bioactive flavonoids in medicinal plants: structure, activity and biological fate. *Asian Journal of Pharmaceutical Sciences* 13 (1):12–23.

War AR, Buhroo AA, Hussain B, Ahmad T, Nair RM, Sharma HC (2020) Plant defense and insect adaptation with reference to secondary metabolites. In *Co-Evolution of Secondary Metabolites*. Mérillon JM and Ramawat K (Eds). Springer, pp. 795–822.

War AR, Paulraj MG, Ahmad T, Buhroo AA, Hussain B, Ignacimuthu S, Sharma HC (2012) Mechanisms of plant defense against insect herbivores. *Plant Signaling & Behavior* 7 (10):1306–1320.

Watanabe K, Takihana N, Aoyagi H, Hanada S, Watanabe Y, Ohmura N, Saiki H, Tanaka H (2005) Symbiotic association in Chlorella culture. *FEMS Microbiology Ecology* 51 (2):187–196.

Williams L, Rodriguez-Saona C, del Conte SCC (2017) Using non-model systems to explore plant-pollinator and plant-herbivore interactions: methyl jasmonate induction of cotton: a field test of the'attract and reward'strategy of conservation biological control. *AoB Plants* 9 (5):plx032.

Wilson J, Newman E (1987) Competition between upland grasses: root and shoot competition between Deschampsia flexuosa and Festuca ovina. *Acta Oecologica Oecologia Generalis* 8 (4):501–509.

Xiao L, Carrillo J, Siemann E, Ding J (2019) Herbivore-specific induction of indirect and direct defensive responses in leaves and roots. *AoB Plants* 11 (1):plz003.

Yong JW, Wei JW, Khew J, Rong SC, San WW (2015) *A Guide to the Common Epiphytes and Mistletoes of Singapore*. Cengage Learning Asia Singapore.

Yoshinaga N, Mori N (2018) Function of the lepidopteran larval midgut in plant defense mechanisms. In *Chemical Ecology of Insects*: Tabata J (Ed.), CRC Press, pp. 28–54.

Zhang A, Liu Z, Lei F, Fu J, Zhang X, Ma W, Zhang L (2017) Antifeedant and oviposition-deterring activity of total ginsenosides against Pieris rapae. *Saudi Journal of Biological Sciences* 24 (8):1751–1753.

Zheng Y, Chiang T-Y, Huang C-L, Gong X (2018) Highly diverse endophytes in roots of Cycas bifida (Cycadaceae), an ancient but endangered gymnosperm. *Journal of Microbiology* 56 (5):337–345.

Zhu-Salzman K, Zeng R (2015) Insect response to plant defensive protease inhibitors. *Annual Review of Entomology* 60:233–252.

Zotz G (2016) *Plants on Plants: The Biology of Vascular Epiphytes*. Springer.

Zunjarrao SS, Tellis MB, Joshi SN, Joshi RS (2020) Plant-insect interaction: the saga of molecular coevolution. In *Co-Evolution of Secondary Metabolites*: Mérillon JM, Ramawat KG (Eds.), Springer, pp. 19–45.

9 CRISPR/Cas9
An Efficient Tool for Improving Biotic Stress in Plants

Parul Sharma and Rajnish Sharma

CONTENTS

9.1 INTRODUCTION

Crop productivity in plants is reduced mostly because of a varied range of surrounding stresses, which are classified according to the nature of the stress, i.e., abiotic (non-living factors) stresses and biotic (living organisms) stresses. Abiotic stress factors include drought, salinity, extreme temperatures, water logging, heavy metals or minerals, etc. Alternatively, biotic stresses involve pathogens (fungi, bacteria, viruses, insects, nematodes, etc.) and attacks. Plant pathogens are solely responsible for the damage caused in terms of pre- and post-harvest losses, thereby affecting crop productivity. Naturally, plants have been bestowed with certain defence mechanisms to overpower some of these threats. On the basis of the capability of plants to sense the surrounding external stress, plants develop apt cellular reactions, thereby making themselves a bit tolerant to stress (Gull et al. 2019).

Traditional plant breeding systems and transgenic approaches have allowed breeders to produce improved crop varieties with agronomic qualitative/quantitative traits.

DOI: 10.1201/9781003213864-9

185

While combating the biotic stress, the former system is centred on the introgression of desirable resistant alleles into the best genotypes through intergeneric crosses or via induced mutagenesis, labelling it as an arbitrary and time-consuming approach (Ran et al. 2017). The latter still continues to contribute toward crop improvement to reduce the rate of biotic and abiotic stresses and increase yield (Tyczewska et al. 2018). For example, over the past few decades, transgenic crops possessing herbicide tolerance and insect resistance, such as soybean and maize, grab the top position with respect to commercialization. Other commercialized transgenics in this category are virus resistant squash hybrids, Hawaiian genetically modified (GM) papaya rainbow and sun-up (ISAAA 2016).

However, this approach received a lot of criticism regarding the commercialization of transgenic crops because of the foreign nature of the transgene (which may have inadvertent effects on targeted and non-targeted organisms), insertion of transgene at random sites in the genome, intellectual property issues, public concern, regulatory burdens, etc. (Langner et al. 2018).

The ever-evolving innovative advancements in molecular biology and biotechnology have immense potential for the future. Multifaceted reciprocity between a plant and a pathogen and the relevant stress response involves quite a few mechanisms responsible for the outbreak of a disease (Boyd and O'Toole 2012; Borelli et al. 2018; Dracatos et al. 2018). While using genome engineering methods, such as zinc-finger nucleases (ZFNs), transcription-activator-like effector nucleases (TALENs), and clustered regularly interspaced short palindromic repeat/CRISPR-associated protein 9 (CRISPR/Cas9), against biotic stress, a deep understanding of the molecular-level mechanisms of host plant–pathogen chemistry that is strengthened through competent molecular technologies is very significant.

These tools mainly exploit targeted endonucleases-based activity to identify sites, thereby generating specific double stranded DNA breaks (DSBs). Subsequently, these breaks are created by cellular DNA repair mechanisms through pathways such as non-homologous end-joining (NHEJ) or homology directed/dependent recombination (HDR). The error-prone NHEJ is the main repair pathway preferred by cells, which leads to knocking out alleles via random small insertions or deletion mutations. The HDR pathway enables precise base changes or gene replacement or insertion at the DSBs only if any homologous DNA or repair template exists (Que et al. 2019).

The CRISPR/Cas9-class II system, derived from bacteria (*Streptococcus pyogenes*), has overpowered other editing technologies in terms of specificity, cost involved, reprogramming, etc. As a result, this has become a widely used platform for various purposes. This system originated from an innate bacterial anti-viral immunity mechanism, in which RNA (tracrRNA + crRNA) guided Cas9 nuclease initiates cleavage at exact sites on foreign nucleic acid(s), triggering their consequent breakdown. By way of a genome editing system, CRISPR/Cas is comprised of a customizable single guide RNA (sgRNA) that fuses two RNA assemblages, e.g., the protospacer-containing CRISPR RNA (crRNA) and transactivating crRNA (tracrRNA) and a Cas protein. The sgRNA binds to target DNA via the Watson-Crick base pairing system. To target an exact sequence, it contains a Cas protein binding site and a well-marked spacer (~20 nucleotide sequence). Two-lobed {recognition (REC) and nuclease (NUC)} Cas9 is a DNA endonuclease that demonstrates nuclease action. The REC lobe is larger, and the NUC lobe is smaller. In addition, the NUC lobe encloses two nuclease domains: the conserved one – HNH and the variable C-terminal domain (CTD) – RuvC that binds the protospacer adjacent motif (PAM). The conserved HNH domain is accountable for slicing the guide RNA (gRNA) complementary target strand, and RuvC domain slices the opposite non-targeted DNA strand. Additionally, CRISPR interference, crRNA maturation, and spacer acquisition exhibit a pivotal role in the Cas9 endonuclease enzyme (Jiang and Doudna 2017).

To target any sequence, the CRISPR/Cas9 system relies on the availability of a nuclease analogous PAM sequence, i.e., G-rich (5-NGG-3), which is usually present in the genomes. However, their necessity during Cas9 action, specifically in the A-T-rich regions, narrows the range of potential genomic target sequences. The chances of off-target mutagenesis may increase because of the possible identification of alternate PAM sequences such as NAG and NGA (where N may be any nucleotide base) via the Cas9-sgRNA unit (Zhang et al. 2014). Improved specificity along with an array of target sequences can be broadened because divergent groups of bacteria with numerous Cas9 proteins exist, so the demand of variable PAM sequences generated variants of the unique *S. pyogenes* based on the CRISPR/Cas9 (SpCas9) system, e.g., *Streptococcus thermophilus* Cas9 (StCas9), *Staphylococcus aureus* Cas9 (SaCas9), and *Neisseria meningitidis* Cas9 (NmCas9). Apart from type II systems, the new genome editing tools are Cas12a (also well-known as Cpf1 belonging to Class 2, type V-A Cas enzyme) comprising of alternative forms such as FnCpf1, AsCpf1, and LbCpf1 from bacterial sources, namely *Francisella novicida*, *Acidaminococcus* sp. and *Lachnospiraceae bacterium*, respectively. Currently, these systems (Cas9 or Cpf1) target only DNA. On the other hand, class 2 type VI systems, such as LwaCas13a and PspCas13b, have been engineered to target or edit RNA (Chen *et al.* 2019). Therefore, this CRISPR/Cas9 approach has been utilized to target the vital genes related to pathogen growth and development, e.g., susceptibility (S) genes or the viral genomes.

9.2   OUTLINE OF CRISPR/CAS9 WORKFLOW

- Target sequence determination (using online programs).
- Design synthetic gRNAs (using online tools to assist with selection and order of CRISPR components).
- Delivery of CRISPR components (an ideal delivery technique suitable for the cell type involved in a particular experiment is significant).
- Gene editing analysis (using a number of techniques, including mismatch cleavage assays (relying on T7 endonuclease I), tracking of indels by decomposition (TIDE) sequencing analysis, next-generation sequencing (NGS), site-seq (cell-based analyses) and fluorescence-activated cell sorting (FACS) to monitor CRISPR editing efficiency).
- Phenotypic evaluation of edited plants.

9.3   CRISPR/CAS FOR DISEASE RESISTANCE IN PLANTS

To attain resistance against plant pathogens through CRISPR/Cas, significant knowledge about the genetics of resistance toward a particular disease and interrelated factors is important. The understanding of molecular-level information of a target gene of a plant or pathogen is of prime importance, as it is a pre-requisite to carry out CRISPR/Cas experiments. This information is accessible because entire genetic components in numerous crops and the linked microbial community have been sequenced (Shelake et al. 2019). Over time, advancements in molecular biology led to an increased level of clarification of the molecular mechanisms involved in host–pathogen interactions, thereby triggering the identification of many host-plant genes that usually participate in these complex processes. CRISPR/Cas9 could target various pathogens such as viruses, fungi, and bacteria by affecting the many vital host-plant elements necessary for pathogen sustainability and, therefore, protecting the plants.

9.4   CRISPR/CAS FOR TARGETING VIRAL PATHOGENS

9.4.1   CRISPR/Cas9-Mediated Interference against DNA Viruses

Plants are rarely affected by double-stranded DNA (dsDNA) viruses, so the more prevalent single-stranded DNA (ssDNA) viruses belonging to the families Geminiviridae and Nanoviridae could be targeted utilizing the CRISPR/Cas9 system.

Among them, Geminiviridae is the biggest, most well-known family, with more than 360 species accountable for noteworthy destruction in agronomically important crops worldwide. It encompasses a broad variety of plant viruses such as African cassava mosaic virus, bean yellow dwarf virus, tomato yellow leaf curl virus, beet curly top virus, etc. (Roossinck 2011, 2015). These viruses have a circular, ssDNA genome, and, like many viruses, their genomes encode only limited proteins, hence their reliance on host-cell factors for replication. The replication (Rep) (for the replication genome) and capsid protein (CP) (forms isometric particles and aids in packing the genome into them) are the main proteins encoded by these viruses. Like bacteriophages (M13) and many plasmids, viruses replicate in the infected host-plant cell by means of a rolling circle amplification (RCA) mechanism involving the formation of dsDNA intermediate. This intermediate script itself is an appropriate target for Cas-based endonucleases. Also, the intergenic (IR) sequence needed for virus replication serves as another hit for creating resistant plants because of IR activity loss caused by indels. These viruses have been targeted using genome editing tools, including CRISPR/Cas9. There is immense potential for this strategy to reveal various plant DNA viruses by utilizing the programmed version of single gRNA and corresponding endonuclease, which triggers a conserved section in various geminiviruses (Table 9.1).

9.4.2   CRISPR/Cas9-Mediated Interference against RNA Viruses

Plant pathogenic RNA viruses comprise many reverse transcription-based viruses, dsRNA viruses, and negative (-) and positive (+) sense ssRNA viruses. So, increasing plant immunity against these RNA viruses becomes essential. Mostly, the Cas9 or

TABLE 9.1
CRISPR/Cas9-Mediated Targeted Gene(s) against DNA Viruses in Plants

Plant	Virus	Target gene(s)	Target gene function	Strategy/vector used	Output/result	Reference
N. benthamiana and *A. thaliana*	BeYDV –Bean Yellow Dwarf Virus	IR[a], Rep[b], CP[d],	Involvement in viral RCA[c] replication mechanism	*Agrobacterium* transformation using Cas9/gRNA vectors	Development of highly resistant plants against virus infection	Ji et al. (2015)
N. benthamiana	BSCTV – Beet Severe Curly Top Virus	IR[a], Rep[b]	Involvement in viral RCA[c] replication mechanism	*Agrobacterium* transformation using Cas9/gRNA vectors	Development of plants with reduced viral load and symptoms	Baltes et al. (2015)
N. benthamiana	BCTV – Beet Curly Top Virus, TYLCV – Tomato Yellow Leaf Curl Virus, MeMV – Merremia Mosaic Virus	IR[a], Rep[b], CP[d],	Involvement in viral RCA[c] replication mechanism	*Agrobacterium* transformation using Cas9/gRNA vectors	Development of strategies for durable virus interference and resistance	Ali et al. (2016)

IR[a]: intergenic sequence; Rep[b]: replication; RCA[c]: rolling circle amplification; CP[d]: capsid protein

Cas12 reliant systems aim to target DNA only. Alternatively, Cas13, also known as C2c2 (i.e., single effector protein C2c2), which belongs to class 2, type-VI Cas endonuclease, targets only RNA not DNA. The CRISPR/Cas13 system from LwaCas13a (*Leptotrichia wadei*) (Abudayyeh et al. 2017) and PspCas13b (*Prevotella* sp. P5-125) (Cox et al. 2017) has been successfully engineered for defined RNA editing. The very specific, efficient, and flexible CRISPR-RNA editing system is REPAIR, i.e., RNA editing for programmable A to I (G) replacement. A few examples of the CRISPR/Cas-based editing that possess prospective use for targeting RNA viruses are depicted in Table 9.2. These studies would surely boost the recognition and advancement of alternate appropriate RNA editing systems.

9.5 CRISPR/CAS FOR TARGETING FUNGAL PATHOGENS

Fungi and fungal-like organisms (FLOs) are the foremost plant pathogens, causing 30% of the evolving diseases that lead to enormous crop losses in comparison to other pathogens worldwide (Giraud et al. 2010). To achieve immunity against fungi, the CRISPR/Cas9-based editing system focuses on the disease-causing genes and their protein products related to resistance. Controlling fungal diseases becomes a main challenge owing to their varied behaviours, such as wide host range (instant new attacks), breaking resistance (ruled by R-resistance gene), and generating new resistance (toward fungicides) (Yin and Qiu 2019). Recently, this challenge has been addressed by modifying host S genes in the crops mentioned in Table 9.3.

9.6 CRISPR/CAS FOR TARGETING BACTERIAL PATHOGENS

About 200 plant pathogenic bacterial species accountable for crop losses have been documented. The chief ones include bacteria from the genera *Agrobacterium, Dickeya, Erwinia, Pectobacterium, Pseudomonas, Ralstonia, Xanthomonas, Xylella,* etc. (Buttimer et al. 2017). The difficulty in controlling epidemic bacterial diseases stem from their numerous existence and their ability to reproduce rapidly. Identification of several host-plant genes, predominantly the S genes, that participate in the complex processes of host–bacteria interactions increased the applicability of the CRISPR/Cas9 system against many bacterial diseases (Table 9.4). Hence, the documentation of more durable S genes in different plant species would be helpful in gaining resistance toward many other pathogenic bacteria.

9.7 REGULATORY MEASURES OVER CRISPR-EDITED CROPS

Many countries have documented regulatory measures for genome-edited crops, e.g., Canada, India, Malaysia, Mexico, New Zealand, South Africa, and Thailand. These regulations have not been framed in countries such as Australia, Japan, and the USA. Also, only case-specific regulations have been formulated in Argentina, Brazil, and Chile. Additionally these countries have regulatory frameworks for the assessment and release of transgenics or genetically modified crops. The European Union still has an unclear regulatory approach for genetically modified crops as well

TABLE 9.2

CRISPR/Cas9-Mediated Targeted Gene(s) against RNA Viruses in Plants

Plant	Virus	Target gene(s)	Target gene function	Vector/ strategy used	Output/result	Reference
C. sativus	CVYV – Cucumber Vein Yellow Virus, ZYMV – Zucchini Yellow Mosaic Virus, PRSV-W – Papaya Ring Spot Mosaic Virus-W	eIF^{a}4E	IF^b associated with virus translation	Agrobacterium transformation using Cas9/gRNA vector	Development of virus-resistant plants	Chandrasekaran et al. (2016)
A. thaliana	TuMV- Turnip mosaic Virus	eIF^a (iso)4E	IF^b associated with virus translation	Agrobacterium transformation using Cas9/gRNA vector	Development of a new method for the generation of potyvirus resistance	Pyott et al. (2016)
N. benthamiana	TuMV – Turnip Mosaic Virus	GFP^{c}1, GFP^{c}2, HC-Pro^d, CP^e	Viral replication	Agrobacterium transformation using TRV vector	Development of CRISPR/Cas13a system for RNA virus interference	Aman et al. (2018)
N. benthamiana and A. thaliana	CMV – Cucumber Mosaic Virus, TMV – Tobacco Mosaic Virus	CP^e (ORF^f 1,2,3, CP^e & 3' UTR^g)	Viral replication	Agrobacterium transformation using FnCas9/gRNA vectors	Development of plants resistant to RNA viruses	Zhang et al. (2018)
O. sativa and L. japonica	RTSV – Rice Tungro Spherical Virus	eIF^a 4G	IF^b associated with virus translation	Agrobacterium transformation using Cas9/gRNA vector	Development of RTSV-resistant plants	Macovei et al. (2018)

eIF^a: eukaryotic translation initiation factor; IF^b: initiation factor; GFP^c: green fluorescent protein; HC-Pro^d: helper component proteinase (plant virus protein); CP^e: coat protein; ORF^f: open reading frame; UTR^g: untranslated region

TABLE 9.3

CRISPR/Cas9-Mediated Targeted Gene(s) against Fungal Pathogens in Plants

Plant	Fungal pathogen	Disease name	Target gene	Target gene function	Vector or agent/strategy used	Output/result	Reference
Triticum aestivum	*Blumeria graminis* f. sp. *tritici*)	Powdery mildew	MLO[a]-A1	S gene associated with powdery mildew	Particle bombardment using Cas9/gRNA vector	Resistant wheat plants with mutations in TaMLO-A1 allele	Wang et al. (2014)
Vitis vinifera cv.	*Erysiphe necator*	Powdery mildew	MLO[a]-7	S gene associated with powdery mildew	PEG-mediated protoplast transformation using CRISPR ribonucleo-proteins (RNPs)	Development and direct delivery of CRISPR RNPs for targeted mutagenesis at respective loci	Malnoy et al. (2016)
	Botrytis cinerea	Gray mould	WRKY[b]-52	Transcription factor linked to stress	*Agrobacterium*-mediated transformation using Cas9/gRNA vector	Increased resistance against *B. cinerea* in first generation plants	Wang et al. (2018)
O. sativa and *L. japonica*	*Magnaporthe oryzae*	Rice blast disease	ERF[c] 922	Transcription factor linked to stress	*Agrobacterium*-mediated transformation using Cas9/gRNA vector	Enhanced blast resistance in rice T_2 generation plants	Wang et al. (2016)
Solanum lycopersicon	*Oidium neo lycopersici*	Powdery mildew	MLO[a] 1	S gene associated with powdery mildew	*Agrobacterium*-mediated transformation using Cas9/gRNA vector	Fully resistant *slmlo1* tomato variety *"Tomelo"* T_1 generation plants	Nekrasov et al. (2017)
Theobroma cacao	*Phytophthora tropicalis*	Black pod disease	NPR[d] 3	Regulatory gene of the plant immune system	*Agrobacterium*-mediated transformation using Cas9/gRNA vector	Increased resistance along with increased activity of downstream defence genes in somatic embryos	Fister et al. (2018)
O. sativa and *L. japonica*	Rice blast disease (*Magnaporthe oryzae*)	Rice blast disease	OsSEC 3A	Part of exocyst complex involved in plant immune signalling	Protoplast transformation using Cas9/gRNA vector	Increased defence response in OsSEC3A mutant plants	Ma et al. (2018)

MLO[a]: mildew resistance locus; WRKY[b]: WRKY transcription factor; ERF[c]: ethylene response factor type transcription factor; NPR[d]: non-expressor of pathogenesis-related; OsSEC[e]: subunit of the exocyst complex in *Oryza sativa*

TABLE 9.4
CRISPR/Cas9-Mediated Targeted Gene(s) against Bacterial Pathogens in Plants

Plant	Bacterial pathogen	Disease name	Target gene(s)	Target gene function	Strategy/vector used	Output/result	Reference
Oryza sativa	Xanthomonas oryzae pv. oryza	Bacterial blight	SWEET[a] 13	Sucrose transporter gene conferring blight susceptibility	Agrobacterium-transformation using Cas9/gRNA vector	Bacterial blight-resistant plants	Zhou et al. (2015)
Citrus paradisi	Xanthomonas citri	Citrus canker	LOB[b] 1	S gene conferring canker susceptibility	Agrobacterium-transformation using Cas9/gRNA vector	Mutant plants showed improved canker resistance by suppressing formation of pustules	Jia et al. (2016)
Malus domestica	Erwinia amylovora	Fire blight	DIPM[c] -1 DIPM[c] -2 DIPM[c] -4	Encodes essential pathogenicity effectors	PEG-mediated protoplast transformation using CRISPR Ribonucleoproteins (RNPs)	Direct delivery of CRISPR RNPs for targeted mutagenesis at respective loci	Malnoy et al. (2016)
Citrus sinensis osbeck	Xanthomonas citri	Citrus canker	LOB[b] 1	S gene conferring canker susceptibility	Agrobacterium-transformation using Cas9/gRNA vector	Development of canker-resistant mutant plants	Peng et al. (2017)

SWEET[a]: sugar will eventually be exported transporter; LOB[b]: lateral organ boundaries; DIPM[c]: Dsp (disease specific) interacting proteins of Malus × domestica as genome-edited crops with some oppositions (Mounadi et al. 2020). To realize the potential of this technology, thorough studies are still needed to bring about its adoption and utilization at a commercial level with no risks. Thus, the associated issues need to be well addressed in a very simplified way in order for the general public to understand it and its applicability.

9.8 ADVANTAGES AND DISADVANTAGES OF CRISPR-EDITED TECHNIQUES

CRISPR-edited techniques possess some advantages and disadvantages.
Advantages:

- Relatively simple, very specific, cost effective, and efficient method compared to other editing tools.
- Simultaneous targeting of multiple sites (multiplex editing), and base editing can be possible.
- Ribonucleoproteins, or RNPs, virus-based, and nanoparticle-linked genome editing systems are the substitutes for tissue-culture based methods.

Disadvantages:

- Restricted target selection because of PAM requirement or specificity.
- Undesired mutation at the target(s), i.e., off-target(s) is one of the major problems.
- Many plants lack a reproducible transformation/regeneration system, which hinders the mutants selection process in plant tissue culture-based *Agrobacterium tumefaciens* or direct gene transfer methods.
- A proper, obligatory monitoring system to assess the risks and concerns linked with genome-edited crops, if any, has to be established globally to exploit the actual potential of this revolutionary technology.

9.9 CONCLUSION

Undoubtedly, until now, CRISPR/Cas9 has been one of the expansively used editing platforms in many aspects of crop improvement such as yield, quality, male sterility, and resistance/immunity for biotic and abiotic stresses owing to its simplicity, efficacy, and flexibility. But, some technical considerations restrict its effectual broader application for disease resistance across many plant species. These include (a) direct targeting of S genes because of their linkage to other desirable growth and development associated genes, (b) off-target mutations caused by misleading gRNA or gRNA-independent mode, (c) safety and commercialization (as the resulting mutants are free from any unknown nucleic acid, they should be tagged as non-GMO, leading to their rapid adoption; however, there are legal obstacles to their acceptance, regulation, and commercialization), (d) non-availability of sequence information, and (e) lack of standard regeneration and transformation protocols in a large number of crops, etc.

Production of targeted disease-resistant plants through CRISPR/Cas could be a method of tackling the obstacles in resistance breeding programmes. Thus, CRISPR/Cas targeted, edited, or mutated plants will have definite positive impressions on meeting the increasing global food demands and security.

REFERENCES

Abudayyeh, O. O., Gootenberg, J. S., Essletzbichler, P., et al. 2017. RNA targeting with CRISPR-Cas13. *Nature* 550 (7675): 280–84.

Ali, Z., Ali, S., Tashkandi, M., Shan, S., Zaidi, A., and Mahfouz, M. M. 2016. CRISPR/Cas9-mediated immunity to geminiviruses: Differential interference and evasion. *Scientific Reports* 6 (August): 1–16. doi:10.1038/srep26912.pdf.

Aman, R., Ali, Z., Butt, H., et al. 2018. RNA virus interference via CRISPR/Cas13a system in plants. *Genome Biology* 19 (1): 1–9. doi:10.1186/s13059-017-1381-1.pdf.

Baltes, N. J., Hummel, A. W., Konecna, E., et al. 2015. Conferring resistance to geminiviruses with the CRISPR–Cas prokaryotic immune system. *Nature Plants* 1 (September): 1–4. doi:10.1038/nplants.2015.145.pdf.

Borrelli, V. M. G., Brambilla, V., Rogowsky, P., Marocco, A., and Lanubile, A. 2018. The enhancement of plant disease resistance using CRISPR/Cas9 technology. *Frontiers in Plant Science* 9 (August): 1–15. doi:10.3389/fpls.2018.01245.pdf.

Boyd, C. D., and O'Toole, G. A. 2012. Second messenger regulation of biofilm formation: Breakthroughs in understanding c-di-GMP effector systems. *Annual Review of Cell and Developmental Biology* 28: 439–62.

Buttimer, C., McAuliffe, O., Ross, R. P., Hill, C., O'Mahony, J., and Coffey, A. 2017. Bacteriophages and bacterial plant diseases. *Frontiers in Microbiology* 8 (January): 1–15. doi:10.3389/fmicb.2017.00034.pdf.

Chandrasekaran, J., Brumin, M., Wolf, D., et al. 2016. Development of broad virus resistance in non-transgenic cucumber using CRISPR/Cas9 technology. *Molecular Plant Pathology* 17 (7): 1140–53.

Chen, K., Wang, Y., Zhang, R., Zhang, H., and Gao, C. 2019. CRISPR/Cas genome editing and precision plant breeding in agriculture. *Annual Review of Plant Biology* 70: 667–97.

Cox, D. B. T., Gootenberg, J. S., Abudayyeh, O. O., et al. 2017. RNA editing with CRISPR-Cas13. *Science* 358 (6366): 1019–27.

Dracatos, P. M., Haghdoust, R., Singh, D., and Fraser, P. 2018. Exploring and exploiting the boundaries of host specificity using the cereal rust and mildew models. *New Phytologist* 218: 453–62.

Fister, A. S., Landherr, L., Maximova, S. N., and Guiltinan, M. J. 2018. Transient expression of CRISPR/Cas9 machinery targeting TcNPR3 enhances defense response in *Theobroma cacao*. *Frontiers in Plant Science* 9 (March): 1–15. doi:10.3389/fpls.2018.00268.pdf.

Giraud, T., Gladieux, P., and Gavrilets, S. 2010. Linking the emergence of fungal plant diseases with ecological speciation. *Trends in Ecology and Evolution* 25 (7): 387–95.

Gull, A., Lone, A. A., and Wani, N. I. 2019. Biotic and abiotic stresses in plants. In *Abiotic and Biotic Stress in Plants*, ed. A. B. Oliveira, 1–6. IntechOpen, London, UK.

The International Service for the Acquisition of Agri-biotech Applications. 2016. Global status of commercialized biotech/GM crops: 2016. ISAAA Brief No. 52. http://www.isaaa.org/resources/publications/briefs/52.

Ji, X., Zhang, H., Zhang, Y., Wang, Y., and Gao, C. 2015. Establishing a CRISPR–Cas-like immune system conferring DNA virus resistance in plants. *Nature Plants* 1 (September): 15144. doi:10.1038/nplants.2015.144.pdf.

Jia, H., Orbovic, V., Jones, J. B., and Wang, N. 2016. Modification of the PthA4 effector binding elements in type I CsLOB1 promoter using Cas9/sgRNA to produce transgenic Duncan grapefruit alleviating Xcc1pthA4:dCsLOB1.3 infection. *Plant Biotechnology Journal* 14 (5): 1291–01.

Jiang, F., and Doudna, J. A. 2017. CRISPR–Cas9 structures and mechanisms. *Annual Review of Biophysics* 46: 505–29.

Langner, T., Kamoun, S., and Belhaj, K. 2018. CRISPR crops: Plant genome editing toward disease resistance. *Annual Review of Phytopathology* 56: 479–12.

Ma, J., Chen, J., Wang, M., et al. 2018. Disruption of OsSEC3A increases the content of salicylic acid and induces plant defense responses in rice. *Journal of Experimental Botany* 69 (5): 1051–64.

Macovei, A., Sevilla, N.R., Cantos, C., et al. 2018. Novel alleles of rice eIF4G generated by CRISPR/Cas9-targeted mutagenesis confer resistance to Rice tungro spherical virus. *Plant Biotechnology Journal* 16 (11): 1918–27.

Malnoy, M., Viola, R., Jung, M. H., et al. 2016. DNA free genetically edited grapevine and apple protoplast using CRISPR/Cas9 ribonucleoproteins. *Frontiers in Plant Science* 7 (December): 1–9. doi:10.3389/fpls.2016.01904.pdf.

Mounadi, K. E., Floriano, M. L. M., and Ruiz, H. G. 2020. Principles, applications, and biosafety of plant genome editing using CRISPR-Cas9. *Frontiers in Plant Science* 11 (February): 1–16. doi: 10.3389/fpls.2020.00056.pdf.

Nekrasov, V., Wang, C., Win, J., Lanz, C., Weigel, D., and Kamoun, S. 2017. Rapid generation of a transgene-free powdery mildew resistant tomato by genome deletion. *Scientific Reports* 7 (March): 1–6. doi: 10.1038/s41598-017-00578-x.pdf.

Peng, A., Chen, S., Lei, T., et al. 2017. Engineering canker resistant plants through CRISPR/Cas9-targeted editing of the susceptibility gene CsLOB1 promoter in citrus. *Plant Biotechnology Journal* 15 (12): 1509–19.

Pyott, D.E., Sheehan, E., and Molnar, A. 2016. Engineering of CRISPR/Cas9-mediated potyvirus resistance in transgene-free Arabidopsis plants. *Molecular Plant Pathology* 17 (8): 1276–88.

Que, Q., Chen, Z., Kelliher T., et al. 2019. Plant DNA repair pathways and their applications in genome engineering. In *Methods in Molecular Biology*, ed. Y. Qi, 3–24. Humana Press, New York.

Ran, Y., Liang, Z., and Gao, C. 2017. Current and future editing reagent delivery systems for plant genome editing. *Science China Life Sciences* 60 (5): 490–05.

Roossinck, M. J. 2011. The big unknown: plant virus biodiversity. *Current Opinion in Virology* 1 (1): 63–7.

Roossinck, M. J. 2015. Metagenomics of plant and fungal viruses reveals an abundance of persistent lifestyles. *Frontiers in Microbiology* 12 (5) (January): 1–3. doi:10.3389/fmicb.2014.00767.pdf.

Shelake, R. M., Pramanik, D., and Kim, J. Y. 2019. Exploration of plant-microbe interactions for sustainable agriculture in CRISPR era. *Microorganisms* 7 (8) (August): 1–32. doi:10.3390/microorganisms7080269.pdf.

Tyczewska, A., Wozniak, E., Gracz, J., Kuczynski, J., and Twardowski, T. T. 2018. Towards food security: Current state and future prospects of agrobiotechnology. *Trends in Biotechnology* 36 (12): 1219–29.

Wang, X., Tu, M., Wang, D., et al. 2018. CRISPR/Cas9-mediated efficient targeted mutagenesis in grape in the first generation. *Plant Biotechnology Journal* 16 (4): 844–55.

Wang, F., Wang, C., Liu, P., et al. 2016. Enhanced rice blast resistance by CRISPR/Cas9-targeted mutagenesis of the ERF transcription factor gene OsERF922. *PLoS One* 11 (4) (April): 1–18. doi:10.1371/journal.pone.0154027.pdf.

Wang, Y., Cheng, X., Shan, Q., et al. 2014. Simultaneous editing of three homoeoalleles in hexaploid bread wheat confers heritable resistance to powdery mildew. *Nature Biotechnology* 32: 947–52.

Yin, K., and Qiu, J. L. 2019. Genome editing for plant disease resistance: Applications and perspectives. *Philosophical Transactions of the Royal Society B* 374 (January): 1–8. doi:10.1098/rstb.2018.0322.pdf.

Zhang, T., Zheng, Q., Yi, X., et al. 2018. Establishing RNA virus resistance in plants by harnessing CRISPR immune system. *Plant Biotechnology Journal* 16 (8): 1415–23.

Zhang, Y., Ge, X., Yang, F., et al. 2014. Comparison of non-canonical PAMs for CRISPR/Cas9-mediated DNA cleavage in human cells. *Scientific Reports* 4 (June): 1–13. doi:10.1038/srep05405.pdf.

Zhou, J., Peng, Z., Long, J., et al. 2015. Gene targeting by the TAL effector PthXo2 reveals cryptic resistance gene for bacterial blight of rice. *The Plant Journal* 82 (4): 632–43.

10 Plant-Microbe Interaction and Recent Trends in Biotechnology for Secondary Metabolite Production in Medicinal Plants

Neha Sharma and Hemant Sood

CONTENTS

10.1 INTRODUCTION

The associations between medicinal plants and microorganisms are highly diverse, which can affect their growth and development and production of secondary metabolites. These interactions can be endophytic or epiphytic, which cause either mutual benefits or infections in plants (Figure 10.1). The exudates from the roots of the plant provide nutrients, thereby colonizing a plethora of microbes in the rhizosphere and, in turn, these microorganisms improve plant health and make the plants resistant to various stress conditions (Mendes et al. 2013). Diversity of microbes in the rhizosphere depends upon the plant species, its health and development stage, soil type, climate, and other biotic and abiotic factors. Some of the beneficial and pathogenic plant-associated microorganisms can be passed down to next generations via seed and pollen grains (Fürnkranz et al. 2012; Hirsch and Mauchline 2012). The positive

DOI: 10.1201/9781003213864-10

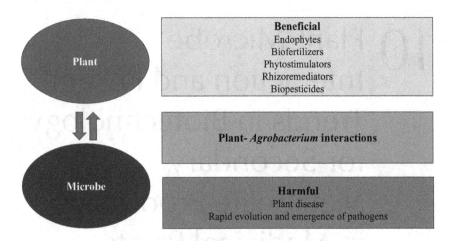

FIGURE 10.1   Different types of plant-microbe interactions.

interactions with microbes change growth capacities of plants by improving the nutrient efficiency and tolerance against various biotic and abiotic stresses. Negative associations with microorganisms cause diseases in plants, and studying these interactions may play role in identification of microbial and plant factors responsible for conception of disease as well as unravelling their molecular functions. Various DNA- and RNA-based tools, such as molecular markers, next generation sequencing, and transcriptome analysis, have proven to be valuable methods of studying plant pathogen interactions.

Medicinal plants shelter distinctive microbes corresponding to the presence of different biologically active secondary metabolites (Köberl et al. 2013). The biochemical products in plant cells are categorized as primary and secondary metabolites. Compounds that are essential and involved in growth and development of the plant are known as primary metabolites, while secondary metabolites are not directly essential for plant cell life but are required for the survival of plants during unfavourable environmental conditions (Sharma 2017). Previous studies on medicinal plants focused mainly on the production and enhancement of secondary metabolites by modulating the secondary metabolite pathways and elicitor treatment or optimizing the media for plant growth (Sharma et al. 2015; Kumar et al. 2016; Sharma et al. 2016a); however, plant-microbe interactions can also trigger the secondary metabolite production in plants. The microorganisms associated with medicinal plants influence the metabolome of the host, thereby effecting the efficacy of herbal medicine (Huang et al. 2018). These microorganisms share common metabolic pathways with the host plant and signify their potential as a storehouse of genes for secondary metabolite production (Shelake et al. 2019). Secondary metabolites' content and quality can vary in the same medicinal plant species depending on their place of cultivation, which could be due to their association with different microbial communities in different locations (Huang et al. 2018). The microorganisms acclimatize to associate with medicinal plants in such a way that they get involved in regulating the production of secondary metabolites, such as artemisinin biosynthesis by *Colletotrichum* sp. and paclitaxel production by *Taxomyces andreanae* (Huang et al. 2018). Köberl et al. (2013) reviewed several Chinese medicinal plants and found that they hosted a specific actinobacterial community that have shown antimicrobial and antitumor properties and act as a unique source for novel antibiotics. Therefore, the current research focus is shifting toward understanding the multidimensional interactions that plants keep with microbes and enable us to understand how these interactions exert beneficial or detrimental effects and plan strategies according to their best utilization and effective disease management measures in future.

In this chapter, we discuss various beneficial and negative plant-microbe interactions with a special focus on medicinal plants. We also review the recent advancements for studying plant-microbe interactions.

10.2   BENEFICIAL PLANT-MICROBE INTERACTIONS

Medicinal plants and their herbal formulations have been part of traditional health care systems for thousands of years worldwide. The therapeutic value of medicinal plants is due to the presence of secondary metabolites, such as terpenoids, phenolics, and alkaloids (Sharma 2017). Studies have shown that microbes associated with medicinal plants, particularly endophytes, are able to produce bioactive compounds chemically similar to secondary metabolites produced by their host plants (Köberl et al. 2013). Endophytes are microorganisms (bacteria or fungi) residing inside living tissues of plants without causing any visible symptoms. Several important secondary metabolites, such as taxol, camptothecin, podophyllotoxin, vinblastine, etc., with high therapeutic properties have been produced by endophytes (Köberl et al. 2013; Naik et al. 2019). Some of the bioactive compounds that are produced by endophytes have been summarized in Table 10.1. Endophytes that produce secondary metabolites similar to those produced by their host plants may be due to the horizontal gene transfer between host and endophyte (Naik et al. 2019). Now the question arises: Why do endophytes biosynthesize these bioactive phytochemicals? This may occur because the secondary metabolites enables the endophytes to create supremacy over other invading endophytes or pathogens or to enhance plant defences, deliberating higher fitness to the host against different pathogens, thereby insuring fitness benefits to themselves indirectly (Soliman et al. 2013, 2015; Mousa et al. 2016). Further, while producing cytotoxic metabolites, endophytes, themselves, must be resistant to the toxic metabolites or may transport such metabolites into the intercellular spaces of plant tissues and thereby assisting the host plants to act against pathogens. Some endophytes sequester toxic metabolites as in intracellular hydrophobic bodies and release them through exocytosis, while some endophytic fungi transport these metabolites through the mycelial cell membranes and the cell wall into intercellular spaces of the host plant (Soliman et al. 2015; Naik et al. 2019). Therefore, endophytes have potential to produce different bioactive phytochemicals; however, there is a strong need to understand their evolution and harness them for a variety of industrial applications.

TABLE 10.1

Some Plant Secondary Metabolites Produced by Endophytes

Secondary metabolite	Therapeutic properties	Host plant	Producing microorganism	Reference
Azadirachtin A and B	Antibacterial, antifungal and anti-inflammatory	Azadirachta indica	Eupenicillium parvum	Kusari et al. 2012
Artemisinin	Antimalarial	Artemisia annua	Colletotrichum sp.	Wang et al. 2001
Artemisinin	Antimalarial	A. annua	Pseudonocardia sp.	Li et al. 2012
Camptothecin	Anticancer, antiviral (HIV)	Nothapodytes foetida	Entrophospora infrequens	Puri et al. 2005; Amna et al. 2006
Camptothecin	Anticancer, antiviral (HIV)	N. foetida	Neurospora sp.	Rehman et al. 2008
Camptothecin	Anticancer, antiviral (HIV)	Camptotheca acuminate	Fusarium solani	Kusari et al. 2009a
Camptothecin	Anticancer, antiviral (HIV)	Apodytes dimidiata	F. solani	Shweta et al. 2010
Camptothecin	Anticancer, antiviral (HIV)	C. acuminata	Trichoderma atroviride LY357, Aspergillus sp. LY355, and Aspergillus sp. LY341	Pu et al. 2013.
Camptothecin	Anticancer, antiviral (HIV)	C. acuminata	Alternaria sp.	Su et al. 2014
Ginkgolide B	Neuroprotective	Ginkgo biloba	Fusarium oxysporum	Cui et al. 2012
Podophyllotoxin	Anticancer	Podophyllum hexandrum	Alternaria sp.	Yang et al. 2003
Podophyllotoxin	Anticancer	Podophyllum peltatum	Phialocephala fortinii	Eyberger et al. 2006
Podophyllotoxin	Anticancer	Juniperus recurva	F. oxysporum	Kour et al. 2008
Podophyllotoxin	Anticancer	Juniperus communis	Aspergillus fumigatus	Kusari et al. 2009b
Podophyllotoxin	Anticancer	P. hexandrum	A. fumigatus	Kusari et al. 2009b
Podophyllotoxin	Anticancer	P. hexandrum	F. solani	Nadeem et al. 2012
Rohitukine	Anticancer	Dysoxylum binectariferum	Fusarium proliferatum	Kumara et al. 2012
Sanguinarine	Antimicrobial, antioxidant, anti-inflammatory, and proapoptotic properties	Macleaya cordata	F. proliferatum	Min et al. 2014

(Continued)

TABLE 10.1 (CONTINUED)
Some Plant Secondary Metabolites Produced by Endophytes

Secondary metabolite	Therapeutic properties	Host plant	Producing microorganism	Reference
Taxol	Anticancer	Taxus chinensis var. mairei	Botrytis sp.	Hu et al. 2006
Taxol	Anticancer	Podocarpus sp.	A. fumigatus	Sun et al. 2008
Taxol	Anticancer	Taxus celebica	F. solani	Chakravarthi et al. 2008
Taxol	Anticancer	Taxus cuspidata	Botrytis sp.	Zhao et al. 2008
Taxol	Anticancer	T. chinensis	F. solani	Deng et al. 2009
Taxol	Anticancer	T. chinensis	Metarhizium anisopliae	Liu et al. 2009
Taxol	Anticancer	T. chinensis	Mucor rouxianus	Miao et al. 2009
Taxol	Anticancer	Taxus x media	Cladosporium cladosporioides MD2	Zhang et al. 2009
Taxol	Anticancer	T. cuspidata	Pestalotiopsis versicolor	Kumaran et al. 2010
Taxol	Anticancer	Taxus globosa	Nigrospora sp.	Ruiz-Sanchez et al.2010
Taxol	Anticancer	T. chinensis var. mairei	Ozonium sp.	Wei et al. 2010
Taxol	Anticancer	Rhizophora annamalayana	F. oxysporum	Elavarasi et al. 2012
Taxol	Anticancer	Taxus baccata L. subsp. wallichiana	Fusarium redolens	Garyali et al. 2013
Taxol	Anticancer	T. baccata	Cladosporium sp.	Kasaei et al. 2017
Vinblastine	Anticancer	Catharanthus roseus	Talaromyces radicus—CrP20	Palem et al. 2016
Vincristine	Anticancer	C. roseus	F. oxysporum	Zhang et al. 2000
Vincristine	Anticancer	C. roseus	Talaromyces radicus—CrP20	Palem et al. 2016

There are several other rhizospheric microorganisms whose association promotes the growth and disease resistance in plants and are classified as biofertilizers, phyto-stimulators, rhizo-remediators, and biopesticides (Singh et al. 2019). Biofertilizers enhance the nutrient availability to plants. They include symbiotic root-nodulating rhizobia and mycorrhizal fungi. The rhizobial bacteria forms a symbiotic relationship with the plant in which the bacteria provide the host with fixed inert nitrogen in the form of ammonium and nitrate while the host plant provides the nutrients and niche to rhizobia. The most effective genera of the biofertilizing bacteria includes *Rhizobium*, *Sinorhizobium*, *Mesorhizobium*, *Bradyrhizobium*, *Azorhizobium*, and *Allorhizobium* (Singh et al. 2019). Mycorrhizal fungi colonize with the plant roots and help them to capture nutrients such as phosphorus, sulfur, nitrogen, and other micronutrients from soil. It also helps the plant to fight against soil-borne pathogens and other environmental stresses such as drought and salinity (Lugtenberg et al. 2002). Phyto-stimulators are the microbes that promote the plant growth by providing plant hormones such as indole-3-acetic acid. Some volatile organic compounds are also released by some rhizobacteria such as *Burkholderia cepacia*, *Staphylococcus epidermidis*, and *Bacillus subtilis*, which have the ability to stimulate plant growth (Köberl et al. 2013). Rhizoremediators degrade the organic pollutants, while biopesticides control diseases by producing antibiotics and antifungal metabolites (Gupta et al. 2016).

Therefore, interactions of medicinal plants with microbes influence their growth and health. Also, the endophytes that produce secondary metabolites of great medicinal value can provide a promising source of metabolite production at commercial level.

10.3   HARMFUL PLANT-MICROBE INTERACTIONS

Plant diseases are caused by detrimental interactions and are usually restricted to a particular combination of host and pathogen. These interactions are complex and facilitated by pathogen- and plant-derived molecules such as proteins, sugars, and lipopolysaccharides (Gupta et al. 2015). In general, plants express pattern recognition receptors (PRRs) that precisely recognize the pathogen-associated molecular patterns (PAMPs), which are unique, structurally conserved, and indispensable to microbes and, thereby, allow the host to recognize and respond to invading pathogens. The PAMP-triggered immunity (PTI) is the first line of defense and is effective against a wide range of pathogenic microorganisms. On the other hand, these invading microorganisms inoculate effector proteins into the plant cell to suppress PTI and effectively thrive on their host causing effector-triggered susceptibility (ETS). Plants have also evolved a second layer of defense in which resistance (R) proteins defend the effector-mediated dysfunction of host components and convert the virulent factors into avirulent factors. These avirulent factors allow the plant to detect previously successful pathogens and trigger the rapid hypersensitive response to restrict their growth. In most plants, PTI restricts diseases against many pathogens; however, effector-triggered immunity (ETI) has also evolved by successful pathogens with specific plant varieties that have gained the competence to recognize the interaction of effector proteins with their targets in the cell (Postel and Kemmerling 2009). This antagonistic interaction between hosts and pathogens frequently results in arms races in which both plants and pathogens evolve in response to each other (Figure 10.2). Marimuthu et al. (2018) discussed various diseases in different medicinal plants such as *Morinda citrifolia* L., *Gloriosa superba, Solanum nigrum, Coleus forskohlii, Cassia angustifolia,* and *Withania somnifera,* which are caused by different bacterial and fungal pathogens such as *Colletotrichum gloeosporioides, Alternaria alternata, Pantoea agglomerans, Macrophomina phaseolina, Leveillula taurica, Fusarium chlamydosporum, Fusarium solani, Ralstonia solanacearum, Phomopsis cassia, Fusarium oxysporum,* and *Phoma putaminum.* These pathogens pose a serious threat to the biomass and secondary metabolite production in medicinal plants. Therefore, there is a strong need to understand and identify emerging pathogens and develop disease management strategies to counter them.

10.4   PLANT–*AGROBACTERIUM* INTERACTIONS

Another class of plant-microbe interactions includes interaction between plant and *Agrobacterium* sp. The genus *Agrobacterium* contains a number of species in which *Agrobacterium tumefaciens, Agrobacterium rhizogenes,* and *Agrobacterium rubi* are causative agents of plant diseases such as crown gall, hairy root, and cane gall, respectively, while *Agrobacterium radiobacter* is an avirulent species. These bacteria transport the transferred-DNA (T-DNA) into the plant nucleus, which is randomly integrated into the plant genome and alters the physiology and morphology of infected host plants. The Ti plasmid of *A. tumefaciens* consists of T-DNA and virulence (*Vir*) region for genetic transformation of the host plant. Imam et al. (2016) found that the *Agrobacterium* pathogenicity and effective transformation mostly depend on various plant-derived signals, such as phenolic compounds, sugars, and acidic signals, that activate the virulence in *Agrobacterium* along with the hormones that lower the defense level in plants. *Agrobacterium*-mediated gene transfer has

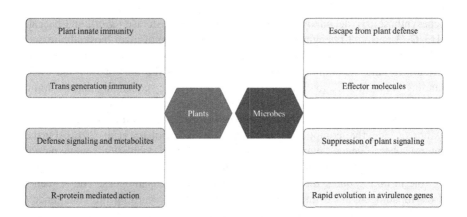

FIGURE 10.2   Plant-microbes/pathogens strategies to counter each other.

emerged as a great tool in modern plant molecular genetics and genetic engineering. It has the potential to increase the biosynthesis of bioactive compounds in plant cells. Several genetic transformation attempts using *A. tumefaciens* in various medicinal plants belonging to different families, including Apocynaceae, Araceae, Araliaceae, Asphodelaceae, Asteraceae, Begoniaceae, Crassulaceae, Fabaceae, Lamiaceae, Linaceae, Papaveraceae, Plantaginaceae, Scrophulariaceae, and Solanaceae, have been made (Bandurska et al. 2016). The hairy roots induced by Ri plasmid of *Agrobacterium rhizogenes* have been proven to be an efficient means of secondary metabolite production in medicinal plants. Abraham and Thomas (2017) reviewed the use of hairy roots for biosynthesis of useful secondary metabolites in various medicinal plants such as *Artemisia annua, Camptotheca acuminate, Catharanthus roseus, Coleus forskohlii, Gentiana dinarica, Gentiana scabra, Inula helenium, Nicotiana tabacum, Linum mucronatum, Phyllanthus odonladenius, Rauvolfia serpentine, Salvia miltiorrhi, Silybum marianum, Valeriana officinalis*, etc. We have also summarized some of the medicinal compounds that are produced using hairy root culture techniques in Table 10.2. Hence, this plant–*Agrobacterium* interaction offers a platform for valuable secondary metabolite production.

10.5   ADVANCES TO STUDY PLANT-MICROBE INTERACTION

Understanding the mechanisms of plant-microbe interaction and elucidating the involved genes are crucial for their future application and planning efficient disease management strategies. For a microorganism or endophyte to produce a particular metabolite, it is essential to acquire all the genes encoding enzymes of the biosynthetic pathway. Several enzymes of biosynthetic pathways of various secondary metabolites, such as taxol, picrosides, podophyllotoxin, aconites, etc., have been cloned and well characterized in different medicinal plants (Kusari et al. 2012; Malhotra et al. 2014; Kumar et al. 2015; Sharma et al. 2016b). The presence of key genes encoding important enzymes of the secondary metabolite pathway serves as a marker to screen the microorganisms for production of bioactive metabolites. Vasundhara et al. (2016) found that gene-encoding enzymes, viz., taxa-4(5),11(12)-dienesynthase (*ts*), debenzoyltaxane-2′-a-*O*-benzoyltransferase (*dbat*), and baccatin III13-*O*-(3-amino-3-phenylpropanoyl) transferase (*bapt*) of taxol biosynthetic pathway, showed amplification in taxol producing fungi. The genomics, transcriptomics, proteomics, and metabolomics tools aid in gathering the information and understanding plant-microbe interactions. The whole-genome sequencing and its mining have enabled the identification of gene clusters of biosynthetic pathways of known as well as undetected metabolites, and their manipulation can improve the yield and quality of bioactive compounds in endophytes (Vasundhara et al. 2016). Further, genomics-based genetic markers also help in generating population genetic data for various plant pathogens, thereby leading to early detection of diseases and playing role in taking quick measures against various phytopathogens. The sequences of phytopathogens have improved our understanding regarding the emergence of new pathogens and their ability to cause diseases in plants so that improved disease management strategies can be developed (Imam et al. 2016). Tools such as CDNA

TABLE 10.2
Medicinal Compounds Produced in Some Plants Using Hairy Root Culture Techniques

Plant name	Medicinal compound	Therapeutic properties	Reference
Aconitum heterophyllum	Aconites	Anticancer, anti-inflammatory, antimicrobial, pesticidal	Giri et al. 1997
Agastache rugosa	Rosmarinate	Antioxidant, anti-inflammatory, antimutagenic, antimicrobial, antiviral, astringent	Lee et al. 2007a
Atropa belladonna	Tropane	Pain reliever, muscle relaxer, anti-inflammatory	Yang et al. 2011
Angelica gigas	Deoursin	Anticancer, antibacterial, antinematodal	Xu et al. 2008
Artemisia annua	Artemisinins	Antimalarial	Liu et al. 2002
Astragalus mongholicus	Cycloartane saponin	Antimicrobial, antifungal	Ionkova et al. 1997
Beeta vulgaris	Betalains	Antioxidant, anti-inflammatory	Pavlov et al. 2005
Brugmansia candida	Tropane alkaloids, hyoscyamine, scopolamine	Antiasthmatic, anticholinergic, narcotic, anaesthetic	Marconi et al. 2008
Camptotheca acuminate	Camptothecin	Anticancer	Ni et al. 2011
Catharanthus roseus	Alkaloids	Anticancer	Hanafy et al. 2016
Centella asiatica	Asiaticoside	Wound healing, memory improvement, cognition and mood modulation	Nguyen et al. 2019
Coleus forskolli	Forskolin and rosmarinic acid	Anti-inflammatory and antipyretic	Li et al. 2005
Echinacea purpurea	Alkamides	Immunostimulatory, anti-inflammatory	Romero et al. 2009
Fagopyrum esculentum	Rutin	Antioxidant, anti-inflammatory, anticancer, antimicrobial, cardioprotective, hypolipidemic, antidiabetic, reno-protective, wound healing, anti-stress	Lee et al. 2007b
Gentiana macrophylla	Gentiopicroside	Antimicrobial	Zhang et al. 2010
Gentiana dinarica	Xanthone	Antioxidant, anti-inflammatory, antimicrobial, antiviral, antidiabetic	Vinterhalter et al. 2015
Gentiana scabra	Iridoids, secoiridoids	Antibacterial and stomachic	Huang et al. 2014
Gingko biloba	Terpenoids	Antioxidant, improve blood flow to brain	Ayadi and Guiller 2003

(Continued)

TABLE 10.2 (CONTINUED)
Medicinal Compounds Produced in Some Plants Using Hairy Root Culture Techniques

Plant name	Medicinal compound	Therapeutic properties	Reference
Gloriosa superba	Colchicine, colchicoside	Anti-inflammatory, antiarthritis, antigout, analgesic, antiperiodic, abortifacient	Glorybai and Agastian 2013
Inula helenium	Insulin, helenin, sesquiterpene lactones	Expectorant, invigorating, emmenagogue, diuretic, cholagogue, stomachic, antidiabetic	Shirazi et al. 2013
Isatis tinctoria	Flavonoids	Antibacterial, anticancer, antiviral	Gai et al. 2015
Linum flavum	Lignans	Anticancer	Renouard et al. 2018
Linum mucronatum	Podophyllotoxin, 6-methoxy podophyllotoxin	Anticancer	Samadi et al. 2014
Lithospermum erythrorhizon	Shikonin	Antimicrobial	Tatsumi et al. 2016
Nicotiana tabacum	Nicolin	Antispasmodic, discutient, diuretic, emetic, expectorant, irritant, narcotic, sedative	Zhao et al. 2013
Panax ginseng	Ginsenosides	Neuroprotection, anticancer, antidiabetic, hepatoprotective, immunomodulatory	Murthy et al. 2017
Papaver somniferum	Morphine, sanguinarine	Analgesic, narcotic, sedative, stimulant	Le Flem-Bonhomme et al. 2004
Picrorhiza kurroa	Picroliv	Hepatoprotective, anti-inflammatory, anticholestatic, antiulcerogenic, antiasthmatic, antidiabetic, immuno-modulatory	Verma et al. 2015
Portulaca oleracea	Dopamine	Muscle relaxant, anti-inflammatory, diuretic	Moghadam et al. 2014
Rauvolfia micrantha	Ajmalicine and ajmaline	Neuroprotective	Sudha et al. 2003
Rauvolfia serpentine	Vomilenine, reserpine	Hypnotic, hypotensive, sedative	Madhusudanan et al. 2008
Rubia akane	Alizarin and purpurin	Anticancer, antimalarial, antimicrobial, antifungal, antioxidant	Lee et al. 2010
Salvia miltiorrhiza	Tanshinone	Antioxidative, neuroprotective, antifibrotic, anti-inflammatory, antineoplastic	Gupta et al. 2011

(Continued)

TABLE 10.2 (CONTINUED)
Medicinal Compounds Produced in Some Plants Using Hairy Root Culture Techniques

Plant name	Medicinal compound	Therapeutic properties	Reference
Silybium marianum	Flavonolignan	Hepatoprotective, anticancer, antihepatitis	Rahnama et al. 2008
Swertia japonica	Amarogenetin	Antibacterial, antihepatitic, anticholinergic, chemo preventive, antileishmanial	Ishimaru et al. 1990
Valeriana officinalis	Valerenic acid	Anxiolytic, sedative	Torkamani et al. 2014
Withania somnifera	Withanoloid A	Aphrodisiac, liver tonic, anti-inflammatory	Murthy et al. 2008

microarrays, super SAGE gene expression profiling, RNA seq, 2D gels, MS/MS, and iTRAQ have been developed to dissect plant-microbe interactions. In the post-genomic era, the aim is to link the sequences to phenotypes and validate the biologically and functionally obtained results (Imam et al. 2016). Recently, clustered regularly interspaced short palindromic repeats (CRISPR)/Cas has emerged as a great genome editing tool and is highly capable of exploring plant-microbe interactions. It can precisely determine the function of genes in plants and microbes by completely knocking down the target gene. In recent years, various studies related to the plant–rhizobia symbiotic association for nitrogen fixation, along with the understanding of secondary metabolite pathways and their production, have been done in various plants and microbes using CRISPR-based techniques (Wang et al. 2016; Shi et al. 2017; Yi et al. 2018; Pyne et al. 2019; Shanmugam et al. 2019). Additionally, this technique has been used to study plant–pathogen interactions. Shelake et al. (2019) stated that candidate plant genes from PTI or ETI or pathogen genes have been targeted using the genome editing techniques to confer the resistance against various pathogens. Therefore, detailed molecular exploration of beneficial or harmful plant-microbe interactions with these techniques will facilitate their application to enhance secondary metabolite production along with the design of proper disease management strategies.

10.6 CONCLUSIONS AND FUTURE PERSPECTIVES

Studying the fundamentals of plant-microbe interactions in medicinal plants enables us to clarify the role of microbes in increasing the plant growth and fitness and the production of commercially important secondary metabolites. Beneficial plant-microbe interactions can lead to better cultivation methods for improving the medicinal quality of herbs as well as conservation of medicinal plants in their natural habitat. The battle between plants and pathogens is very challenging and, in order to identify as well as manage the new emerging and re-emerging phytopathogens, there is a strong need to understand the signaling pathways involved in the interactions and disease progression as well as to identify key factors playing a role in plant immune responses. Molecular biology, omics tools (genomics, transcriptomics, proteomics, metabolomics), and next generation sequencing (NGS) technologies have provided more detailed insight into the identification of gene transcripts, proteins, or metabolites and aid in understanding the mechanisms behind beneficial and harmful plant-microbe interactions. These techniques, along with computational biology, can be further explored for detecting early, taking quick action against various phytopathogens, and enhancing secondary metabolite production in medicinal plants.

ACKNOWLEDGMENTS

The authors are thankful to the Division of Crop Improvement, the Central Potato Research Institute, and the Department of Biotechnology and Bioinformatics, Jaypee University of Information Technology for providing the opportunity to work on this project.

REFERENCES

Abraham, J., and T. D. Thomas. 2017. Hairy root culture for the production of useful secondary metabolites. In *Biotechnology and Production of Anti-cancer Compounds*, Malik S. (eds). Springer, Cham, pp. 201-230. https://doi.org/10.1007/978-3-319-53880-8_9.
Amna, T., Puri, S. C., Verma, V., Sharma, J. P., Khajuria, R. K, Musarrat, J., Spiteller, M., and G. N. Qazi. 2006. Bioreactor studies on the endophytic fungus *Entrophospora infrequens* for the production of an anticancer alkaloid camptothecin. *Can. J. Microbiol.* 52:189–196.
Ayadi, R. and J. T. Guiller. 2003. Root formation from transgenic calli of *Ginkgo biloba*. *Tree Physiol.* 23:713–718.
Bandurska, K., Berdowska, A., and M. Król. 2016. Transformation of medicinal plants using *Agrobacterium tumefaciens*. *Postepy Hig. Med. Dosw.* 70:1220–1228.
Chakravarthi, B. V. S. K., Das, P., Surendranath, K., Karande, A. A., and C. Jayabaskaran. 2008. Production of paclitaxel by *Fusarium solani* isolated from *Taxus celebica*. *J. Biosci.* 33:259–267.
Cui, Y., Yi, D., Bai, X., Sun, B., Zhao, Y., and Y. Zhang. 2012. Ginkgolide B produced endophytic fungus (*Fusarium oxysporum*) isolated from *Ginkgo biloba*. *Fitoterapia* 83:913–920.
Deng, B. W., Liu, K. H., Chen, W. Q., Ding, X. W., and X. C. Xie. 2009. *Fusarium solani*, Tax-3, a new endophytic taxol-producing fungus from *Taxus chinensis*. *World J. Microbiol. Biotechnol.* 25:139–143.
Elavarasi, A., Rathna, G. S., and M. Kalaiselvam. 2012. Taxol producing mangrove endophytic fungi *Fusarium oxysporum* from *Rhizophora annamalayana*. *Asian Pac. J. Trop. Biomed.* 2:1081–1085.
Eyberger, A. L., Dondapati, R., and J. R. Porter. 2006. Endophyte fungal isolates from *Podophyllum peltatum* produce podophyllotoxin. *J. Nat. Prod.* 69(8):1121–1124.
Fürnkranz, M., Lukesch, B., Müller, H., Huss, H., Grube, M., and G. Berg. 2012. Microbial diversity inside pumpkins: microhabitat-specific communities display a high antagonistic potential against phytopathogens. *Microb. Ecol.* 63:418–428.

Gai, Q. Y., Jiao, J., Luo, M., Wei, Z. F., Zu, Y. G. and W. Ma. 2015. Establishment of hairy root cultures by *Agrobacterium rhizogenes* mediated transformation of *Isatis tinctoria* L. for the efficient production of flavonoids and evaluation of antioxidant activities. *PLoS One* 10:e0119022.

Garyali, S., Kumar, A., and M. S. Reddy. 2013. Taxol production by an endophytic fungus, *Fusarium redolens*, isolated from Himalayan yew. *J. Microbiol. Biotechnol.* 23(10), 1372–80.

Giri, A., Banerjee, S., Ahuja, P. S. and C. C. Giri. 1997. Production of hairy roots in *Aconitum heterophyllum* Wall. using *Agrobacterium rhizogenes*. *In Vitro Cell Dev. Biol. Plant* 33:280–284.

Glorybai, A. L. and P. Agastian. 2013. *Agrobacterium rhizogenes* mediated hairy root induction for increased colchicine content in *Gloriosa superba* L. *J. Acad. Ind. Res.* 2:68–78.

Gupta, R., Lee, S. L., Agrawal, G. K., Rakwal, R., Park S., Wang, Y., and S. T. Kim. 2015. Understanding the plant-pathogen interactions in the context of proteomics-generated apoplastic proteins inventory. *Front. Plant Sci.* 6:352.

Gupta, S. K., Liu, R. B., Liae, S. Y., Chan, H. S. and H. S. Tsay. 2011. Enhanced tanshinone production in hairy roots of *Salvia miltiorrhiza* Bunge under the influence of plant growth regulators in liquid culture. *Bot. Stud.* 52:435–443.

Gupta, S., Seth, R., and A. Sharma. 2016. Plant growth-promoting rhizobacteria play a role as phytostimulators for sustainable agriculture. In *Plant-microbe interaction: an approach to sustainable agriculture*:475–493. https://doi.org/10.1007/978-981-10-28 54-0_22

Hanafy, M. S., Matter, M. A., Asker, M. S. and M. R. Rady. 2016. Production of indole alkaloids in hairy root cultures of *Catharanthus roseus* L. and their antimicrobial activity. *S. Afr. J. Bot.* 105:9–18.

Hirsch, P. R., and T. H. Mauchline. 2012. Who's who in the plant root microbiome? *Nat. Biotechnol.* 30:961–962.

Hu, K., Tan, F., Tang, K., et al. 2006. Isolation and screening of endophytic fungi synthesizing taxol from *Taxus chinensis* var. mairei. *J. Southwest China Normal Univ. Nat. Sci. Edit.* 31:134–137.

Huang, S. H., Vishwakarma, R. K., Lee, T. T., Chan, H. S. and H. S. Tsay. 2014. Establishment of hairy root lines and analysis of iridoids and secoiridoids in the medicinal plant *Gentiana scabra. Bot. Stud. Int. J.* 55:17.

Huang, W., Long, C., and E. Lam. 2018. Roles of plant-associated microbiota in traditional herbal medicine. *Trends Plant Sci.* 23(7):559–562.

Imam, J., Singh, P. K., and P. Shukla. 2016. Plant microbe interactions in post genomic era: perspectives and applications. *Front. Microbiol.* 7:1488.

Ionkova, I., Karting, T. and W. Alfermann. 1997. Cycloartane saponin production in hairy root cultures of *Astragalus mongholicus. Phytochem* 45:1597–1600.

Ishimaru, K., Sudo, H., Salake, M., Malsugama, Y., Hasagewa, Y., Takamoto, S. and K. Shimomura. 1990. Amarogenetin and amaroswertin and four xanthones from hairy root cultures of *Swertia japonica. Phytochem* 29:1563–1565.

Kasaei, A., Mobini, D. M., F. Mahjoubi, et al. 2017. Isolation of taxol producing endophytic fungi from Iranian yew through novel molecular approach and their effects on human breast cancer cell line. *Curr. Microbiol.* 74:702–709.

Köberl, M., Schmidt, R., Ramadan, E. M., Bauer, R., and G. Berg. 2013. The microbiome of medicinal plants: diversity and importance for plant growth, quality, and health. *Front. Microbiol.* 4:400.

Kour, A., Shawl, A. S., Rehman, S., Sultan, P., Qazi, P. H., Suden, P., Khajuria, R. K., and V. Verma. 2008. Isolation and identification of an endophytic strain of *Fusarium oxysporum* producing podophyllotoxin from *Juniperus recurva. World J. Microbiol. Biotechnol.* 24:1115–1121.

Kumar V., Sharma, N., Sood, H., and R. S. Chauhan. 2016. Exogenous feeding of immediate precursors reveals synergistic effect on picroside-I biosynthesis in shoot cultures of *Picrorhiza kurroa* Royle ex Benth. *Sci. Rep.* 6:29750.

Kumar, P., Pal, T., Sharma, N., Kumar V., Sood H., and R. S. Chauhan. 2015. Expression analysis of biosynthetic pathway genes vis-à-vis podophyllum content in *Podophyllum hexandrum* Royle. *Protoplasma* 252:1253–1262.

Kumara, P. M., Zuehlke, S., Priti, V., et al. 2012. *Fusarium proliferatum*, an endophytic fungus from *Dysoxylum binectariferum* Hook.f, produces rohitukine, a chromane alkaloid possessing anti-cancer activity. *Antonie Van Leeuwenhoek* 101:323–329.

Kumaran, R. S., Kim, H. J., and B. K. Hur. 2010. Taxol promising fungal endophyte, *Pestalotiopsis* species isolated from *Taxus cuspidata*. *J. Biosci. Bioeng.* 110:541–546.

Kusari, S., Lamshoft, M., and M. Spiteller. 2009b. *Aspergillus fumgigatus* Fresenius, an endophytic fungus from *Juniperus communis* L. Horstmann as a novel source of the anticancer pro-drug deoxypodophyllotoxin. *J. Appl. Microbiol.* 107:1019–1030.

Kusari, S., Verma, V. C., Lamshoeft, M., and M. Spiteller. 2012. An endophytic fungus from *Azadirachta indica* A. Juss that produces azadirachtin. *World J. Microbiol. Biotechnol.* 28:1287–1294.

Kusari, S., Zühlke, S., and M. Spiteller. 2009a. An endophytic fungus from *Camptotheca acuminata* that produces camptothecin and analogues. *J. Nat. Prod.* 72:2–7.

Le Flem-Bonhomme, V., Laurain-Mattar, D. and M. A. Fliniaux. 2004. Hairy root induction of *Papaver somniferum* var. album, a difficult-to-transform plant, by *A. rhizogenes* LBA 9402. *Planta* 218(5):890–893.

Lee, S. U., Kim, S. U. G., Song, W. S., Kim, Y. K., Park, N. I. and S. U. Park. 2010. Influence of different strains of *Agrobacterium rhizogenes* on hairy root induction and production of alizarin and purpurin in *Rubia akane* Nakai. *Rom. Biotechnol. Lett.* 15(4):5405–5409.

Lee, S. Y., Cho, S. J., Park, M.H., Kim, Y. K., Choi, J. I. and S. U. Park. 2007b. Growth and rutin production in hairy root culture of buck wheat (*Fagopyruum esculentum*). *Prep. Biochem. Biotechnol.* 37:239–246.

Lee, S. Y., Xu, H., Kim, Y. K. and S. U. Park. 2007a. Rosmarinic acid production in hairy root cultures of *Agastache rugosa* Kuntze. *World J. Microbiol. Biotechnol.* 20:969–972.

Li, J., Zhao, G. Z., Varma, A., Qin, S., Xiong, Z., Huang, H.Y., et al. 2012. An endophytic *Pseudonocardia* species induces the production of artemisinin in *Artemisia annua*. *PLoS ONE* 7(12): e51410.

Li, W., Koike, K., Asada, Y., et al. 2005. Rosmarinic acid production by *Coleus forskohlii*-hairy root cultures. *Plant Cell Tiss. Organ. Cult.* 80:151–155.

Liu, C. Z., Guo, C., Wang, Y. C. and F. Ouyang. 2002. Effect of light irradiation on hairy root growth and artemisinin biosynthesis of *Artemisia annua* L. *Process Biochem.* 38(4):581–585.

Liu, K., Ding, X., Deng, B., and W. Chen. 2009. Isolation and characterization of endophytic taxol-producing fungi from *Taxus chinensis*. *J. Ind. Microbiol. Biotechnol.* 36:1171–1177.

Lugtenberg, B., Chin-A-Woeng, T. F., and G. V. Bloemberg. 2002. Microbe-plant interactions: principles and mechanisms. *Antonie Van Leeuwenhoek* 81, 373–383.

Madhusudanan, K. P., Banerjee, S., Khanuja, S. P. S. and S. K. Chattopadhyay. 2008. Analysis of hairy root culture of *Rauvolfia serpentine* using direct analysis in real time mass spectrometric technique. *Biomed. Chromatogr.* 22:596–600

Malhotra, N., Kumar, V., Sood, H., Singh, T. R., and R. S. Chauhan. 2014. Multiple genes of mevalonate and non-mevalonate pathways contribute to high aconites content in an endangered medicinal herb, *Aconitum heterophyllum* Wall. *Phytochemistry* 108:26–34.

Marconi, P. L., Selten, L. M., Cslcena, E. N., Alvarez, M. A. and S. I. Pitta-Alvarez. 2008. Changes in growth and tropane alkaloid production in long term culture of hairy roots of *Brugmansia candida*. *Elect. J. Integrative Biosci.* 3:38–44.

Marimuthu, T., Suganthy, M., and S. Nakkeeran. 2018. Common pests and diseases of medicinal plants and strategies to manage them. *New Age Herbals* 289–312. https://doi.org/10.1007/978-981-10-8291-7_14

Mendes, R., Garbeva, P., and J. M. Raaijmakers. 2013. The rhizosphere microbiome: Significance of plant beneficial, plant pathogenic, and human pathogenic microorganisms. *FEMS Microbiol. Rev.* 37(5):634–663.

Miao, Z., Wang, Y., Yu, X., Guo, B., and K. Tang. 2009. A new endophytic taxane production fungus from *Taxus chinensis*. *Appl. Biochem. Microbiol.* 45:81–86.

Min, C. L., Wang, X. J., Zhao, M. F., and W. W. Chen. 2014. Isolation of endophytic fungi from *Macleaya cordata* and screening of sanguinarine producing strains. *Zhongguo Zhong Yao Za Zhi* 39:4288–4292.

Moghadam, Y. A., Piri, K. H., Bahramnejad, B. H. and T. Ghiasvand. 2014. Dopamine production in hairy root cultures of *Portulaca oleracea* (Purslane) using *Agrobacterium rhizogenes*. *J. Agric. Sci. Technol.* 16:409–420.

Mousa, W. K., Shearer, C., Limay-rios, V., et al. 2016. Root-hair endophyte stacking in finger millet creates a physicochemical barrier to trap the fungal pathogen *Fusarium graminearum*. *Nat. Microbiol.* 1:16167.

Murthy, H. N., Dijkstra, C., Anthony, P., White, D. A., Davey, M. R., Powers, J. B., Hahn, E. J and K. Y. Paek. 2008. Establishemnt of *Withania somnifera* hairy root cultures for the production of Withanoloid A. *J. Integr. Plant Biol.* 50:915–981.

Murthy, H. N., Park, S. Y. and K. Y. Paek. 2017. Production of ginsenosides by hairy root cultures of *Panax ginseng*. In:Malik S. (ed.) *Production of plant derived natural compounds through hairy root culture*. Springer, Cham. https://doi.org/10.1007/978-3-319-69769-7_11

Nadeem, M., Mauji, R., Pravej, A. et al. 2012. *Fusarium solani*, P1, a new endophytic podophyllotoxin-producing fungus from roots of *Podophyllum hexandrum*. *Afr. J. Microbiol. Res.* 6:2493–2499.

Naik, S., Shaanker, R. U., Ravikanth, G., and S. Dayanandan. 2019. How and why do endophytes produce plant secondary metabolites? *Symbiosis*. doi:10.1007/s13199-019-00614-6.

Nguyen, K. V., Pongkitwitoon, B., Pathomwichaiwat, T., et al. 2019. Effects of methyl jasmonate on the growth and triterpenoid production of diploid and tetraploid *Centella asiatica* (L.) Urb. hairy root cultures. *Sci. Rep.* 9:18665.

Ni, X., Wen, S., Wang, W., Wang, X., Xu, H. and G. Kai. 2011. Enhancement of camptothecin production in *Camptotheca acuminate* hairy roots by over expressing ORCA3 gene. *J. Appl. Pharm. Sci.* 1:85–88.

Palem, P. P. C., Kuruakose, G. C., and C. Jayabaskaran. 2016. An endophytic fungus *Talaromyces radicus*, isolated from *Catharanthus roseus*, produces vincristine and vinblastine, which induce apoptotic cell death. *PLoS ONE* 11(4):e0153111.

Pavlov, A., Gorgiev, V. and M. Ilieva. 2005. Betalain biosynthesis by red beet (*Beta vulgaris* L.) hairy root culture. *Process Biochem.* 40(5):1531–1533.

Postel, S., and B. Kemmerling. 2009. Plant systems for recognition of pathogen-associated molecular patterns. *Semin. Cell Dev. Biol.* 20(9):1025–1031.

Pu, X., Qu, X., Chen, F., Bao, J., Zhang, G., and Y. Luo. 2013. Camptothecin-producing endophytic fungus *Trichoderma atroviride* LY357: isolation, identification, and fermentation conditions optimization for camptothecin production. *Appl. Microbiol. Biotechnol.* 97:9365–9375.

Puri, S. C., Verma, V., Amna, T., Qazi, G. N., and M. Spiteller. 2005. An endophytic fungus from *Nothapodytes foetida* that produces camptothecin. *J. Nat. Prod.* 68:1717–1719.

Pyne, M. E., Narcross, L., and V. J. J. Martin. 2019. Engineering plant secondary metabolism in microbial systems. *Plant Physiol.* 179:844–861.

Rahnama, H., Hasanloo, T., Shams, M. R. and R. Sepehrifar. 2008. Silymarin production by hairy root culture of *Silybium marianum* (L.) Gaertn. *Iran. J. Biotechnol.* 6:113–118.

Rehman, S., Shawl, A. S., Verma, V., Kour, A., Athar, M., Andrabi, R., P. Sultan, et al. 2008. An endophytic *Neurospora* sp. from *Nothapodytes foetida* producing camptothecin. *Appl. Biochem. Microbiol.* 44, 203–209.

Renouard, S., Corbin, C., Drouet, S., Medvedec, B., Doussot, J., Colas, C., Maunit, B., Bhambra, A. S., Gontier, E., Jullian, N., Mesnard, F., Boitel, M., Abbasi, B. H., Arroo, R. R. J., Lainé, E. and C. Hano. 2018. Investigation of *Linum flavum* (L.) hairy root cultures for the production of anticancer aryltetralin lignans. *Int. J. Mol. Sci.* 19(4):990.

Romero, F.R., Delate, K., Kraus, G.A., et al. 2009. Alkamide production from hairy root cultures of *Echinacea. In Vitro Cell Dev. Biol. Plant* 45:599.

Ruiz-Sanchez, J., Flores, B. Z. R., Dendooven, L., et al. 2010. A comparative study of taxol production in liquid and solid- state fermentation with *Nigrospora* sp. a fungus isolated from *Taxus globosa. J. Appl. Microbiol.* 109:2144–2150.

Samadi, A., Jafari, M., Nejhad, N. M. and F. Hossenian. 2014. Podophyllotoxin and 6-methoxy podophyllotoxin production in hairy root cultures of *Linum mucronatum ssp. Mucronatum. Pharmacogn. Mag.* 10:154–160.

Shanmugam, K., Ramalingam, S., Venkataraman, G., and G. N. Hariharan. 2019. The CRISPR/Cas9 system for targeted genome engineering in free-living fungi: advances and opportunities for lichenized fungi. *Front. Microbiol.* 10:62.

Sharma, N. 2017. *Investigation of morphogenetic differences and seaweed extract stimulated increase in biomass and picroside-I content in Picrorhiza species.* PhD diss., Jaypee University of Information Technology, Solan, India.

Sharma, N., Chauhan, R. S., and H. Sood. 2016b. Discerning picroside-I biosynthesis via molecular dissection of in vitro shoot regeneration in *Picrorhiza kurroa. Plant Cell. Rep.* 35:1601–1615.

Sharma, N., Chauhan, R.S., and H. Sood. 2015. Seaweed extract as a novel elicitor and medium for mass propagation and picroside-I production in an endangered medicinal herb *Picrorhiza kurroa. Plant Cell Tissue Organ Cult.* 122:57–65.

Sharma, N., Kumar, V., Chauhan, R. S., and H. Sood. 2016a. Modulation of picroside-I bio-synthesis ingrown elicited shoots of *Picrorhiza kurroa*in vitro. *J. Plant Growth Regul.* 35:965–973.

Shelake, R. M., Pramanik, D., and J. Y. Kim. 2019. Exploration of plant-microbe interactions for sustainable agriculture in CRISPR era. *Microorganisms* 7:269.

Shi, T. Q., Liu, G. N., Shi, K., Song, P., Ren, L. J., Huang, H., and X. J. Ji. 2017. CRISPR/Cas9-based genome editing of the filamentous fungi: The state of the art. *Appl. Microbiol. Biotechnol.* 101:7435–7443.

Shirazi, Z., Piri, K., Asl, A. M. and T. Hasanloo. 2013. Establishment of *Inula helenium* hairy root culture with the use of *Agrobacterium rhizogenes. Int. J. Sci. Basic Appl. Res.* 4:1034–1038.

Shweta, S., Zuehlke, S., Ramesha, B. T., Priti, V., Kumar, P. M., Ravikanth, G., Spiteller, M., Vasudeva, R., and R. Uma Shaanker. 2010. Endophytic fungal strains of *Fusarium solani*, from *Apodytes dimidiata* E.Mey. ex Arn (Icacinaceae) produce campto-thecin, 10-hydroxycamptothecin and 9-methoxycamptothecin. *Phytochemistry* 71: 117–122.

Singh, P. P., Kujur, A., Yadav, A., Kumar, A., Singh, S. K., and B. Prakash. 2019. Mechanisms of plant-microbe interactions and its significance for sustainable agriculture. *PGPR Amelioration in Sustain. Agric.* 17–39. https://doi.org/10.1016/B978-0-12-815879-1.00 002-1

Soliman, S. S. M., Greenwood, J. S., Bombarely, A., Mueller, L. A., Tsao, R., Mosser, D. D., and M. N. Raizada. 2015. An endophyte constructs fungicide-containing extracellular barriers for its host plant. *Curr. Biol.* 25:2570–2576.

Soliman, S. S. M., Trobacher, C. P., Tsao, R., Greenwood, J. S., and M. N. Raizada. 2013. A fungal endophyte induces transcription of genes encoding a redundant fungicide pathway in its host plant. *BMC Plant Biol.* 13:93.

Su, H., Kang, J., Cao, J., et al. 2014. Medicinal plant endophytes produce analogous bioactive compounds. *Chiang. Mai. J. Sci.* 41:1–13.

Sudha, C. G., Obul Reddy, B., Ravishankar, G. A. and S. Seeni. 2003. Production of ajmalicine and ajmaline in hairy root cultures of *Rauvolfia micrantha* Hook f., a rare and endemic medicinal plant. *Biotechnol. Lett.* 25(8):631–636.

Sun, D., Ran, X., and J. Wang. 2008. Isolation and identification of a taxol producing endophytic fungus from Podocarpus. *Acta Microbiol. Sin.* 48:589–595.

Tatsumi, K., Yano, M., Kaminade, K., Sugiyama, A., Sato, M., Toyooka, K., Aoyama, T., Sato, F. and K. Yazaki, 2016. Characterization of shikonin derivative secretion in *Lithospermum erythrorhizon* hairy roots as a model of lipid-soluble metabolite secretion from Plants. *Front. Plant Sci.* 7:1066.

Torkamani, M. R. D., Jafari, M., Abbaspour, N., Heidary, R. and N. Safaie, 2014. Enhanced production of valerenic acid in hairy root culture of *Valeriana officinalis* by elicitation. *Cent. Eur. J. Biol.* 9:853.

Vasundhara, M., Kumar, A. and M. S. Reddy, 2016. Molecular approaches to screen bioactive compounds from endophytic fungi. *Front. Microbiol.* 7:1774.

Verma, P. C., Singh, H., Negi, A. S., Saxena, G., Rahman, L. U. and S. Banerjee. 2015. Yield enhancement strategies for the production of picroliv from hairy root culture of *Picrorhiza kurroa* Royle ex Benth. *Plant Signal Behav.* 10(5):e1023976.

Vinterhalter, B., Milosevic, D. K., Jankovic, T., Plejevljakusic, D., Ninkovic, S., Smigocki, A. and D. Vinterhalter. 2015. *Gentiana dinarica* Beck. hairy root cultures and evaluation of factors affecting growth and xanthone production. *Plant Cell Tissue Organ Cult.* 121:667–679.

Wang, J. W., Zhang, Z., and R. X. Tan. 2001. Stimulation of artemisinin production in *Artemisia annua* hairy roots by the elicitor from the endophytic *Colletotrichum* sp. *Biotechnol. Lett.* 23, 857–860.

Wang, L., Wang, L., Tan, Q., Fan, Q., Zhu, H., Hong, Z., Zhang, Z., and D. Duanmu. 2016. Efficient inactivation of symbiotic nitrogen fixation related genes in *Lotus japonicus* using CRISPR-Cas9. *Front. Plant Sci.* 7:76.

Wei, Y., Zhou, X., Liu, L., et al. 2010. An efficient transformation system of taxol- producing endophytic fungus EFY-21 (*Ozonium* sp.). *Afr. J. Biotechnol.* 9:1726–1733.

Xu, H., Kim, Y. K., Suh, S. Y., Udin, M. R., Lee, S. Y. and S. U. Park. 2008. Deoursin production from hairy root culture of *Angelica gigas*. *J. Korea Soc. Appl. Biol. Chem.* 51:349–351.

Yang, C., Chen, M., Zeng, L., Zhang, L., Liu, X., Tang, K. and Z. Liao. 2011. Improvement of tropane alkaloids production in hairy root cultures of *Atropa belladonna* by overexpressing pmt and h6h genes. *Plant Omics J.* 4:29–33

Yang, X., Guo, S., Zhang, L. et al. 2003. Selection of producing podophyllotoxin endophytic fungi from podophyllin plant. *Nat. Prod. Res. Dev.* 15:419–422.

Yi, Y., Li, Z., Song, C., and O. P. Kuipers. 2018. Exploring plant-microbe interactions of the rhizobacteria *Bacillus subtilis* and *Bacillus mycoides* by use of the CRISPR-Cas9 system. *Environ. Microbiol.* 20:4245–4260.

Zhang, H. L., Xue, S. H., Pu, F., et al. 2010. Establishment of hairy root lines and analysis of gentiopicroside in the medicinal plant *Gentiana macrophylla*. *Russ. J. Plant Physiol.* 57:110–117.

Zhang, L., Guo, B., Li, H., et al. 2000. Preliminary study on the isolation of endophytic fungus of *Catharanthus roseus* and its fermentation to produce products of therapeutic value. *Chin. Tradit. Herbal Drug.* 31:805–807.

Zhang, P., Zhou, P. P., and L. J. Yu. 2009. An endophytic taxol-producing fungus from *Taxus media*, Cladosporium cladosporioides MD2. *Curr. Microbiol.* 59:227–232.

Zhao, B., Agblevor, F. A., Ritesh, K. C., and J. G. Jelesko. 2013. Enhanced production of the alkaloid nicotine in hairy root cultures in *Nicotiana tabacum* L. *Plant Cell Tissue Organ Cult.* 113:121–129.

Zhao, K., Zhao, L., Jin, Y., et al. 2008. Isolation of a taxol-producing endophytic fungus and inhibiting effect of the fungus metabolites on HeLa cell. *Mycosystema* 27:735–744.

11 Overview of Bioactive Compounds from Endophytes

Parikshana Mathur, Payal Mehtani,
Charu Sharma, and Pradeep Bhatnagar

CONTENTS

11.1 INTRODUCTION

Endophyte is a family of microorganisms that grows in the intra- and intercellular regions of plant tissues without causing any damage to the host plant. In 1999, Taylor et al. discovered endophytic fungi from the living fossil plant *Trachycarpus fortunei*. A long relationship between the microorganism and the host plant leads to mutual exchange of information and development of a common genetic stem that has helped the microbe to adapt to environmental stresses more competently. This has resulted in the increased compatibility of the endophyte with the host plant. Endophytes interact with other microbial groups that colonize plant tissues, such as mycorrhiza, pathogens, saprotrophs, and epiphytes, and are important components of plant microbiomes (Jain and Pundir 2015). The production of bioactive

DOI: 10.1201/9781003213864-11

compounds/secondary metabolites by endophytes is believed to be directly linked to the evolution of the host plant. Endophytes may have incorporated genetic information from the host, which has helped them to adapt better within the host and in performing defensive roles (Strobel 2003).

Conventionally, plants were considered to be the only source of bioactive compounds, but now, the endophytic microbes associated with plants are also reported to produce a variety of bioactive metabolites. Once these endophytes are isolated from the plant and characterized, they can serve as an admirable source of novel bioactive compounds. These novel compounds can have applications in medicine, agriculture, and food industries and can be a source of new drugs for treatment against a variety of diseases (Shukla et al. 2014) (Table 11.1 and 11.2). Medicinal plants are especially known to allow growth of endophytic microorganisms that add to the value of the plant by producing compounds of pharmaceutical importance. In addition to these applications, endophytes are useful in developing a sustainable environment, improving crop productivity, enhancing plant growth, and increasing biotic and abiotic stress tolerance in host plants (Fadiji and Babalola 2020).

11.2 ENDOPHYTES

Endophytes are an endosymbiotic group of microorganisms that inhabit the roots, stem, leaves, buds, fruits, and seeds of the plant (Stepniewska and Kuzniar 2013). All or some part of the life cycle of these microbes occurs within their hosts without causing any disease, and they exhibit complex interactions with their hosts. They may show mutualistic, antagonistic, and sometimes parasitic relations (Nair and Padmavathy 2014). These are also reported to improve nutrient gain and host growth but their main focus is on helping the plant's defense mechanism by enhancing its ability to endure biotic and abiotic stresses. They also improve the resistance of plants to insects and pests. Endophytes synthesize bioactive compounds that are of biotechnological interest, including the production of pharmaceutical drugs and enzymes (Parthasarathi et al. 2012). In addition, endophytes also aid in nitrogen fixation for plants. Plant hormones produced by certain endophytes seem to be necessary for bryophyte development (Hornschuh et al. 2002).

Both plants and microorganisms benefit from the synthesis of secondary metabolites, as they help in many physiological functions that are common to the host as well as the endophyte. Hence, the purpose of utilizing botanical diversity to discover novel chemicals has led to the discovery of new microbial strains that are able to produce compounds that were previously considered to be normal plant products (Nicoletti and Fiorentino 2015). All plants are inhabited by endophytic microorganisms, which can synthesize secondary metabolites. Shukla et al. (2014) suggested that extraction of bioactive compounds from endophytic microorganisms is affected by a number of factors such as climatic condition, geographical location of the plant, and the season of sample collection. Therefore, research is now more focused on considering these products as a major factor that influence the evolution and establishment of mutualistic interrelations (Schulz et al. 2002).

TABLE 11.1

Common Bioactive Compounds Isolated from Different Plants and Their Bioactivity

Bioactive compound	Plant source of endophyte	Endophyte	Bioactivity	Reference
Lignans (cathartics, emetics and cholagogue)	*Podophyllum hexandrum*	–	Anticancerous activity	Konuklugil 1995
Resins (etoposide and teniposide)	*Podophyllum emodi*	–	Anticancerous activity	Konuklugil 1995
Taxols	*Taxus brevifolia*	*Taxomyces andreanae*	Anticancerous activity	Wani et al. 1971
Ergoflavin	*Mimosops elengi*	Fungal culture PM0651480	Anticancerous activity	Deshmukh et al. 2009
Huperzine A	*Huperzia serrata*	*Shiraia* sp.	Cholinesterase inhibitor	Wang et al. 2011
Pestalotheol C	–	*Pestalotiopsis theae*	Anti-HIV activity	Li et al. 2008
7-amino-4-methylcoumarin	*Ginkgo biloba*	*Xylaria* sp. YX-28	Food preservative	Liu et al. 2007
Chaetomugilin A and D	*Ginkgo biloba*	*Chaetomium globosum*	Antifungal activity	Qin et al. 2009
Pestacin	*Terminalia morobensis*	*Pestalotiopsis microspora*	Antioxidant activity	Harper et al. 2003
Graphislactone A	*Trachelospermum jasminoides*	*Cephalosporium* sp.	Antioxidant activity	Song et al. 2005

TABLE 11.2

Bioactive Compounds Isolated from Endophytes and Their Use Against Microorganisms

Bioactive compound	Compound type	Endophyte	Inhibits microbe	References
Cryptocandin A	Cyclic lipopeptide	*Cryptosporiopsis quercina*	*Candida albicans, Trichophyton* sp.	Strobel et al. 1999
Beauvericin	Protein	*Fusarium proliferatum*	*Clostridium botulinum*	Meca et al. 2010
Javanicin	Napthoquinones	*Chloridium* sp.	*Pseudomonas* sp.	Kharwar et al. 2008
Tyrosol	Phenol and Quinone	*Diaporthe helianthi*	*Enterococcus hirae, Salmonella* sp.	Specian et al. 2012
Hypercin and Emodin	Naphthodianthrone	*Hypericum perforatum*	*Staphylococcus aureus, Klebsiella pneumoniae*	Kusari et al. 2008

11.3 MAJOR CLASSES OF SECONDARY METABOLITES AND THEIR APPLICATION

Secondary metabolites are biosynthetic derivatives of primary metabolites and are classified according to their biosynthetic origin. The following section discusses some bioactive compounds gained from endophytes and their applications.

11.3.1 ALKALOIDS

Alkaloids are considered an important source of drugs consisting of nitrogenous organic compounds. They have pharmaceutical and industrial importance owing to their wide-ranging biological properties, such as antifungal, anticancer, and antiviral activities (Mathur et al. 2021). The discovery of taxol, an anticancer drug isolated from *Taxus brevifolia,* has been a trailblazing success of natural compounds obtained from endophytes (Wani et al. 1971). This compound, which is reported to cure ovarian, lung, and breast cancer, has been later isolated from various other endophytic species such as *Metarhizium anisopliae* and *Colletotrichum gloeosporioides* isolated from plants *Taxus chinensis* and *Justicia gendarussa*, respectively (Liu et al. 2009; Gangadevi and Muthumary 2008).

Another effective anticancer agent, camptothecin, was obtained from the endophytic fungus *Fusarium solani* and was isolated from the *Camptotheca acuminate* tree (Kusari et al. 2009). Antimicrobial compounds asperfumoid and asperfumin were isolated from an endophytic fungus *Aspergilus fumigatus* CY018 from the plant *Cynodon dactylon* (Liu et al. 2004). Cholinesterase inhibitor huperzine A (HupA) was obtained from endophytic fungi *Shiraia* sp., which was isolated from the plant *Huperzia serrata* (Wang et al. 2011). Spiroquinazoline alkaloids are produced by *Eupenicillium* sp., isolated from *Murraya paniculate* (Barros and Rodrigues-Filho 2005). The production of ergot alkaloids is reported by endophytic fungi *Neotyphodium* sp. (Panaccione et al. 2006) and endophytic fungus isolated from *Acremonium coenophialum* (Bacon 1988).

11.3.2 TERPENOIDS

Triterpenoids are widely distributed in edible and ethno-medicinal plants (Souza et al. 2011). They are composed of three terpene units and are biosynthetically derived from the acyclic C30 hydrocarbon, squalene. Zhao et al. (2020) reported sesquiterpenoids as the most important kind of terpenoids and *Trichoderma* and *Aspergillus* as major terpenoid-producing fungal endophytes. Antimicrobial, cytotoxic, anti-inflammatory, antipathogenic, antitumor agents, and enzyme inhibition properties have been reported in terpenoids. Triterpenoids are further classified into sub-classes.

11.3.2.1 Steroids/Sterols

Plant sterols and steroids are compounds that yield a broad range of biological activities such as reproduction, plant growth, and dealing with biotic and abiotic stresses.

The primary use of natural steroids in health care is to reduce inflammation and other disease symptoms (Rasheed et al. 2013). Antifungal agent Penicisteroid A is isolated from *Penicillium chrysogenum* (Gao et al. 2011). *Colletotrichum*, a fungal endophyte residing in *Artemisia annua*, produces steroids 3β-hydroxy-5α,8α epidioxy-ergosta-6,22-dien, 3β-hydroxy-ergosta-5-ene, 3-oxo-ergosta-4,6,8(14),22-tetraene, 3β,5α-dihydroxy-6β-acetoxy-ergosta-7,22-diene, and 3β,5α-dihydroxy-6β-phenylacetyloxy-ergosta-7,22-diene (Lu et al. 2000).

11.3.2.2 Saponins

Saponins are compounds secreted by plants and endophytic microorganisms. Saponins are glycosides that have a sugar moiety bonded with glycosidic linkage to a non-sugar portion. They have diverse structural, pharmacological, physicochemical, and biological properties, which has led to their commercial importance. Saponins are reported to exhibit activities such as antimicrobial, cell membrane perturbing, hemolytic, neutraceutical hypolipidemic, immune modulating, and surface active and cell cytotoxicity (Netala et al. 2015). The fungal endophyte *Fusarium* sp., isolated from *Panax ginseng*, exhibited the capacity to synthesize ginsenoside saponins (Wu et al. 2013).

11.3.3 Phenolic Compounds

Phenols comprise the largest class of secondary metabolites. The term phenolic compounds includes a variety of water-soluble plant substances made up of an aromatic ring having hydroxyl substituents (Mathur et al. 2021). Antifungal phenolic acids obtained from endophyte *Epichloe typhina*, isolated from *Phleum pratense* include p-hydroxyphenylacetic acid, tyrosol, p-coumaric acids, and p-hydroxybenzoic acid (Koshino et al. 1988). *Pezicula* strain 553, a fungal endophyte, synthesizes 2-methoxy-4-hydroxy-6-methoxymethylbenzaldehyde and has antifungal properties against *Cladosporium cucumerinum* (Schulz et al. 1995). Altenusin, a chemotherapeutic agent, which has the potential to treat trypanosomiasis and leishmaniasis, was isolated from *Alternaria sp.* residing in *Trixis vauthieri* (Cota et al. 2008). Tricin and related flavone glycosides were isolated from *Poa ampla*, a perennial plant infected with a symbiotic fungus *Neotyphodium typhnium* (Ju et al. 1998). Anti-cancerous activity of lignans and resins isolated from the endophyte *Podophyllum hexandrum* has been reported (Konuklugil 1995). Ergoflavin, synthesized by fungi coded as PM0651480 and isolated from the leaves of the plant *Mimosops elengi*, has shown good anti-cancerous activity (Deshmukh et al. 2009). Antifungal properties were reported by Oocydin A, which was isolated from bacterial endophyte *Serratia marcescens* recovered from *Rhyncholacis penicillata* (Strobel et al. 1999). Marinho et al. (2005) reported strong antibacterial activity in polyketide citrinin produced by endophytic fungus *Penicillium janthinellum* that inhabits *Melia azedarach*. Antimicrobial activity has been reported in p-amino-acetophenonic acids isolated from *Streptomyces griseus* that inhabits *Kandelia candel* (Guan et al. 2005). Cytonic acids A and D, viral inhibitors produced by endophyte *Cytomaema sp.* and isolated from *Quercus sp.*, have been reported to inhibit the human cytomegalovirus (Guo et al. 2000). Antiviral drug brefeldin A was isolated from endophyte *Cladosporium sp.* that inhabits *Quercusvariabilis* (Wang et al. 2007). Pestacin and Graphislactone A, synthesized by endophyte *Pestalotiopsis microspora* and recovered from plants *Terminalia morobensis* and *Trachelospermum jasminoides*, were reported to have antioxidant activities (Harper et al. 2003; Song et al. 2005). Phenolics synthesized by *Xylaria* sp. that inhabits *Ginkgo biloba* has shown antioxidant properties (Liu et al. 2007). Phenols are further classified into subclasses.

11.3.3.1 Flavonoids

Flavonoids are the largest group of phenolic compounds and are mainly water-soluble. They are structurally derived from parent-substance flavones. In leguminous plants, flavonoids act as signaling molecules during the process of nodulation (Kikuchi et al. 2007). *Aspergillus* strains isolated from *Ginkgo biloba* L. were found to be able to produce phenolic and flavonoid compounds that may serve as a potential source of natural medicines or prodrugs (Qiu et al. 2010). *Aspergillus japonicus* strain isolated from *Euphorbia indica* L. has been reported to enhance the production of flavonoids in soyabean and sunflower seedlings and can also improve stress tolerance of the host along with increased biomass production (Ismail et al. 2018). *Alternaria alternata* and *Fusarium proliferatum* isolated from *Salvia miltiorrhiza* Bge. f. alba has been reported to produce more phenol and flavonoid than those of the host roots and also exhibit strong antioxidant activities (Li et al. 2015). Phytotoxins such as hydroxypestalopyrone, pestaloside, and pestalopyrone have antifungal property and were obtained from *Pestalotiopsis microspora* isolated from *Taxus taxifolia* (Lee et al. 1995).

11.3.3.2 Tannins

Tannins are water-soluble phenolic compounds of plant origin that are able to transform raw animal skins into leather because of their cross-linking ability with protein. It has been suggested that many tannin components are antimicrobial and anticarcinogenic (Chung et al. 1998).

11.3.4 POLYPEPTIDES

Polypeptides are high molecular weight polymers of amino acids. A protein may have a simple chain of polypeptides or several identical chains. Endophytic fungus *Acremonium sp.* produces anticancerous compound Leucinostatin A that colonizes in the plant *Taxus baccata* (Strobel et al. 1997). Echinocandin A, which showed antifungal activity, was obtained from endophytic fungi *Cryptosporiopsis* sp. and *Pezicula* sp. inhabiting *Pinus sylvestris* and *Fagus sylvatica* respectively (Noble et al. 1991). Antifungal compound cryptocandin was isolated from the fungus *Candida quercina*, residing in *Tripterigeum wilfordii* (Strobel et al. 1999). Antifungal and antimalarial properties were reported in Coronamycin isolated from endophyte *Streptomyces sp.* recovered from *Monstera* sp. (Ezra et al. 2004). Antibiotics such as *Munumbicins* and *Kakadumycins* were isolated from *Streptomyces sp.* inhabiting in *Kennedia nigriscans* and *Kennedia nigriscans* (Castillo et al. 2002, 2003). Endophytic *Xylaria* sp. isolated from *Abies holophylla* produced antifungal polyketide Griseofulvin (Park et al. 2005). Many hydrolytic enzymes such as cellulase, protease, and lipase were obtained from multiple endophytic bacteria isolated from sea grass *Halodule uninervis* (Bibi et al. 2018).

11.3.5 OILS AND FATTY ACIDS

Essential oils of plants have a large variety of compounds that can be extracted through distillation. They contribute to the fragrance of plants and have antibiotic and antitumor properties (Bakkali et al. 2008). Oils and fatty acids obtained as secondary metabolites of endophytic fungi may have a role in crop resistance/susceptibility toward pests (Nicoletti and Fiorentino 2015). *Pseudomonas viridiflava*, isolated from grass, is reported to synthesize antimicrobial lipopeptides named Ecomycins B and C, which have the ability to inhibit human pathogens *Cryptococcus neoformans* and *Candida albicans* respectively (Miller et al. 1998).

11.4 OTHER APPLICATIONS OF ENDOPHYTES

Endophytes not only synthesize bioactive compounds but also show other defensive features as well. Endophyte-associated crops show resistance to biotic stress conditions such as insect herbivory and nematocidal attack by acting as biocontrol agents in various plants. They also help the host plants in dealing with abiotic stresses such as high temperature, drought, salinity, and heavy metals and, thereby, aid in plant growth (Kaur 2020). Isolation, characterization, and evaluation of the bioactive products of various endophytic bacteria have been reported in black pepper (*Piper nigrum*) against *Phytophthora capsici*. Endophytes such as *Pseudomonas aeruginosa*, *Pseudomonas putida*, and *Bacillus megaterium* show antagonistic properties against *Phytophthora*, which causes foot rot in black pepper (Aravind et al. 2009). Endophyte *Beauveria bassiana* has been reported to control the borer insects in sorghum (Tefera and Vidal 2009) and coffee seedlings (Posada and Vega 2006). Strong antagonistic properties of endophytic bacteria *Bacillus subtilis* inhabiting *Speranskia tuberculata* against *Botrytis cinerea* have been reported by Wang et al. (2009). Strains of *Burkholderia pyrrocinia* JK-SH007 and *Burkholderia cepacia* act as potential biocontrol agents against poplar canker (Ren et al. 2011). Organic extracts of *Talaromyces pinophilus* isolated from *Arbutus unedo* has displayed lethal effects against pea aphid, which indicates the possibility of the involvement of the compound in defensive mutualism (Vinale et al. 2017) (Figure 11.1).

The endophytic strain of *Pseudomonas pseudoalcaligenes* reportedly induces the accumulation of higher concentrations of glycine betaine-like compounds that helped in improving salinity stress tolerance in rice (Jha et al. 2011). *Burkholderia phytofirmans* has been reported to enhance cold tolerance in grapevine plants by altering photosynthetic activity and metabolism of carbohydrates involved in cold stress tolerance (Fernandez et al. 2012) (Figure 11.2).

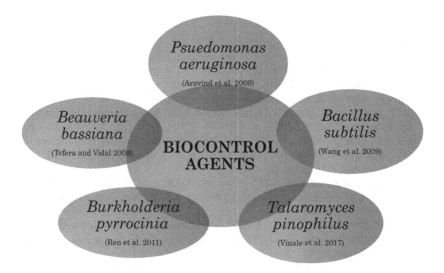

FIGURE 11.1 Endophytes acting as biocontrol agents in various plants.

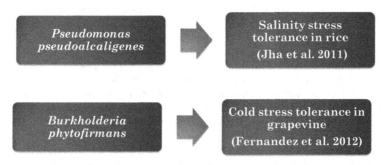

FIGURE 11.2 Endophytes exhibiting stress tolerance in agricultural crops.

A study on the role of endophytes in bioremediation in *Nicotiana tabaccum* plants shows enhanced biomass production under cadmium stress, which establishes the beneficial effects of endophytes on metal toxicity (Russell et al. 2011).

11.5 CONCLUSION

The interactions between the host plant and its inhabiting endophyte both contribute to the co-production of bioactive molecules. Endophytes are a less investigated group of microorganisms that have the ability to synthesize bioactive compounds and have a variety of applications. They are the source of new compounds for novel drug discovery and can act as antimicrobial, antifungal, antiviral, antioxidant, anti-inflammatory agents. Hence, it is of vital importance to bring the attention of researchers to this promising field and its exploration in various fields such as medical, pharmaceutical, food, and cosmetics.

REFERENCES

Aravind, R., Kumar, A., Eapen, S.J., Ramana, K.V. 2009. Endophytic bacterial flora in root and stem tissues of black pepper (*Piper nigrum* L.) genotype: isolation, identification and evaluation against *Phytophthora capsici*. *Lett. Appl. Microbiol.* 48(1):1472–765.

Bacon, C.W. 1988. Procedure for isolating the endophyte from tall fescue and screening isolates for ergot alkaloids. *Appl. Environ. Microbiol.* 54(11):2615–8.

Bakkali, F., Averbeck, S., Averbeck, D., Idaomar, M. 2008. Biological effects of essential oils: A review. *Food Chem. Toxicol.* 46:446–475.

Barros, F.A.P., Rodrigues-Filho, E. 2005. Four spiroquinazoline alkaloids from *Eupenicillium* sp. isolated as an endophytic fungus from leaves of *Murraya paniculata* (Rutaceae). *Biochem. Syst. Ecol.* 33(3):257–68.

Bibi, F., Naseer, M.I., Hassan, A.M., Yasir, M., Al-Ghamdi, A.A.K., Azhar, E.I. 2018. Diversity and antagonistic potential of bacteria isolated from marine grass *Halodule uninervis*. *3 Biotech.* 8:48.

Castillo, U.F., Strobel, G.A., Ford, E.J. 2002. Munumbicins, wide-spectrum antibiotics produced by *Streptomyces* NRRL 30562, endophytic on *Kennedia nigriscans*. *Microbiol.* 148(9):2675–85.

Castillo, U., Harper, J.K., Strobel, G.A. 2003. Kakadumycins, novel antibiotics from *Streptomycessp.* NRRL 30566, an endophyte of *Grevillea pteridifolia*. *FEMS Microbiol. Lett.* 224(2):183–90.

Chung, K.T., Wong, T.Y., Wei, C.I., Huang, Y.W., Lin, Y. 1998. Tannins and human health: A review. *Crit. Rev. Food Sci. Nutr.*38(6):421–64.

Cota, B.B., Rosa, L.H., Caligiorne, R.B., Rabello, A.L.T., Almeida Alves, T.M., Rosa, C.A. 2008. Altenusin, a biphenyl isolated from the endophytic fungus *Alternaria sp.*, inhibits trypanothione reductase from *Trypanosoma cruzi*. *FEMS Microbiol. Lett.* 285:177–82.

Deshmukh, S.K., Mishra, P.D., Kulkarni, A. 2009. Anti-inflammatory and anticancer activity of ergoflavin isolated from an endophytic fungus. *Chem. Biodivers.* 6(5):784–9.

Ezra, D., Castillo, U.F., Strobel, G.A. 2004. Coronamycins, peptide antibiotics produced by a verticillate *Streptomyces* sp. (MSU-2110) endophytic on *Monstera* sp. *Microbiol.* 150(4):785–93.

Fadiji, A.E., Babalola, O.O. 2020. Elucidating mechanisms of endophytes used in plant protection and other bioactivities with multifunctional prospects. *Front. Bioeng. Biotechnol.* 8:467. doi:10.3389/fbioe.2020.00467

Fernandez, O., Theocharis, A., Bordiec, S., Feil, R., Jacquens, L., Clement, C., Fontaine, F., Barka, E.A. 2012. *Burkholderiaphytofirmans* PsJN acclimates grapevine to cold by modulating carbohydrate metabolism. *Mol. Plant-Microbe Interact.* 25(4):496.

Gangadevi, V., Muthumary, J. 2008. Isolation of *Colletotrichum gloeosporioides*, a novel endophytic taxol-producing fungus from the leaves of a medicinal plant, *Justicia gendarussa*. *Mycologia Balcanica* 5:1–4.

Gao, S.S., Li, X.M., Li, C.S., Proksch, P., Wang, B.G. 2011. Penicisteroids A and B, antifungal and cytotoxic polyoxygenated steroids from the marine alga-derived endophytic fungus *Penicillium chrysogenum* QEN-24S. *Biorg. Med. Chem. Lett.* 21:2894–7.

Guan, S.H., Sattler, I., Lin, W.H., Guo, D.A., Grabley, S. 2005. p-Aminoacetophenonic acids produced by a mangrove endophyte: *Streptomyces griseus* subspecies. *J. Nat. Prod.* 68(8):1198–200.

Guo, B., Dai, J.R., Ng, S., Huang, Y., Leong, C., Ong, W., Carte, B.K. 2000. Cytonic acids A and B: novel tripeptide inhibitors of hCMV protease from the endophytic fungus *Cytonaema species*. *J. Nat. Prod.* 63(5):602–4.

Harper, J.K., Arif, A.M., Ford, E.J. 2003. Pestacin: A 1,3- dihydro isobenzofuran from *Pestalotiopsis microspora* pos possessing antioxidant and antimycotic activities. *Tetrahedron.* 59(14):2471–6.

Hornschuh, M., Grotha, R., Kutschera, U. 2002. Epiphytic bacteria associated with the bryo-phyte *Funaria hygrometrica*: Effects of methylobacterium strains on protonema devel-opment. *Plant Biol.* 4(6):682–7.

Ismail, Hamayun, M., Hussain, A., Iqbal, A., Khan, S.A., Lee, I. 2018. Endophytic fungus Aspergillus japonicus mediates host plant growth under normal and heat stress condi-tions. *BioMed. Res. Int.* 2018. 1–11. doi:10.1155/2018/7696831.

Jain, P., Pundir, R.K. 2015. Diverse Endophytic Microflora of Medicinal Plants. Plant Growth Promoting Rhizobacteria and Medicinal Plants. In: Egamberdieva D., Shrivastava S., Varma A. (eds) *Plant-Growth-Promoting Rhizobacteria (PGPR) and Medicinal Plants. Soil Biology.* Springer, Cham 42:341–57.

Jha, Y., Subramanian, R.B., Patel, S. 2011. Combination of endophytic and rhizospheric plant growth promoting rhizobacteria in *Oryza sativa* shows higher accumulation of osmo-protectant against saline stress. *Acta Physiologiae Plantarum.* 33:797.

Ju, Y., Sacalis, J.N., Still, C.C. 1998. Bioactive flavonoids from endophyte-infected blue grass (*Poa ampla*). *J. Agric. Food Chem.* 46:3785–8.

Kaur, T. 2020. Fungal endophyte-host plant interactions: Role in sustainable agriculture. In M. Hasanuzzaman, M.C.M.T. Filho, M. Fujita and T.A.R. Nogueira (eds.) *Sustainable Crop Production.* IntechOpen, London. doi: 10.5772/intechopen.92367.

Kharwar, R.N., Verma, V.C., Kumar, A., Gond, S.K., Harper, J.K., Hess, W.M., Lobkovosky, E., Ma, C., Ren, Y., Strobel, G.A. 2008. Javanicin, an antibacterial naphthaquinone from an endophytic fungus of neem, *Chloridium* sp. *Current Microbiol.* 58:233–8.

Kikuchi, K., Matsushita, N., Suzuki, K. et al. 2007. Flavonoids induce germination of basidiospores of the ectomycorrhizal fungus *Suillus bovinus.* *Mycorrhiza* 17:563–70. doi:10.1007/s00572-007-0131-8

Konuklugil, B. 1995. The importance of Aryl tetralin (*Podophyllum*) lignins and their distri-bution in the plant kingdom. *Ankara Univ. Eczacilik Fak. Derg.* 24(2):109–25.

Koshino, H., Terada, S.I., Yoshihara, T., Sakamura, S., Shimanuki, T., Sato, T. 1988. Three phenolic acid derivatives from stromata of *Epichloe typhina* on *Phleum pratense.* *Phytochemistry.* 27:1333–8.

Kusari, S., Lamshoft, M., Zuhlke, S., Spiteller, M. 2008. An endophytic fungus from *Hypericum perforatum* that produces hypericin. *J. Nat. Prod.* 71: 1059–63.

Kusari, S., Zuhlke, S., Spiteller, M. 2009. An endophytic fungus from *Camptotheca acumi-nata* that produces camptothecin and analogues. *J. Nat. Prod.* 72:2–7.

Lee, J.C., Yang, X., Schwartz, M., Strobel, G., Clardy, J. 1995. The relationship between an endangered North American tree and an endophytic fungus. *Chem. Biol.* 2(11):721–7.

Li, E., Tian, R., Liu, S., Chen, X., Guo, L., Che, Y. 2008. Pestalotheols A-D, bioactive metab-olites from the plant endophytic fungus *Pestalotiopsis theae.* *J. Nat. Prod.* 71(4):664–8. doi:10.1021/np700744t

Li, Y.L., Xin, X.M., Chang, Z.Y., Shi, R.J., Miao, Z.M., Ding, J., Hao, G.P. 2015. The endo-phytic fungi of *Salvia miltiorrhiza* Bge.f. alba are a potential source of natural antioxi-dants. *Bot. Stud.* 56(1):5. doi: 10.1186/s40529-015-0086-6

Liu, J.Y., Song, Y.C., Zhang, Z., Wang, L., Guo, Z.J., Zou, W.X., Tan, R.X. 2004. *Aspergillus fumigatus* CY018, an endophytic fungus in *Cynodon dactylon* as a versatile producer of new and bioactive metabolites. *J. Biotechnol.* 114:279–287.

Liu, X., Dong M., Chen, X., Jiang M., Lv, X., Yan, G. 2007. Antioxidant activity and pheno-lics of an endophytic *Xylaria* sp. from *Ginkgo biloba.* *Food Chem.* 105(2):548.

Liu, K., Ding, X., Deng, B., Chen, W. 2009. Isolation and characterization of endophytic taxol producing fungi from *Taxuschinensis.* *J. Ind. Microbiol. Biotechnol.* 36(9):1171–7.

Lu, H., Zou, W.X., Meng, J.C., Hu, J., Tan, R.X. 2000. New bioactive metabolites pro-duced by *Colletotrichum sp.*, an endophytic fungus in *Artemisia annua.* *Plant Sci.* 151:67–73.

Marinho, M.R., Rodrigues-Filho E., Moitinho, R., Santos L.S. 2005. Biologically active polyketides produced by *Penicillium janthinellum* isolated as an endophytic fungus from fruits of *Melia azedarach. J. Braz. Chem. Soci.* 16:280–3.

Mathur, P., Mehtani, P., Sharma, C. 2021. Leaf endophytes and their bioactive compounds. In: Shrivastava N., Mahajan S., Varma A. (eds) *Symbiotic Soil Microorganisms. Soil Biology*, 60. Springer, Cham. doi:10.1007/978-3-030-51916-2_9

Meca, G., Sospedra, I., Soriano, J.M., Ritieni, A., Moretti, A., Manes, J. 2010. Antibacterial effect of the bioactive compound beauvericin produced by *Fusarium proliferatumon* solid medium of wheat. *Toxicon* 56(3):349–54.

Miller, C.M., Miller, R.V., Garton-Kenny, D. 1998. Ecomycins, unique antimycotics from *Pseudomonas viridiflava. J. Appl. Microbiol.* 84(6):937–44. doi:10.1046/j.1365-2672.1 998.00415.x

Nair, D.N., Padmavathy, S. 2014. Impact of endophytic microorganisms on plants, environment and humans. *Sci. World J.* 2014, 250693. doi:10.1155/2014/250693

Netala, V.R., Ghosh, S.B., Bobbu, P., Anitha, D., Tartte, V. 2015. Triterpenoid saponins: A review on biosynthesis, applications and mechanism of their action. *Int. J. Pharm. Pharm. Sci.* 7(1):24–8.

Nicoletti, R., Fiorentino, A. 2015. Plant bioactive metabolites and drugs produced by endophytic fungi of spermatophyta. *Agri.* 5:918–70.

Noble, H.M., Langley, D., Sidebot- tom, P.J., Lane,S.J., Fisher, P.J. 1991. Anechinocandin from an endophytic *Cryptosporiopsis* sp. and *Pezicula* sp. in *Pinus sylvestris* and *Fagus sylvatica. Mycol. Res.* 95:1439–40.

Pannacione, D.G., Cipoletti, J.R., Sedlock, A.B., Blemings, K.P., Schardi, C.L., Machado, C., Seidel, G.E. 2006. Effects of ergot alkaloids on food preference and satiety in rabbits, as assessed with gene-knockout endophytes in Perennial Ryegrass (*Lolium perenne*). *J. Agric. Food Chem.* (54)13: 4582–7.

Park, J.H., Choi, G.J., Lee, H.B. 2005. Griseofulvin from *Xylaria* sp. strain F0010, an endophytic fungus of *Abies holophylla* and its antifungal activity against plant pathogenic fungi. *J. Microbiol. Biotechnol.* 15(1):112–7.

Parthasarathi, S., Sathya, S., Bupesh, G., Samy, D.R., Mohan, M.R., Selva, G.K. 2012. Isolation and characterization of antimicrobial compound from marine *Streptomyces hygroscopicus* BDUS 49. *World J. Fish Mar. Sci.* 4(3):268–77.

Posada, F., Vega, F.E. 2006. Inoculation and colonization of coffee seedlings (*Coffea arabica* L.) with the fungal entomopathogen *Beauveria bassiana* (Ascomycota: Hypocreales). *Mycoscience* 47(5):284–9.

Qin, J.C., Zhang, Y.M., Gao, J.M., Bai, M.S., Yang, S.X., Laatsch, H., Zhang, A.L. 2009. Bioactive metabolites produced by *Chaetomiumglobosum* an endophyte isolated from *Ginkgo biloba. Bioorg. Med. Chem. Lett.* 19:1572–4.

Qiu, M., Xie, R., Shi, Y. 2010. Isolation and identification of two flavonoid-producing endophytic fungi from *Ginkgo biloba* L. *Ann Microbiol.* 60:143–50. doi:10.1007/ s13213-010-0016-5

Rasheed, A., Qasim, Md. 2013. Review of natural steroids and their applications. *Int. J. Pharm.Sci. Res.* 4(2):520–31.

Ren, J.H., Ye, J.R. Liu, H., Xu, X.L., Wu, X.Q. 2011. Isolation and characterization of a new *Burkholderia pyrrocinia* strain JK-SH007 as a potential biocontrol agent. *World J. Microbiol. Biotechnol.* 27(9):2203–15.

Russell, J.R., Huang, J., Anand, P. 2011. Biodegradation of polyester polyurethane by endophytic fungi. *Appl. Environ. Microbiol.* 77(17):6076–84.

Schulz, B., Sucker, J., Aust, H.J., Krohn, K., Ludewig, K., Jones, P.G. 1995. Biologically active secondary metabolites of endophytic *Pezicula* species. *Mycol. Res.* 99:1007–15.

Schulz, B., Boyle, C., Draeger, S., Römmert, A.K., Krohn, K. 2002. Endophytic fungi: A source of novel biologically active secondary metabolites. *Mycol. Res.* 106:996–1004.

Shukla, S.T., Habbu, P.V., Kulkarni, V.H., Jagadish, K.S., Pandey, A.R., Sutariya, V.N. 2014. Endophytic microbes: A novel source for biologically/pharmacologically active secondary metabolites. *Asian J. Pharmacol. Toxicol.* 2(3):1–16.

Song, Y.C., Huang, W.Y., Sun, C., Wang, F.W., Tan, R.X. 2005. Characterization of graphislactone A as the antioxidant and free radical-scavenging substance from the culture of *Cephalosporium* sp. IFB-E001, an endophytic fungus in *Trachelospermum jasminoides*. *Biol. Pharm. Bull.* 28(3):506–9.

Souza, J., Vieira, I., Rodrigues-Filho, E., Braz-Filho, R. 2011. Terpenoids from Endophytic Fungi. *Molecules* 16:10604–18. 10.3390/molecules161210604.

Specian, V., Sarragiotto, M.H., Pamphile, J.A., Clemente, E. 2012. Chemical characterization of bioactive compounds from the endophytic fungus *Diaporthe helianthi* isolated from *Luehea divaricata*. *Braz. J. Microbiol.* 43(3):1174–82.

Stepniewska, Z., Kuzniar, A. 2013. Endophytic microorganisms-promising applications in bioremediation of greenhouse gases. *Appl. Microbiol. Biotechnol.* 97(22):9589–9596.

Strobel, G.A. 2003. Endophytes as sources of bioactive products. *Microbes Infect.* 5(6):535–44.

Strobel, G.A., Hess, W.M., Li, J.Y., Ford, E., Sears, J., Sidhu, R.S., Summerell, B. 1997. *Pestalotiopsis guepinii*, a taxol producing endophyte of the Wollemi Pine, *Wollemia nobilis*. *Aust. J. Bot.* 45(6):1073–82.

Strobel, G., Li, J.Y., Sugawara, F., Koshino, H., Harper, J., Hess, W.M. 1999a. Oocydin A, a chlorinated macrocyclic lactone with potent anti-oomycete activity from *Serratia marcescens*. *Microbiology* 145:3557–64. doi:10.1099/00221287-145-12-3557

Strobel, G.A., Miller, R.V., Martinez-Miller, C., Condron, M.M., Teplow, D.B., Hess, W.M. 1999b. Cryptocandin, a potent antimycotic from the endophytic fungus Cryptosporiopsis cf. quercina. *Microbiology* 145(8):1919–26. doi:10.1099/13500872-145-8-1919

Taylor, J.E., Hyde, K.D., Jones, E.B.G. 1999. Endophytic fungi associated with the temperate palm *Trachycarpus fortunei* within and outside its natural geographical range. *New Phytol.* 142(2):335–46.

Tefera, T., Vidal, S. 2009. Effect of inoculation method and plant growth medium on endophytic colonization of sorghum by the entomopathogenic fungus *Beauveria bassiana*. *BioControl.* 54:663–9. doi:10.1007/s10526-009-9216-y

Vinale, F., Nicoletti, R., Lacatena, F., Marra, R., Sacco, A., Lombardi, N., d'Errico, G., Digilo, M.C., Lorito, M., Woo, S.L. 2017.Secondary metabolites from the endophytic fungus *Talaromyces pinophilus*. *Nat. Prod. Res.* 31(15):1778–85.

Wang, F.W., Jiao, R.H., Cheng, A.B., Tan, S.H., Song, Y.C. 2007. Antimicrobial potentials of endophytic fungi residing in *Quercus variabilis* and brefeldin A obtained from *Cladosporium* sp. *World J. Microbiol. Biotechnol.* 23(1):79–83.

Wang, S., Hu, T., Jiao, Y., Wei, J., Cao, K. 2009. Isolation and characterization of *Bacillus subtilis* EB-28, an endophytic bacterium strain displaying biocontrol activity against *Botrytis cinerea* Pers. *Front. Agric. China.* 3(3):247–52.

Wang, Y., Zeng, Q.G., Zhang, Z.B., Yan, R.M, Wang, L.Y., Zhu, D. 2011. Isolation and characterization of endophytic huperzine A-producing fungi from *Huperzia serrata*. *J. Ind. Microbiol. Biotechnol.* 38(9):1267–78. doi:10.1007/s10295-010-0905-4

Wani, M.C., Taylor, H.L., Wall, M.E., Coggon, P., McPhail, A.T. 1971. The isolation and structure of taxol, a novel antileukemic and antitumor agent from *Taxus brevifolia*. *J. Am. Chem. Soc.* 93(9):2325–7.

Wu, H., Yang, H.Y., You, X.L., Li, Y.H. 2013. Diversity of endophytic fungi from roots of Panax ginseng and their saponin yield capacities. *SpringerPlus.* 2:107.

Zhao, Y., Cui, J., Liu, M., Zhao, L. 2020. Progress on terpenoids with biological activities produced by plant endophytic fungi in China between 2017 and 2019. *Nat. Prod. Comm.* 15(7):1–18. doi:10.1177/1934578X20937204

12 Nanomaterials Augmenting Plant Growth-Promoting Bacteria and Their

Potential Application in Sustained Agriculture

Mohammed Azharuddin Savanur, Mallappa M.,
Syeda Ulfath Tazeen Kadri, and Sikandar I. Mulla**

CONTENTS

12.1 INTRODUCTION

Limited resources and abrupt changes in climatic conditions are major challenges for the global agriculture sector. Currently, synthetic agrochemicals are employed to contain phytopathogens and increase crop yield (Damalas and Eleftherohorinos

---

* Corresponding authors.

DOI: 10.1201/9781003213864-12

2011). However, indiscriminate and excessive use of these chemicals further aggravates global warming and causes hostile consequences such as increased resistance in pathogens and their expansion and adverse conditions for favorable microbes, which pose risks for human health and the environment (Cavicchioli et al. 2019; Chauhan et al. 2014). Moreover, abiotic stress, such as pollution, drought, salinity, frequent infections by phytopathogens, and paucity of nutrients, also causes crop production to decline, exacerbating food security issues (Pandey et al. 2017). Changing environmental conditions will significantly vary the reliability and effectiveness of currently employed conventional agriculture practices, including chemical and biological controls (Barratt et al. 2018; Xu 2016). Hence, ingenious and sustainable approaches that combine agricultural and environmental strategies are essential to successfully circumvent the adverse influence of climate change and low crop yields.

Plant growth-promoting bacteria (PGPB) are root colonizing, heterogeneous bacteria that enhance plant growth by either direct or indirect phyto-stimulatory mechanisms. They provide protection to plants against biotic and abiotic stress, augment phytoremediation, and increase soil biodiversity (Olanrewaju, Glick, and Babalola 2017b). Usually, PGPB are inoculated as suspensions on seed, root, and soil surfaces (de Souza, Ambrosini, and Passaglia 2015). Sometimes, bacterial populations decline rapidly because of unfavorable conditions, thus, reducing their efficacy in the rhizosphere. The major problem in exploring PGPB efficacy in the field is their variation in performance under diverse climatic conditions, different parameters, and varied characteristics of soil (Timmusk et al. 2017a). Therefore, to ensure PGPB efficacy in diverse conditions, there is a need to develop efficient and effective methods that will provide a favorable microenvironment, physical protection, and reliable, sustainable application in the natural environment (Timmusk et al. 2017a). A detailed study of the way PGPB interacts with the host and surrounding environment is required to develop a PGPB formulation that will ensure their survival, availability to crops, and provide security against pathogens.

Nanotechnology is an emerging field in sustainable agriculture, which enhances biotic and abiotic stress tolerance and disease prevention and provides refined nutrients to plants (Fu et al. 2019). Plant root exudates that are secreted in the rhizosphere attract many microorganisms and dissolve minerals, thus, providing a mechanism for producing various mineral nanoparticles in the soil that are biocompatible (Timmusk, Seisenbaeva, and Behers 2018). Mineral nanoparticles have a large surface area and absorb various molecules; these nano-bio interaction systems are known to promote plant growth when used in combination with PGPB (Mahawar and Prasanna 2018). Thus, nanoparticle (NP)–PGPB interactions can be explored to deliver beneficial bacteria and their active compounds in a regulated manner. It is essential to understand the value of potential NP-PGPB interaction for successful use of nanomaterials (NMs) to achieve their optimal activity in agriculture formulations, as these formulations can be integrated into developing agriculture practices to ensure sustainable crop production.

12.2   PERSPECTIVE OF PGPB IN SUSTAINED AGRICULTURE

Many bacteria found in the rhizosphere harbor plant roots and promote their growth; such species are consequently termed plant growth-promoting bacteria or PGPB.

Various strains of PGPB can colonize on the root surface and form nodules, while others exist within tissues as endophytes (Glick 2012). These bacteria can be categorized as biofertilizers because they can mineralize and boost the availability of essential nutrients through nitrogen fixation, solubilization, and mobilization of phosphates. They can also be categorized as biocontrol agents or biopesticides because they suppress or control phytopathogens and provide resistance against biotic and abiotic stress (Timmusk et al. 2017b). Because of their positive interactions with host plants, PGPB may be a promising and sustainable tool for improving agriculture production. The use of environmentally unsafe chemical fertilizers that contaminate and leach nutrients from the soil can be overcome by using environmentally friendly, low cost, and highly productive alternatives, such as PGPB (Glick 2012), which will help to achieve the goal of sustainable agriculture worldwide. Various strategies for consortium formulations of PGPB aim to enhance the desired benefits and agricultural production.

12.2.1 Insights of Plant–PGPB Interactions

The release of photosynthetically assimilated carbon in the form of lipids, carbohydrates, amino acids, enzymes, growth factors, and organic acids via plant roots provide a substantial amount of nutrients to soil microbes, thus, harboring diverse microorganisms that may colonize surface areas, areas adjacent to the surface, and apoplastic spaces (Wille et al. 2019). In response, these microbes secrete metabolites that may benefit plant growth, provide resistance against biotic and abiotic stress, or cause detrimental effects (Schirawski and Perlin 2018). The communication between plant roots and associated microbes is an intricate process involving chemical interactions that can be stimulative for biofilm formation or suppressive for pathogens (Rudrappa, Biedrzycki, and Bais 2008). These cross-talks between plants and microbes are not just a differentiation between friend/foe, but rather, the phenotypic outcome relies on the presence of particular strain and plant genotype (Ntoukakis and Gifford 2019). The association of microbes with plant roots and the availability of root exudates are also affected by the physicochemical properties of soil, including the availability of nutrients, content of organic matter, structure, pH, and texture (Ho 2017).

Copious amounts of signaling molecules are secreted by host plants and associated rhizobacteria, which regulates the expression of genes in both species. The exuded signaling metabolites facilitate chemical communications and establish mutual relationships (Smith et al. 2015; Venturi and Keel 2016). Plants recognize microbes by employing specific intracellular and extracellular receptors termed pattern recognition receptors (PRRs) and resistance genes (Ntoukakis and Gifford 2019). Pathogen entry is prevented by identifying small conserved molecular motifs within microbes known as microbe/pathogen-associated molecular pattern (PAMP) using PRRs and, thus, triggering immune responses in plants. PAMP-triggered immunity (MTI) is downregulated by the emission of effector molecules, which instigate effector-triggered susceptibility (ETS) (Jones and Dangl 2006). The ETS is reduced via secretion of resistance proteins by the host to identify effector molecules of pathogens and initiate a vigorous and rapid second line of immune response referred to as effector-triggered immunity (ETI), which is also coupled with hypersensitive response (De Coninck et al. 2015). Nuclear or cytosolic proteins from damaged, necrotic, and stressed cells, labeled as danger/damage-associated molecular patterns (DAMPs) or alarmin, are also recognized by plant PPRs and are countered by immune signaling cascades (Mhlongo et al. 2018).

For PGPB to create an efficient symbiotic relationship with plants, it is obligatory to circumvent the aforementioned immune barriers by means of synergic chemical interactions. Analogous molecular strategies are deployed by both pathogens and symbionts to reduce and succeed the immune attack. The symbiotic interaction between PGPB and the host plant also employs the same PRR-MAMP detection system that aids in harboring beneficial microbes (Hacquard et al. 2017). Acyl homoserine lactones (AHSL), which is involved in the quorum sensing process, plays a vital role in regulating the phenotype and genotype of bacteria that colonize plant roots. Flavonoids such as 2 phenyl-1,4-benzopyrone derivatives found in exudates of plant roots are able to imitate quorum sensing molecules of bacteria, thus, influencing their metabolism and helping to establish symbiotic relationships with plants (Hassan and Mathesius 2012). These findings suggest that susceptibility, resistance, and mutualism play a significant role in establishing symbiotic relationships between PGPB and host plants.

12.2.2 MECHANISMS OF ACTIONS OF PGPB

PGPB can elevate plant growth directly either by regulating plant hormones or by assisting in resource acquisition and indirectly by inhibiting phytopathogenic bacterial growth or their consequences. The direct mechanisms include synthesis of phytohormones, such as auxin, which play a principal role in synchronization of many growth and behavioral processes, such as hydrotropism, geotropism, phototropism, differentiation of vascular tissue, apical dominance, cell division, and lateral and adventitious root initiation (Abel and Theologis, 2010). Plants' auxin pools are regulated by PGPB indole-3-acetic acid (IAA) and can have more than one biosynthetic pathway (Goswami, Thakker, and Dhandhukia 2016; Çakmakçı et al. 2020). PGPB produced cytokines regulate cell division and shoot formation, delay tissue senescence, counter apical dominance, influence development of fruits and flowers and plant–pathogen interactions, and improve drought tolerance (Schaller, Street, and Kieber 2014; Hassen, Bopape, and Sanger 2016). Gibberellins, along with other hormones, act as transducers of elicitor signals and are similar to those of plants. They stimulate and activate important growth-related processes, which include germination of the seed, elongation of the stem, flowering, and enhance chlorophyll content and rate of photosynthesis (Vejan et al. 2016). PGPB downregulates 1-aminocyclopropane-1-carboxylic acid (ACC) deaminase and ethylene production, thereby reducing the detrimental effects caused by biotic stress in plants (Glick 2014). Root-colonizing PGPB fix atmospheric nitrogen using nitrogenase enzymes, thus, providing essential nutrients to plants (Geddes et al. 2015). Organic acids, such as gluconic or keto gluconic acids, secreted by phosphate solubilizing PGPB, along with carboxyl and hydroxyl ions, convert inorganic phosphate to phosphate ions, thus, making minerals available to plants (Alori, Glick, and Babalola 2017).

PGPB may also promote plant growth by indirect means such as synthesis of anti-biotics, including 2,4-diacetyl phloroglucinol, pyoluteorin, phenazine-1-carboxyclic acid, oomycin, pyrrolnitrin, pantocin, kanosamine, and zwittermycin-A, via regu-lation of endogenous signals such as AHSL, sensor kinases, and sigma factors (Olanrewaju, Glick, and Babalola 2017). PGPB are known to synthesize cell-wall degrading enzymes such as protease, chitinase, and lipase that will lyse pathogen cells to prevent infection (Veliz, Martínez-Hidalgo, and Hirsch 2017). Siderophores producing PGPB can reduce the availability of iron for pathogens and avert their growth (Ferreira, Soares, and Soares 2019). In addition, some PGPB can fight the pathogens by competing for nutrients and limiting their binding sites (Olanrewaju, Glick, and Babalola 2017). Hydrogen cyanide synthesis by PGPB can act synergisti-cally with antibiotics and prevent the development of resistance in pathogens (Saraf, Pandya, and Thakkar 2014). PGPB can prime the host plants against consequent pathogen attacks by urging induced systemic resistance and alleviating detrimental effects (Rashid and Chung 2017). PGPB can disrupt the quorum sensing phenomena by producing lactonase, which degrades inducers of quorum sensing, thus, impeding the virulent properties of pathogens (Bauer and Mathesius 2004).

12.3 PERSPECTIVE OF NMS IN SUSTAINED AGRICULTURE

Sustained crop production demands the availability of optimal nutrients and is an indispensable area of research in agriculture. This essentiality can be achieved by exploring the field of nano nutrients, which provide adequate nutrition for crops and support sustainable agriculture. The versatile physicochemical properties of NMs make them able to operate in different fields such as electronics, life sciences, and chemical engineering (Jeevanandam et al. 2018). Recently, agriculture research has explored nanotechnology to develop minuscule competent systems for delivering vital molecules to plants, which will improve seed germination, plant growth, and protection against biotic and abiotic stress conditions (Sanzari et al. 2019). For the NMs to be potentially effective in agriculture applications following aspects such as their biocompatibility, the nature of interaction with other molecules, the mechanism of uptake and transport in plants, and methodology of implementation need to be considered carefully (Salme and Timmusk 2017). These features enhance the effi-cacy of NMs in producing formulations for agriculture that will facilitate controlled delivery of agricultural inputs at the target site.

Designing and synthesizing NMs that have unique and ideal characteristics is the major challenge proposed in modern nanotechnology research. The miniature form of nanoparticles provide a large surface area and higher activity, and thereby provide excellent potential for developing advanced nanotools (Koh 2007). The current prog-ress in NM engineering has yielded different morphology and particle sizes that have a wide range of applications in industries such as health care, agriculture, environ-ment, and food processing (Bandala and Berli 2019). In agriculture, NMs are used as fertilizers through the controlled and slow release of essential elements for crop growth and development and to prevent nutrient deficiency in the soil (Solanki et al. 2015; Sanzari et al. 2019). Nano pesticides are used to control pests and diseases, suppressing the activity of pests and infections caused by pathogens (Elmer and White 2018). In addition, NMs also induce plant-based systemic resistance against pests. NMs are implicated as carriers for the delivery of agrochemicals, organic molecules, and fertilizers to the plants (Adisa et al. 2019). They are also employed in plant genetic engineering in which they deliver foreign DNA and chemicals that modify target genes. Nano-sensors are used to measure and monitor many parameters such as soil condition, nutrient deficiency, crop growth, toxicity, entry of chemicals in the environment, and diseases, which will assist in ensuring the health of the soil and plants, quality of the product, and overall safety of the environmental systems (Srivastava, Dev, and Karmakar 2018). NMs provide protection against the effects of climate change, including drought, temperature, salinity, and pollution by activation of specific enzymes, regulation of hormones, expression of stress genes, and balancing metal uptake (Zhao et al. 2020). Thus, for sustained agriculture, NMs are used to ameliorate the growth and production of crops and to increase the quality without affecting environmental conditions.

12.3.1   INSIGHTS OF NMS–PGPB INTERACTION

The plant metabolism's ability to dissolve the surrounding minerals using root exudates is imperative and essential for their adaptation in stress conditions. These exudates have diverse roles in communication between soil microorganisms and plants. They are also involved in generating biocompatible nanoparticles through the release of chelating organic acids and anions, which enhance the aggregation of single cells and act as glue between plant and microbes (Timmusk, Seisenbaeva, and Behers 2018). Therefore, for the PGPBs and their bioactive compounds to be more effective, they need to be delivered at the target site by favorable means. Biocompatible nanoparticles can be employed for reproducible and stable application of PGPBs in the field. In order to use nanoparticles as carriers, it is indispensable to understand their interactions with root colonizing planktonic bacterial cells and their biofilms. These interactions can be described as the attachment of nanoparticles to the microbial cell surface and their transport within cells and biofilms, which aids in developing agricultural formulations (Shakiba et al. 2020).

The physicochemical characteristics of both nanoparticles and PGPBs determine their fate of interactions. Physicochemical properties, such as size, surface charge, morphology, hydrophobicity, and functional groups of nanoparticles, influence their interaction with both bacterial surface molecules and the extracellular polymeric substances (polysaccharides, proteins, lipids, and extracellular DNA) of biofilms (Singh et al. 2019). Upon interaction, different organic molecules are adsorbed on the surface of the nanoparticles, forming a corona-like coating. The properties of the corona that is formed determine their physiological behavior, such as solubility, aggregation, extent of particle uptake, mechanism of toxicity, and interactions with microbes, signifying that nanoparticle remains pristine for a very short time in biological environments (Fulaz et al. 2019; Ikuma, Decho, and Lau 2015). Because these interactions take place in a very active physiological environment, leading to a cascade of collateral phenomena, their exact role and mechanisms are not yet fully uncovered.

In addition to chemical and biological interactions, the electrostatic, hydrophobic, and steric forces are important physical factors that dictate the interactions between nanoparticles and bacteria (Westmeier et al. 2018). Generally, electrostatic interactions are involved in the adhesion of bacteria to the nanoparticle surface, in which positively charged nanoparticles are known to interact with negatively charged bacteria and their extracellular polymeric substances (Huangfu et al. 2019; Ikuma, Decho, and Lau 2015). Hydrophobic interactions were shown to be much stronger than hydrophilic, in which hydrophobic extracellular polymeric components contributed to stability (Renner and Weibel 2011). Steric interactions prevent aggregation of nanoparticles even in higher ionic strength solutions, adding to the colloidal stability of nanoparticles (Huangfu et al. 2019; Ikuma, Decho, and Lau 2015). These findings suggest that not just the net charge but also the synergistic action of both functional groups and other physical forces influence the nanoparticle–bacteria interactions.

12.3.2 Impact of NMs on PGPB

Various nanoparticles that enhance PGPB and their effects on plants are illustrated in Table 12.1, and the impact of nanoparticles on PGPB and their combined physiological effects on plants are shown in Figure 12.1.

12.3.2.1 TiO₂NPs

Studies on the interaction of titanium dioxide nanoparticles (TiO_2NPs) and plants have gained significant importance in the field of nanotechnology and multifarious agricultural applications. The proposed agricultural applications include TiO_2NPs as fertilizers, pesticides, bio-stimulants, and others such as assisting seed germination and growth (Tan, Peralta-Videa, and Gardea-Torresdey 2018). The exudates of plant root systems can biodegrade TiO_2NPs in the presence of chelating carboxylate ligands, such as citrate, oxalate, and lactate, which help convert nanoparticles into naturally abundant soil minerals (Seisenbaeva et al. 2013). Research studies reporting an interaction between nanoparticles and bacteria are mainly restricted to toxicity. Recently, the idea of nanoparticle–biofilm interactions elicited the use of TiO_2NPs as agents that synergizes the PGPR–plant interactions.

Physicochemical properties, such as zeta potential, surface charge, size, and morphology of nanoparticles, greatly influence their functional and biological effects (Singh et al. 2019). The zeta potential of TiO_2NPs varies (positive to negative) when there is an increase in pH and the bacterial surface has a net negative charge, highlighting the importance of other factors, such as hydrophobic and steric forces, apart from electrostatic, as they also play a vital role in the interaction of both (Palmqvist et al. 2015). Bacteria–TiO_2NP interaction studies have demonstrated a higher microbial cell attachment and increased metabolism in TiO_2NPs, which have a greater surface area (Park et al. 2008). Furthermore, the natural organic matters, such as humic acid and bacteria in effluent aquatic systems, contribute to the aggregation

TABLE 12.1

Nanoparticles Augmenting PGPB and Their Effects on Plants

Nano materials	PGPB	Plants	Effects: Bacteria	Effects: Plants	Reference
TiO_2	*Bacillus amyloliquefaciens* UCMB5113	Oilseed rape plants (*Brassica napus*)	Supported bacterial adhesion, promoted root colonization, and increased survival and fitness	Protected plants against fungal pathogen *Alternaria brassicae*	Palmqvist et al. 2015
	Bacillus thuringiensis AZP2, *Paenibacillus polymyxa* A26, *Alcaligenes faecalis* AF	Wheat (*Triticum aestivum* cv. *Stava*)	Supported bacterial adhesion, promoted root colonization, and increased performance	Increased seedling and plant biomass, enhanced drought and stress tolerance and protection against *Fusarium culmorum* pathogens	Timmusk et al. 2018
ZnO	*Pseudomonas chlororaphis* O6 (PcO6)	Bean (*Phaseolus vulgaris*)	Increased siderophore production, reduced phenazine and AHSLs production and quorum sensing	Limited growth of plant pathogens, reduced accumulation of toxic metals, provided nutritional benefits from iron uptake, and stimulated systemic resistance	Anderson et al. 2018
CuO	*Pseudomonas chlororaphis* O6 (PcO6)	Wheat (*Triticum aestivum* cv. *Stava*)	Increased IAA and nitric oxide production, reduced siderophore production and promoted biofilm formation	Plant growth promotion, enhanced metal phytoremediation, modified root morphology, and enhanced lignifications and drought resistance	Anderson et al. 2018
Ag	*Pseudomonas aeruginosa* PAO1 *Pseudomonas fluorescence, Bacillus cereus*	Maize (*Zea mays*)	Promoted biofilm formation, lipopolysaccharide synthesis, and upregulation of quorum sensing	Increased root area and length, phytohormone production, oxidative stress resilience, and bioremediation potential	Yang and Alvarez 2015; Khan and Bano 2016

(Continued)

TABLE 12.1 (CONTINUED)
Nanoparticles Augmenting PGPB and Their Effects on Plants

Nano materials	PGPB	Plants	Effects Bacteria	Plants	Reference
SiO$_2$	Pseudomonas fluorescens VUPF5 Bacillus subtilis VRU1	Pistachio UCB-1	Increased auxin (IAA) production	Enhanced the root and shoot length, increased chlorophyll content, enhanced phytohormone and metabolite absorption	Pour 2019
	Bacillus megaterium, Bacillus brevis, Pseudomonas fluorescens, Azotobacter Vinelandii	Maize (Zea mays)	Promoted growth and total soil bacterial population	Enhanced seed germination	Karunakaran et al. 2013
	Citrobacter freundii ATHM38	Tomato (Solanum lycopersicum)	Production of polysaccharides and phytohormones, and reduced ethylene content	Reduced proline content and antioxidant enzyme activities and increased plant growth under salinity stress.	Moshabaki Isfahani et al. 2019
Au	Pseudomonas monteilii	Cowpea (Vigna unguiculata)	Enhanced IAA production and prebiotic effect	Plant growth promotion	Panichikkal et al. 2019

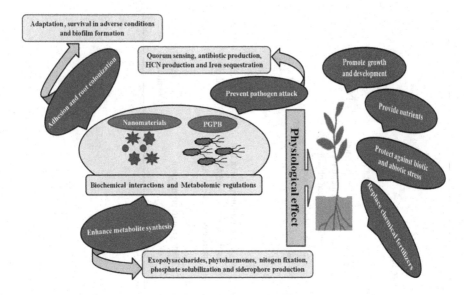

FIGURE 12.1   Impact of nanomaterials on PGPB and their combined physiological effects on plants.

and deposition of TiO_2NPs (Chowdhury, Cwiertny, and Walker 2012). Such properties of TiO_2NP are explored to support PGPB which enhance their strength as well as resistivity.

Palmqvist et al., (2015) examined the novel application of TiO_2NPs in the nano–interface interaction between PGPB (*Bacillus amyloliquefaciens* UCMB5113) and oilseed rape plants (*Brassica napus*) for protection against a fungal pathogen. The study shows the use of two different types (CaptiGel and TiBALDH) of biocompatible, negatively charged, and identically sized (3–3.5 nm) TiO_2NPs that had been synthesized using the Sol Gel approach (Groenke et al. 2012; Seisenbaeva et al. 2013). The surface negative charge of TiBALDH- and CaptiGel-derived particles were stabilized by lactate ions and triethanolamine ligands, which are involved in the formation of inner-sphere complexes with phosphates of exposed membrane phospholipids in bacteria and plant roots, thus, explaining the mechanism of interaction. Experimental evidence demonstrated that TiO_2NPs first facilitated adhesion and then colonized bacteria to the roots, subsequently protecting against fungal infections.

In accordance with the above results, another study strongly correlates and highlights the role of TiO_2NPs as agents mediating nano–interface interactions between winter wheat plants (*Triticum aestivum* cv. Stava) and PGPB (*Bacillus thuringiensis* AZP2 and *Paenibacillus polymyxa* A26) (Timmusk, Seisenbaeva, and Behers 2018). CaptiGel TiO_2NPs produced by the Sol Gel method, as discussed earlier, were used to formulate PGPB strains, and these nanoparticles were shown to be fully biocompatible without any negative bioeffects in tobacco at high concentrations (100 µg/ml). Double inoculants of PGPB–TiO_2NPs formulations in peat soil significantly increased seedling biomass under stress conditions, demonstrating the importance of nanoparticles in forming stable bacterial microniche and their root attachment. The significance of the method of inoculum delivery in soil systems using TiO_2NPs was highlighted, considering the existence of natural soil nanoparticles when applying TiO_2NPs formulations.

From these findings it is apparent that TiO_2NPs provide an efficacious platform for plant growth-promoting rhizobacteria (PGPR) application via the root microbial community and, thus, serves as a tool for the design of biofilms that aid in resolving food security issues. However, delineating signaling pathways will enhance the efficiency of TiO_2NPs and help to improve plant–PGPB interactions, allowing their regulated delivery at specific tissues or cells at specific times.

12.3.2.2 ZnONPs

Zinc is an essential micronutrient that plays a vital role in many physiological processes such as growth, photosynthesis, metabolism, and immunity. It is an integral part of many biomolecules use to facilitate important functions and stability (Singh et al. 2018). The nanoforms of zinc (ZnONPs) have unique optoelectrical, physical, and chemical properties because of the large surface-area-to-volume ratio, offering various potential applications in the food, pharmaceutical, and agriculture sector (Mishra and Adelung 2018; Sabir, Arshad, and Chaudhari 2014). ZnONPs are used as fertilizers at lower concentrations, as they provide essential elements for cellular functions, and as pesticides at higher doses due to dose dependent toxicity. They also help in seed germination and pigmentation and boost stress resilience and, thus, have tremendous potential to improve the yield and growth of food crops (Reddy Pullagurala et al. 2018). Although antimicrobial properties of ZnONPs have been extensively analyzed, their beneficial interactions with root-colonizing bacteria have gained attention recently and require further understanding of how rhizosphere bacteria modulate the effects of nanoparticles that are favorable for plants.

Research studies pertaining to the interaction of ZnONPs with bacteria illustrate that, at the cellular, biochemical, and transcriptome levels, these nanoparticles influence key metabolites, enzymes, and signaling molecules that are involved in communication between plant and bacterial cells for functioning in the rhizosphere (Goodman et al. 2016). ZnONPs at lower concentrations (≥ 10 mg metal/kg) interact with beneficial root-associated microbes and alter the synthesis of phenazines (elicitors that induce resistance in plants against pathogens), siderophores (iron-chelating compounds that increase iron bioavailability), acyl homoserine lactones (signaling molecules involved in quorum sensing), and IAA (plant growth regulator) as well as the activity of phosphate mobilizing enzymes, such as phosphates and phytase, impacting growth and resistance in plants (Anderson et al. 2018).

Pseudomonas chlororaphis O6 (PcO6) is a pseudomonad that colonizes many plant roots and confers protection against pathogens such as viruses, bacteria, and fungi by producing antibiotics such as phenazines (Loper et al. 2012; Weller 1988). The synthesis of phenazine is regulated by AHSL, which are produced by activating the Gac/Rsm signal transduction pathway. The AHSL molecules reprogram bacteria upon reaching high cell densities to synthesize phenazine, a phenomenon termed as quorum sensing (Papenfort and Bassler 2016). The interaction of ZnONPs with PcO6 affects the quorum-sensing process by reducing the production of phenazine and AHSL, altering the metabolic activities. Furthermore, the changes in the production of quorum-sensing molecules are linked to iron (Fe) metabolism and increased pyoverdine-like siderophore secretion (Dimkpa et al. 2019; Goodman et al. 2016). Siderophores are ferric (Fe^{3+}) ions binding compounds that are secreted by bacteria to transport iron across cell membranes, thus, serving in biofilm formation (Visca, Imperi, and Lamont 2007). The increase in siderophore production by PcO6 upon ZnONPs treatment may be attributed to negative regulation by the Gac/Rsm pathway. The transcriptional regulation of pyoverdine biosynthetic genes is correlated with the intracellular levels of the signaling compound c-di-GMP that are governed by the translation inhibitory protein RsmA, explaining the role of the controlled network in the observed effect (Dimkpa et al. 2015; Anderson et al. 2018).

Thus, the alterations of key metabolites elicited by ZnONPs in PcO6 bacteria may potentially benefit plants in the following ways. The enhanced siderophore production increases chelation efficiency for Fe, which will limit the growth of other pathogenic microbes by reducing the Fe availability. Fe-loaded pyoverdine-like siderophores supplement Fe accessibility to plants, thus, providing nutritional benefits. Siderophores are also known to trigger the plant's defense system by increasing immunity through the induced systemic resistance mechanism. Plant growth-promoting rhizobacteria synthesize phosphatases and phytase, as these enzymes play an important role in the mineralization and acquisition of organic phosphate (Singh and Satyanarayana 2011; Shulse et al. 2019). ZnONPs were shown to enhance phosphatases and phytase production, as zinc acts as a cofactor for these enzymes and aids in their activity, reinforcing nutritional efficacy (Raliya, Tarafdar, and Biswas 2016). Collectively, these findings demonstrate that ZnONPs modify the cell signaling process of bacteria by altering the production of key metabolites that ensure beneficial effects of PGPB on plants. However, further study is required to determine the observed effect of ZnONPs on other antibiotics and metabolites produced by bacteria as well as their impact on plants.

12.3.2.3 CuONPs

Like zinc, copper is also an indispensable micronutrient that plays a key role in the growth and development of plants. Being a constituent of various enzymes, it is involved in many physiological processes, such as electron transport, respiration, metabolism, protein trafficking, and iron mobilization (Pilon et al. 2006; Festa and Thiele 2011). The chemical and physical structure of soil, including pH and inorganic matter, decides the solubility and availability of copper. The paucity of copper in soil causes an obstruction to growth and productivity as well as sterility in plants (Purves and Ragg, 1962; White and Greenwood, 2013). A deficiency of copper is overcome by using the nanoform of copper (CuONP). CuONPs are being used as fertilizers to supply this essential element, as pesticides at higher doses, and to induce resistance against drought stress (Adisa et al. 2019). Properties of CuONPs, such as chemical reactivity, electrical conductivity, physical resistance, and optical activity, are related to their size and surface area (Chen 2018). At higher concentrations, these nanoparticles produce stress and toxicity due to the generation of reactive oxygen species (ROS), perturbing cellular metabolism. At optimal dosage, they generate metabolic effects that signal transcriptional regulation and production of antioxidants (glutathione, vitamin C, peroxidase, superoxide dismutase, and catalase), leading to stress tolerance in plants (Dimkpa, McLean, Britt, and Anderson 2012; Sarkar et al., 2020). CuONPs are also known to enhance the production of secondary bioactive metabolites, such as flavonoids, phytoalexins, gymnemic acid II, and phenolic compounds, which play an important role in communication, functioning, adaptation of plants, and symbiotic interactions with beneficial microbes (López-Vargas et al. 2018; Chung et al. 2019; Sarkar et al., 2020). They also increase lignification in roots by altering peroxidase activity and manipulate root morphology that enhances surface area, providing more space for microbe colonization (Jacobson et al. 2018). These beneficial effects can be harnessed to increase the yield of desired metabolites by using CuONPs as elicitors in molecular pharming.

Although direct interactions of CuONPs with plants are being progressively studied, their influence on root-colonizing probiotic bacteria is also an important aspect to be considered, as they play a major role in plant growth and immunity. Like in plants, CuONPs are also known to influence the metabolism of root-colonizing bacteria by modifying the production of signaling metabolites. These nanoparticles reduced the production of pyoverdine-like siderophores in PcO6 bacteria by lowering the levels of proteins and cytoplasmic transporters involved in maturation of pyoverdine precursors (Dimkpa, McLean, Britt, Johnson, et al. 2012; Dimkpa, Mclean, et al. 2012). Contrarily, CuONPs enhanced the production of IAA (plant hormone involved in growth and development) by negative regulation of Gac/Rsm pathway (Dimkpa, Zeng, et al. 2012; Anderson et al. 2018). This elevation in IAA production by PcO6 bacteria modifies root morphology, promotes growth, and supplements phytoremediation of metals (Adams et al. 2017).

Anderson et al. (2018) demonstrated that PcO6-colonized wheat roots displayed enhanced patterns of lignification in root and shoot tissues upon CuONP treatment. Increased lignification was attributed to increased production of lignin and pectin, which may lower water flow in the apoplast and decrease shoot water content, priming the plants for drought tolerance (Jacobson et al. 2018). CuONP-induced lignification of sclerenchyma cells is related to improved resilience against pathogen invasion and sturdy shoots that display higher resistance to wind and prevent lodging (Jacobson et al. 2018). Increased production of nitric oxide was also observed in PcO6 upon CuONP treatment, which signifies enhanced lignification and induces drought protection and copper tolerance. The interaction of PcO6 with CuONPs resulted in increased bacterial size and alginate production by regulation of c-di GMP levels, signaling cells to switch to a biofilm lifestyle and intracellular granule formation (speculated to be polyhydroxybutyrate – storage products formed when there is an excess of carbon) (Bonebrake et al. 2018). These findings show that CuONPs induced increased lignification and enhanced physical strength, nutrition to tissues, and drought tolerance, and synergism with probiotic bacteria involves the beneficial aspects of these nanoparticles, which will improve plant growth and development.

12.3.2.4   AgNPs

Silver nanoparticles (AgNPs) have both positive and negative consequences on plants and associated bacteria, providing diverse agricultural applications. Physiochemical properties, such as size, shape, stability, surface area, reactivity concentration, and type of surface coating, determine the impacts of these nanoparticles. Above optimal concentrations, these nanoparticles tend to generate ROS, alter photosynthesis, and damage DNA, leading to stress and cell death in plants (Yan and Chen 2019). Contrastingly, lower doses of AgNPs confer beneficial effects on plants, such as elevating seed germination, enhancing shoot and root length, accumulating nutrient solutes, increasing chlorophyll content, augmenting the photosynthetic process, ameliorating the antioxidant defense system, and imparting resilience against stress conditions (Sami, Siddiqui, and Hayat 2020). Understanding the interaction of AgNPs with plants will elevate their potential use as an elicitor or biostimulant in crop production.

AgNPs are generally considered to be antimicrobial agents because of their lethal effects on bacteria, fungi, and viruses, thus, they serve as a potential pesticide in agriculture. The proposed toxicity is mainly attributed to the release of silver ions that cause disruption to the cell membrane and obstruct the cellular functions and generate intracellular ROS (Yan and Chen 2019). However, the antimicrobial properties of AgNPs may change when encountering a complex array of biotic and abiotic processes (Sheng et al. 2018). Nevertheless, few studies have focused on the interaction of PGPR with AgNPs, especially in the perspective of plant growth and development. These studies also show how nanoparticles benefit plant bacteria by promoting their growth and sustainability. Yang and Alvarez (2015) reported that sublethal doses of AgNPs (10.8 and 21.6 µg/L) promoted biofilm formation, lipopolysaccharide synthesis, and upregulated quorum sensing in *Pseudomonas aeruginosa* PAO1 by increasing sugar and protein contents and transcriptional upregulation of antibiotic resistance (efflux pump) genes as evidenced by the biofouling and biocorrosion processes. Another study showed that AgNPs improved the PGPR-induced increase in root area and length; root–shoot ratio; activities of peroxidase and catalase; and production of proline, abscisic acid, IAA, and gibberellic acid as well as confronted oxidative stress and bioremediation potential (Khan and Bano 2016). At a sublethal concentration, AgNPs were also shown to activate ROS scavenging systems, increase the production of proteins related to quorum sensing, and enhance the production of extracellular polysaccharides and inorganic phosphate solubilization in biofilms of *Bacillus subtilis* as a synchronized response to stress (Gambino et al. 2015).

12.3.2.5   Other NMs

Silicon nanoparticles are known to influence plant metabolic activities and promote growth, owing to their unique physiochemical properties. Studies of silica nanoparticles demonstrated that they are not toxic to the bacterial community in soil as compared to other sources; rather they elevate PGPB growth by maintaining the pH of soil, enhancing nutrient values and supporting seed germination (Kannan 2013). Nanoencapsulation of PGPB and their metabolites using silica nanoparticles are shown to improve pistachio micropropagation by absorbing phytohormones and other metabolites (Pour 2019). Another study reports that combined treatment of silicon nanoparticles and PGPB enhanced plant growth under salt stress by lowering proline content and modulating the activities of redox enzymes (Al-Garni, Khan, and Bahieldin 2019; Moshabaki Isfahani et al. 2019). Gold nanoparticles and their nano-formulations are gaining huge attention in agricultural applications because of their biocompatible traits. Interaction studies of gold nanoparticles with PGPB demonstrated that these particles significantly enhance the growth of beneficial bacteria and IAA production (Shukla 2015; Panichikkal et al. 2019). The mechanism by which gold nanoparticles enhance PGPB growth remains obscured. Nanochitosan combined with PGPB has been shown to enhance seed germination and promote growth and organic acid production in plants under stress conditions (Khati et al. 2017). Preseed treatment of molybdenum nanoparticles combined with PGPB is known to improve the physiology of host plants and the structural diversity of root-associated microbes by inducing changes in root exudates (Shcherbakova et al. 2017).

12.4 CONCLUSION AND FUTURE DIRECTIONS

Sustainable agriculture imparts a medium for the coordinated existence of biotic and abiotic factors in the natural environment. The symbiotic association of PGPB with plants signifies a promising and sustainable tool to improve agriculture production because these bacteria promote plant growth by imparting several beneficial features. Formulations consisting of PGPB, and other methodologies are being investigated to boost their agriculture applications and maintain the natural ecosystem. In view of combining different methodologies with PGPB, nanotechnology provides an efficient and potential platform for developing bacterial-based nano-formulations that will transform traditional agriculture practices into precision agriculture, thus, ensuring food and environment security. In order to develop successful and effective bacterial nano-formulations, it is imperative to investigate their biochemical interactions and physiological impacts. The physiochemical properties of NMs and bacteria govern their biochemical interactions and physiological effects. The composition and chemistry of soil and plant root exudates also play a major role in determining these interactions.

NMs–PGPB interactions enhance root colonization of bacteria, synthesis of metabolites and phytohormones, overall performance of PGPB, and aid in replacing chemical fertilizers. To optimize biofunctionalization and integration of these agricultural formulations, NM engineering needs to consider aspects such as their biocompatibility, surface coating, incorporation of biochemicals, mechanism of interaction, and multidisciplinary approaches. Available techniques and paradigms of nanotechnology and microbiology need to be collectively used in experimental systems. Specific chemical recognition of molecules that drive the NM–PGPB complex formation need to be thoroughly studied. The biomolecular coronas must be considered when dealing with their application in different physiological environments. Distinguishing between direct and indirect effects of NM–PGPB interactions will provide a complete picture of the mechanism of action. Additionally, research on NM–PGPB biofilm interactions need to be focused, as bacteria occur naturally in biofilms rather than planktonic or sessile cells.

REFERENCES

Abel, S., and Theologis, A. 2010. Odyssey of Auxin. *Cold Spring Harbor Perspectives in Biology* 2: a04572. Cold Spring Harbor Laboratory Press, New York, USA.

Adams, J., Wright, M., Wagner, H., Valiente, J., Britt, D., and Anderson, A. 2017. Cu from Dissolution of CuO Nanoparticles Signals Changes in Root Morphology. *Plant Physiology and Biochemistry* 110: 108–117.

Adisa, I. O., Pullagurala, V. L. R., Peralta-Videa, J. R., Dimkpa, C. O., Elmer, W. H., Gardea-Torresdey, J. L., and White, J. C. 2019. Recent Advances in Nano-Enabled Fertilizers and Pesticides: A Critical Review of Mechanisms of Action. *Environmental Science: Nano* 6: 2002–2030.

Al-Garni, S. M. S., Khan, M. A., and Bahieldin, A. 2019. Plant Growth-Promoting Bacteria and Silicon Fertilizer Enhance Plant Growth and Salinity Tolerance in *Coriandrum sativum*. *Journal of Plant Interactions* 14(1): 386–396.

Alori, E. T., Glick, B. R., and Babalola, O. 2017. Microbial Phosphorus Solubilization and Its Potential for Use in Sustainable Agriculture. *Frontiers in Microbiology* 8: 971–993.

Anderson, A. J., McLean, J. E., Jacobson, A. R., and Britt, D. W. 2018. CuO and ZnO Nanoparticles Modify Interkingdom Cell Signaling Processes Relevant to Crop Production. *Journal of Agricultural* 11: 6513–6524.

Bandala, E. R., and Berli, M. 2019. Engineered Nanomaterials ENMs and Their Role at the Nexus of Food, Energy, and Water. *Materials Science for Energy Technologies* 2: 29–40.

Barratt, B. I. P., Moran, V. C., Bigler, F., and van Lenteren, J. C. 2018. The Status of Biological Control and Recommendations for Improving Uptake for the Future. *BioControl* 63: 155–167.

Bauer, W. D., and Mathesius, U. 2004. Plant Responses to Bacterial Quorum Sensing Signals. *Current Opinion in Plant Biology* 69: 429–433.

Bonebrake, M., Anderson, K., Valiente, J., Jacobson, A., McLean, J. E., Anderson, A., and Britt, D. W. 2018. Biofilms Benefiting Plants Exposed to ZnO and CuO Nanoparticles Studied with a Root-Mimetic Hollow Fiber Membrane. *Journal of Agricultural and Food Chemistry* 66(26): 6619–6627.

Çakmakçı, R., Mosber, G., Milton, A. H., Alatürk, F., and Ali, B. 2020. The Effect of Auxin and Auxin-Producing Bacteria on the Growth, Essential Oil Yield, and Composition in Medicinal and Aromatic Plants. *Current Microbiology* 22: 1239–1245.

Cavicchioli, R., Ripple, W. J., Timmis, K. N., Azam, F., Bakken, L. R., Baylis, M., and Behrenfeld, M. J. 2019. Scientists' Warning to Humanity: Microorganisms and Climate Change. *Nature Reviews Microbiology* 30: 569–586.

Chauhan, B. S., Prabhjyot-Kaur, G. M., Randhawa, R. K., Singh, H., and Kang, M. S. 2014. In I. D. L. B Ed., *Global Warming and Its Possible Impact on Agriculture in India*. Academic Press 123: 65–121.

Chen, H. 2018. Metal Based Nanoparticles in Agricultural System: Behavior, Transport, and Interaction with Plants. *Chemical Speciation and Bioavailability* 30: 123–134.

Chowdhury, I., Cwiertny, D. M., and Walker, S. L. 2012. Combined Factors Influencing the Aggregation and Deposition of Nano-TiO_2 in the Presence of Humic Acid and Bacteria. *Environmental Science and Technology* 46: 6968–6976.

Chung, I. M., Rajakumar, G., Subramanian, U., Venkidasamy, B., and Thiruvengadam, M. 2019. Impact of Copper Oxide Nanoparticles on Enhancement of Bioactive Compounds Using Cell Suspension Cultures of *Gymnema* Sylvestre Retz. *Applied Sciences* 26: 1–31.

Coninck, B. D., Timmermans, P., Vos, C., Cammue, B. P. A., and Kazan, K. 2015. What Lies beneath: Belowground Defense Strategies in Plants. *Trends in Plant Science* 31: 91–101.

Damalas, C. A., and Eleftherohorinos, I. G. 2011. Pesticide Exposure, Safety Issues, and Risk Assessment Indicators. *International Journal of Environmental Research and Public Health* 3390: 1402–1419.

Dimkpa, C. O., McLean, J. E., Britt, D. W., and Anderson, A. J. 2012a. Bioactivity and Biomodification of Ag, ZnO, and CuO Nanoparticles with Relevance to Plant Performance in Agriculture. *Industrial Biotechnology* 14: 344–357.

Dimkpa, C. O., Mclean, J. E., Britt, D. W., and Anderson, A. J. 2012b. CuO and ZnO Nanoparticles Differently Affect the Secretion of Fluorescent Siderophores in the Beneficial Root Colonizer, *Pseudomonas chlororaphis*. *Nanotoxicology* 6: 635–642.

Dimkpa, C. O., McLean, J. E., Britt, D. W., Johnson, W. P., Bruce Arey, A. S. L., and Anderson, A. J. 2012c. Nanospecific Inhibition of Pyoverdine Siderophore Production in Pseudomonas Chlororaphis O6 by CuO Nanoparticles. *Chemical Research in Toxicology* 25: 1066–1074.

Dimkpa, C. O., Zeng, J., McLean, J. E., Britt, D. W., Zhan, J., and Anderson, A. J. 2012d. Production of Indole-3-Acetic Acid via the Indole-3-Acetamide Pathway in the Plant-Beneficial Bacterium *Pseudomonas chlororaphis* O6 Is Inhibited by ZnO Nanoparticles but Enhanced by CuO Nanoparticles. *Applied and Environmental Science* 86: 1404–1410.

Dimkpa, C. O., Hansen, T., Stewart, J., McLean, J. E., Britt, D. W., and Anderson, A. J. 2015. ZnO Nanoparticles and Root Colonization by a Beneficial Pseudomonad Influence Essential Metal Responses in Bean. *Nanotoxicology* 3: 271–278.

Dimkpa, C. O., Singh, U., Bindraban, P. S., Elmer, W. H., Gardea-Torresdey, J. L., and White, J. C. 2019. Zinc Oxide Nanoparticles Alleviate Drought-Induced Alterations in Sorghum Performance, Nutrient Acquisition, and Grain Fortification. *Science of the Total Environment* 688: 926–934.

Elmer, W., and White, J. C. 2018. The Future of Nanotechnology in Plant Pathology. *Annual Review of Phytopathology* 47: 111–133.

Ferreira, C. M. H., Soares, H. M. V. M., and Soares, E. V. 2019. Promising Bacterial Genera for Agricultural Practices: An Insight on Plant Growth-Promoting Properties and Microbial Safety Aspects. *Science of The Total Environment*, 682: 779–799.

Festa, R. A., and Thiele, D. J. 2011. Copper: An Essential Metal in Biology. *Current Biology* 41: 877–883.

Fu, L., Wang, Z., Dhankher, O. P., and Xing, B. 2019. Nanotechnology as a New Sustainable Approach for Controlling Crop Diseases and Increasing Agricultural Production. *Journal of Experimental Botany* 33: 507–519.

Fulaz, S., Vitale, S., Quinn, L., and Casey, E. 2019. Nanoparticle–Biofilm Interactions: The Role of the EPS Matrix. *Trends in Microbiology* 27: 915–926.

Gambino, M., Marzano, V., Villa, F., Vitali, A., Vannini, C., Landini, P., and Cappitelli, F. 2015. Effects of Sublethal Doses of Silver Nanoparticles on *Bacillus subtilis* Planktonic and Sessile Cells. *Journal of Applied Microbiology* 118: 1103–1115.

Geddes, B. A., Ryu, M.-H., Mus, F., Costas, A. G., Peters, J. W., Voigt, C. A., and Poole, P. 2015. Use of Plant Colonizing Bacteria as Chassis for Transfer of N2-Fixation to Cereals. *Current Opinion in Biotechnology* 32: 216–222.

Glick, B. R. 2012. Plant Growth-Promoting Bacteria: Mechanisms and Applications. 2015. Modulating Phytohormone Levels - Beneficial Plant-Bacterial Interactions. In B. R. Glick, Ed. Springer International Publishing. 3: 65–96

Glick, B. R. 2014. Bacteria with ACC Deaminase Can Promote Plant Growth and Help to Feed the World. *Microbiological Research* 169: 30–39.

Goodman, J., McLean, J. E., Britt, D. W., and Anderson, A. J. 2016. Sublethal Doses of ZnO Nanoparticles Remodel Production of Cell Signaling Metabolites in the Root Colonizer: *Pseudomonas chlororaphis* O6. *Environmental Science. Nano* 3: 1103–1113.

Goswami, D. 2016. In M. Tejada Ed., Janki N Thakker, and Pinakin C Dhandhukia. Portraying Mechanics of Plant Growth Promoting Rhizobacteria PGPR: A Review. *Moral. Cogent Food and Agriculture.* 11: 127–150

Groenke, N., Seisenbaeva, G. A., Kaminskyy, V., Zhivotovsky, B., Kost, B., and Kessler, V. G. 2012. Structural Characterization, Solution Stability, and Potential Health and Environmental Effects of the Nano-TiO2 Bioencapsulation Matrix and the Model Product of Its Biodegradation TiBALDH. *RSC Adv* 33: 4228–4235.

Hacquard, S., Spaepen, S., Garrido-Oter, R., and Schulze-Lefert, P. 2017. Interplay Between Innate Immunity and the Plant Microbiota. *Annual Review of Phytopathology* 46: 565–589.

Hassen, A. I., Bopape, F. L., and K, L. 2016. Microbial Inoculants as Agents of Growth Promotion and Abiotic Stress Tolerance in Plants BT: Microbial Inoculants in Sustainable Agricultural Productivity. In Emerging Investigator Series: Polymeric Nanocarriers for Agricultural Applications: Synthesis, Characterization, and Environmental and Biological Interactions, D. P. Singh, H. B. Singh, and R. Prabha. Vol. 1: *Research Perspectives.* Springer India, 23–36.

Hassan, S., and Mathesius, U. 2012. The Role of Flavonoids in Root–Rhizosphere Signalling: Opportunities and Challenges for Improving Plant–Microbe Interactions. *Journal of Experimental Botany* 17: 3429–3444.

Ho, Y.-N. 2017. Plant-Microbe Ecology: Interactions of Plants and Symbiotic Microbial Communities Ch. 6 In D. C. Mathew, Ed. IntechOpen.

Huangfu, X., Xu, Y., Liu, C., and He, Q., Ma, C. M., and Ruixing H. 2019. A Review on the Interactions between Engineered Nanoparticles with Extracellular and Intracellular Polymeric Substances from Wastewater Treatment Aggregates. *Chemosphere* 219: 766–783.

Ikuma, K., Decho, A. W., and Lau, B. L. T. 2015. When Nanoparticles Meet Biofilms— Interactions Guiding the Environmental Fate and Accumulation of Nanoparticles. *Frontiers in Microbiology* 6: 591–603.

Isfahani, M., Faranak, A. T., Hoodaji, M., Ataabadi, M., and Mohammadi, A. 2019. Influence of Exopolysaccharide-Producing Bacteria and SiO_2 Nanoparticles on Proline Content and Antioxidant Enzyme Activities of Tomato Seedlings *Solanum lycopersicum* L. under Salinity Stress. *Polish Journal of Environmental Studies* 3: 153–163.

Jacobson, A., Doxey, S., Potter, M., Adams, J., Britt, D., McManus, P., McLean, J., and Anderson, A. 2018. Interactions Between a Plant Probiotic and Nanoparticles on Plant Responses Related to Drought Tolerance. *Journal of Environmental Studies* 89: 148–156.

Jeevanandam, J., Barhoum, A., Chan, Y. S., Dufresne, A., and Danquah, M. K. 2018. Review on Nanoparticles and Nanostructured Materials: History, Sources, Toxicity and Regulations. *Beilstein Journal of Nanotechnology* 13: 1050–1074.

Jones, J. D. G., and Dangl, J. L. 2006. The Plant Immune System. *Nature* 444: 323–329.

Kannan, N. 2013. Effect of Nanosilica and Silicon Sources on Plant Growth Promoting Rhizobacteria, Soil Nutrients and Maize Seed Germination. *IET Nanobiotechnology* 3: 70–77.

Karunakaran, G., Suriyaprabha, R., Manivasakan, P., Yuvakkumar, R., Rajendran, V., Prabu, P., and Kannan, N. 2013. Effect of Nanosilica and Silicon Sources on Plant Growth Promoting Rhizobacteria, Soil Nutrients and Maize Seed Germination. *IET Nanobiotechnology* 7: 70–77.

Khan, N., and Bano, A. 2016. Role of Plant Growth Promoting Rhizobacteria and Ag-Nano Particle in the Bioremediation of Heavy Metals and Maize Growth under Municipal Wastewater Irrigation. *International Journal of Phytoremediation* 18: 211–221.

Khati, P., Chaudhary, P., Gangola, S., Bhatt, P., and Sharma, A. 2017. Nanochitosan Supports Growth of *Zea mays* and Also Maintains Soil Health Following Growth *Environmental Nanotechnology* 4: 81–102.

Koh, S. 2007. Strategies for Controlled Placement of Nanoscale Building Blocks. *Nanoscale Research Letters* 2: 519–545.

Loper, J. E., Hassan, K. A., Mavrodi, D. V., Davis, E. W., II, Lim, C. K., Shaffer, B. T., and Elbourne, L. D. H. 2012. Comparative Genomics of Plant-Associated *Pseudomonas* Spp.: Insights into Diversity and Inheritance of Traits Involved in Multitrophic Interactions. *PLOS Genetics* 8: p 1002784.

López-Vargas, E. R., Ortega-Ortíz, H., Cadenas-Pliego, G., de Alba Romenus, K., de la Fuente, M. C., Benavides-Mendoza, A., and Juárez-Maldonado, A. 2018. Foliar Application of Copper Nanoparticles Increases the Fruit Quality and the Content of Bioactive Compounds in Tomatoes. *Applied Sciences Switzerland* 77: 807–820.

Mahawar, H., and Prasanna, R. 2018. Prospecting the Interactions of Nanoparticles with Beneficial Microorganisms for Developing Green Technologies for Agriculture. *Environmental Nanotechnology, Monitoring and Management* 10: 477–485.

Mhlongo, M. I., Piater, L. A., Madala, N. E., and Labuschagne, N. and Dubery, I. A. 2018. The Chemistry of Plant–Microbe Interactions in the Rhizosphere and the Potential for Metabolomics to Reveal Signaling Related to Defense Priming. In Induced Systemic Resistance. *Frontiers in Plant Science* 9: 112–135.

Mishra, Y. K., and Adelung, R. 2018. ZnO Tetrapod Materials for Functional Applications. *Materials Today* 13: 631–651.

Ntoukakis, V., and Gifford, M. L. 2019. Plant–Microbe Interactions: Tipping the Balance. *Journal of Experimental Botany*, 113: 4583–4585.

Olanrewaju, O. S., Glick, B. R., and Babalola, O. O. 2017. Mechanisms of Action of Plant Growth Promoting Bacteria. *World Journal of Microbiology and Biotechnology* 33: 1–16.

Palmqvist, N. G. M., Bejai, S., Meijer, J., Seisenbaeva, G. A., and Kessler, V. G. 2015. Nano Titania Aided Clustering and Adhesion of Beneficial Bacteria to Plant Roots to Enhance Crop Growth and Stress Management. *Scientific Reports* 5: 1–12.

Pandey, P., Irulappan, V., Bagavathiannan, M. V., and Senthil-Kumar, M. 2017. Impact of Combined Abiotic and Biotic Stresses on Plant Growth and Avenues for Crop Improvement by Exploiting Physio-Morphological Traits. *Frontiers in Plant Science* 8: 537–566.

Panichikkal, J., Thomas, R., John, J. C., and Radhakrishnan, E. K. 2019. Biogenic Gold Nanoparticle Supplementation to Plant Beneficial *Pseudomonas monteilii* Was Found to Enhance Its Plant Probiotic Effect. *Current Microbiology* 76: 503–509.

Papenfort, K., and Bassler, B. L. 2016. Quorum Sensing Signal–Response Systems in Gram-Negative Bacteria. *Nature Reviews Microbiology* 13: 576–588.

Park, M. R., Banks, M. K., Applegate, B., and Webster, T. J. 2008. Influence of Nanophase Titania Topography on Bacterial Attachment and Metabolism. *International Journal of Nanomedicine* 3: 497–504.

Pérez-de-Luque, A. 2017. Interaction of Nanomaterials with Plants: What Do We Need for Real Applications in Agriculture? *Frontiers in Environmental Science* 5: 12.

Pilon, M., Abdel-Ghany, S. E., Cohu, C. M., Gogolin, K. A., and Ye, H. 2006. Copper Cofactor Delivery in Plant Cells. *Current Opinion in Plant Biology.* 9: 256–263.

Pour, M. M. A. H. 2019. Nano-Encapsulation of Plant Growth-Promoting Rhizobacteria and Their Metabolites Using Alginate-Silica Nanoparticles and Carbon Nanotube Improves UCB1 Pistachio Micropropagation. *Frontiers in Plant Science* 63: 1096–1103.

Pullagurala, R., Venkata, L., Adisa, I. O., Rawat, S., Kim, B., Barrios, A. C., Medina-Velo, I. A., Hernandez-Viezcas, J. A., Peralta-Videa, J. R., and Gardea-Torresdey, J. L. 2018. Finding the Conditions for the Beneficial Use of ZnO Nanoparticles towards Plants: A. Review. *Environmental Pollution* 241: 1175–1181.

Purves, D., and Ragg, J. M. 1962. Copper-Deficient Soils in South-East Scotland. *Journal of Soil Science* 7: 241–246.

Raliya, R., Tarafdar, J. C., and Biswas, P. 2016. Enhancing the Mobilization of Native Phosphorus in the Mung Bean Rhizosphere Using ZnO Nanoparticles Synthesized by Soil Fungi. *Journal of Agricultural and Food Chemistry* 64: 3111–3118.

Rashid, M. H.-O., and Chung, Y. R. 2017. Induction of Systemic Resistance against Insect Herbivores in Plants by Beneficial Soil Microbes. *Frontiers in Plant Science* 76: 801–816.

Renner, L. D., and Weibel, D. B. 2011. Physicochemical Regulation of Biofilm Formation. *MRS Bulletin* 1557(10): 347–355. 36.

Rudrappa, T., Biedrzycki, M. L., and Bais, H. P. 2008. Causes and Consequences of Plant-Associated Biofilms. *FEMS Microbiology Ecology* 64: 153–166.

Sabir, S., Arshad, M., and Chaudhari, S. K. 2014. Zinc Oxide Nanoparticles for Revolutionizing Agriculture: Synthesis and Applications. *The Scientific World Journal* 201: 4925–4934.

Sami, F., and Siddiqui, H. and Hayat, Shamsul. 2020. Impact of Silver Nanoparticles on Plant Physiology: A Critical Review BT - Sustainable Agriculture Reviews M. F. In *Nanotechnology for Plant Growth and Development* by Shamsul Hayat John Pichtel and Qazi Fariduddin. Springer International Publishing 41: 77–111

Sanzari, I., Leone, A., and Ambrosone, A. 2019. Nanotechnology in Plant Science: To Make a Long Story Short. *Front. Bioeng. Biotechnol* 296: 59–63.

Saraf, M., Pandya, U., and Thakkar, A. 2014. Role of Allelochemicals in Plant Growth Promoting Rhizobacteria for Biocontrol of Phytopathogens. *Bioresource Technology* 3: 18–29.

Sarkar, J., Chakraborty, N., Chatterjee, A., and Bhattacharjee, A. 2020. Green Synthesized Copper Oxide Nanoparticles Ameliorate Defence and Antioxidant Enzymes in Lens culinaris. *Nanomaterials* 10: 312–373

Schaller, G. E., Street, I. H., and Kieber, J. J. 2014. Cytokinin and the Cell Cycle. *Current Opinion in Plant Biology* 21: 7–15.

Schirawski, J., and Perlin, M. H. 2018. Plant–Microbe Interaction 2017-The Good, the Bad and the Diverse. *International Journal of Molecular Sciences* 158: 507–516.

Seisenbaeva, G. A., Daniel, G., Nedelec, J. M., and Kessler, V. G. 2013. Solution Equilibrium behind the Room-Temperature Synthesis of Nanocrystalline Titanium Dioxide. *Nanotoxicology* 71: 3330–3336.

Shakiba, S., Astete, C. E., Paudel, S., Sabliov, C. M., Rodrigues, D. F., and Louie, S. M. 2020. Emerging Investigator Series: Polymeric Nanocarriers for Agricultural Applications: Synthesis, Characterization, and Environmental and Biological Interactions. *Environmental Science: Nano* 7: 37–67.

Shcherbakova, E. N., Shcherbakov, A. V., Andronov, E. E., Gonchar, L. N., Kalenskaya, S. M., and Chebotar, V. K. 2017. Combined Pre-Seed Treatment with Microbial Inoculants and Mo Nanoparticles Changes Composition of Root Exudates and Rhizosphere Microbiome Structure of Chickpea *Cicer arietinum* L. *Plant. Symbiosis* 73: 57–69.

Sheng, Z., Nostrand, J. D. V., Zhou, J., and Liu, Y. 2018. Contradictory Effects of Silver Nanoparticles on Activated Sludge Wastewater Treatment. *Journal of Hazardous Materials* 341, 448–456.

Shukla, S. K. 2015. Prediction and Validation of Gold Nanoparticles GNPs on Plant Growth Promoting Rhizobacteria PGPR: A Step toward Development of Nano-Bio-fertilizers. *Nanotechnology Reviews* 4: 439–448.

Shulse, C. N., Chovatia, M., Agosto, C., Wang, G., Hamilton, M., Deutsch, S., Yoshikuni, Y., and Blow, M. J. 2019. Engineered Root Bacteria Release Plant-Available Phosphate from Phytate. *Enzyme and Microbial Technology* 30: 01210–01219.

Singh, A., Singh, N. B., Afzal, S., Singh, T., and Hussain, I. 2018. Zinc Oxide Nanoparticles: A Review of Their Biological Synthesis, Antimicrobial Activity, Uptake, Translocation and Biotransformation in Plants. *Journal of Materials Science* 53: 185–201.

Singh, B., and Satyanarayana, T. 2011. Microbial Phytases in Phosphorus Acquisition and Plant Growth Promotion. *Physiology and Molecular Biology of Plants* 23: 93–103.

Singh, J., Vishwakarma, K., Ramawat, N., Rai, P., Singh, V. K., Mishra, R. K., Kumar, V., Tripathi, D. K., and Sharma, S. 2019. Nanomaterials and Microbes' Interactions: A Contemporary Overview. *3 Biotech* 3: 68–92.

Smith, D., Subramanian, S., Lamont, J., and Bywater-Ekegärd, M. 2015. Signaling in the Phytomicrobiome: Breadth and Potential. *Frontiers in Plant Science* 6: 709–733.

Solanki, P., Bhargava, A., Chhipa, H., Jain, N., and Panwar, J. 2015. Nano-fertilizers and Their Smart Delivery System. In L. M. Mahendra Rai Caue Ribeiro and N. Duran Eds., *Nano-Fertilizers and Their Smart Delivery System: Nanotechnologies in Food and Agriculture*. Springer International Publishing. 81–101.

Souza, R. de, Ambrosini, A., and Passaglia, L. M. P. 2015. Plant Growth-Promoting Bacteria as Inoculants in Agricultural Soils. *Genetics and Molecular Biology* 13: 401–419.

Srivastava, A. K., Dev, A., and Karmakar, S. 2018. Nanosensors and Nanobiosensors in Food and Agriculture. *Environmental Chemistry Letters* 17: 161–182.

Tan, W., Peralta-Videa, J. R., and Gardea-Torresdey, J. L. 2018. Interaction of Titanium Dioxide Nanoparticles with Soil Components and Plants: Current Knowledge and Future Research Needs: A Critical Review. *Environmental Nanatechnology* 13: 257–278.

Timmusk, S., Behers, L., Muthoni, J., Muraya, A., and Aronsson, A. C. 2017a. Perspectives and Challenges of Microbial Application for Crop Improvement. *Frontiers in Plant Science* 33: 1–10.

Timmusk, S., Behers, L., Muthoni, J., Muraya, A., and Aronsson, A.-C. 2017b. Perspectives and Challenges of Microbial Application for Crop Improvement. *Frontiers in Plant Science* 33: 819–849.

Timmusk, S., Seisenbaeva, G., and Behers, L. 2018. Titania TiO2 Nanoparticles Enhance the Performance of Growth-Promoting Rhizobacteria. *Scientific Reports* 37: 578–617.

Vejan, P., Abdullah, R., Khadiran, T., Ismail, S., and Nasrulhaq Boyce, A. 2016. Role of Plant Growth Promoting Rhizobacteria in Agricultural Sustainability: A Review. *Molecules* 33: 215–573.

Veliz, E. A., Martínez-Hidalgo, P., and Hirsch, A. M. 2017. Chitinase-producing bacteria and their role in biocontrol. *AIMS Microbiology* 33: 689–705.

Venturi, V., and Keel, C. 2016. Signaling in the Rhizosphere. *Trends in Plant Science* 213: 187–198.

Visca, P., Imperi, F., and Lamont, I. L. 2007. Pyoverdine Siderophores: From Biogenesis to Biosignificance. *Trends in Microbiology* 15: 22–30.

Weller, D. M. 1988. Biological Control of Soilborne Plant Pathogens in the Rhizosphere with Bacteria. *Annual Review of Phytopathology* 17: 379–407.

Westmeier, D., Hahlbrock, A., Reinhardt, C., Fröhlich-Nowoisky, J., Wessler, S., Vallet, C., Pöschl, U., Knauer, S. K., and Stauber, R. H. 2018. Nanomaterial-Microbe Cross-Talk: Physicochemical Principles and PathoBiological Consequences. *Chemical Society Reviews* 47: 5312–5337.

White, P. J., and Greenwood, D. J. Eds. 2013. Properties and Management of Cationic Elements for Crop Growth. *Soil Conditions and Plant Growth* 10: 978–1002

Wille, L., Messmer, M. M., Studer, B., and Hohmann, P. 2019. Insights to Plant–Microbe Interactions Provide Opportunities to Improve Resistance Breeding against Root Diseases in Grain Legumes. *Plant Cell and Environment* 42: 20–40.

Xu, Y. 2016. Envirotyping for Deciphering Environmental Impacts on Crop Plants. *Theoretical and Applied Genetics* 129: 653–673.

Yan, A., and Chen, Z. 2019. Impacts of Silver Nanoparticles on Plants: A Focus on the Phytotoxicity and Underlying Mechanism. *International Journal of Molecular Sciences* 11: 379–407.

Yang, Y., and Alvarez, P. J. J. 2015. Sublethal Concentrations of Silver Nanoparticles Stimulate Biofilm Development. *Environmental Science and Technology Letters* 2: 221–226.

Zhao, L., Lu, L., Wang, A., Zhang, H., Huang, M., Wu, H., Xing, B., Wang, Z., and Ji, R. 2020. Nano-Biotechnology in Agriculture: Use of Nanomaterials to Promote Plant Growth and Stress Tolerance. *Journal of Agricultural and Food Chemistry 68*: 1935–1947.

13 Siderophores and Their Applications in Heavy Metal Detoxification

K. Thakur, H. Pandey, I. Sharma,
V.K. Dhiman, and D. Pandey

CONTENTS

DOI: 10.1201/9781003213864-13

13.1　INTRODUCTION

Metals are a major constituent of soil, including micronutrients that are essential for various physiological parameters such as the growth and development of plants (Sheng et al. 2008). Copper, zinc, iron, manganese, nickel, and cobalt are some of the important elements, whereas cadmium, silver, lead, and mercury are regarded as harmful and are toxic to the existing flora and fauna (Williams et al. 2000). Anthropogenic activities, including metal mining, fuel burning, inorganic pesticides, sewage waste, insecticide, inorganic fertilizers, municipal wastes, harmful pigments, used batteries, and exhaust from vehicles, lead to environmental pollution (Madhaiyan et al. 2007; Vivas et al. 2003; Leyval et al. 1997; Denton 2007). Further, through the process of biomagnification, these toxic metals enter the food chain and, thereby, pollute the ecosystem (Khan et al. 2000). Metals with atomic numbers above 20(Jing et al. 2007) or metals with a mass specifically of more than $5g/cm^3$ are known as heavy metals (Gadd and Griffiths 1977). Cadmium, chromium, copper, mercury, lead, and nickel are considered to be harmful (Jing et al. 2007). Currently, environmental problems caused by toxic metals is of major concern (Gamalero et al. 2009; Abou-Shanab et al. 2006). Heavy metal contamination has a major impact on the ecosystem (terrestrial and marine) (Khan et al. 2000). These hazardous metals can be utilized by the plant as nutrients in relatively low levels; however, above a certain critical limit, it is harmful for plant growth and development. When absorbed in large quantities, these heavy metals affect the metabolism of the plant and hamper their normal growth and development process (Jing et al. 2007). The presence of excess metal in the soil reduces the soil microbial diversity, which also decreases the soil productivity as well as the production of agricultural crops (McGrath et al. 1995).

Various mechanisms that are of physical and chemical origin have been employed for the degradation of xenobiotics. These costly physicochemical techniques often result in the production of harmful gases and other undesirable compounds. Eco-friendly approaches such as bioremediation have been utilized to remove these toxic elements from the environment. Microbial bioremediation depends on the degradation of xenobiotic contaminants present in the environment into biodegradable products. Enzymatic transformation and microbial ingestion of harmful contaminants are the key methods involved in the process of microbial biomagnification and can also help in cleaning up the environment. Various naturally occurring microorganisms in the environment have the potential to transform, degrade, or form a complex with different toxic metal ions. These ions could not be degraded in their less harmful compounds by any other physiochemical or biological means. However, they can be converted into less toxic compounds by changing their oxidation number or through the formation of an organic complex (Banik et al. 2014). A recent technological advancement, known as "microbial-remediation," utilizes microorganisms for bioremediation. The "microbial-remediation," technique makes use of microbes and plant growth-promoting rhizobacteria (PGPR) for the accumulation, degradation, and transformation of soil pollutants through different biochemical strategies (Prasad et al. 2010). Root rhizosphere bacteria provide nutrients to the plant and inhibit the growth of harmful pathogens. PGPR also exhibits metal-binding ability that is useful in providing tolerance to the plant against these heavy metals and could be utilized for the purpose of soil reclamation. Siderophores are a certain class of low molecular weight peptides synthesized in various bacteria, especially in PGPR, that could provide an alternative method for remediation of soil affected by harmful metals, and it can also act as a chelator for the various metal ions (Jing et al. 2007). Recently, studies have been conducted on these low molecular weight (≤ 10 kDa) molecules, known as siderophores, that form the important part of secretions from various plants and microbial sources into the root zone of the plant (Hider and Kong 2010; Ahmed and Holmstrom 2014; Johnstone and Nolana 2015). In comparison to plant-based siderophores, which are produced in lower quantity, siderophores from bacteria have high iron-binding capacity (Kraemer 2004; Kraeme et al. 2006; Glick 2012). Different biochemical properties, such as pH, complex forming sites, redox potential, etc., directly affect the metal-chelating potential of such compounds (Akafia et al. 2014). These peptides scavenge for metal ions, especially for iron, in the root zone (Aznar and Dellagi 2015). Other important characteristics of siderophores include acting against various kinds of harmful bacteria, through the production of sideromycins (Braun et al. 2009). Siderophores also have a high affinity toward metal ions such as vanadium (Vn), molybdenum (Mo), copper (Cu), and zinc (Zn) (Hood and Skaar 2012). Zincophores are Zn-binding siderophores that are synthesized in some bacteria (Prentice et al. 2007), including *Pseudomonas*, which have a high manganese-binding ability (Harrington et al. 2012; Duckworth et al. 2014). Elements such as Mo and Vn also form complexes with other types of siderophores (Deicke et al. 2013). Similarly, Cu containing complex compounds include methanobactin, yersiniabactin, and coproporphyrin (Chaturvedi et al. 2014). Yersiniabactin, enterobactin, and aerobactin can synthesize gold (Au) nanoparticles (Wyatt et al. 2014). They play a vital role as boron and silicon transporter side moieties and are well characterized in both biotic and abiotic stress (Chaturvedi and Henderson 2014; Butler and Theisen 2010). The existence of the persistent kind of siderophore, which binds to non-metals, is due to the presence of marine siderophores, such as mvibrioferrein, that are isolated from the bacteria *Marinobacter* (Amin et al.2007). Siderophores, such as citrate as well as catecholate, form a complex with Bo to generate a vital signalling system (Sandy and Butler 2009). Furthermore, their importance in clinical studies is now being considered as they play a key role in pathogenesis (Ali and Vidhale 2013). Antioxidant activity of these siderophores is important to molecules studied in different biological fields (Aznar et al. 2015). In studies conducted on siderophores, a new class of protein, known as siderocalin, which has a mammalian origin and belongs to the family of lipocalin, specifically interacts with actinide and lanthanide (Allred et al. 2015). Despite the distinctness observed in the characteristics of these peptides in different biological systems, this review deals with microbial siderophores and their potential utilization in bioremediation.

13.2 CLASSIFICATION OF SIDEROPHORES

Siderophores are produced by various types of bacteria with varying structural properties. Functional characteristics or complex forming groups that interact with Fe3+

ions are criteria for its classification. These are generally classified as catecholates, hydroxamates, and carboxyl (Sah and Singh 2015; Gupta et al. 2015). Mixed ligands, which are better known as pyoverdines, form a small class of siderophores that are classified under the fourth major group of siderophores, which is further divided into different classes based on the functional group attached to it. Various types of siderophores have been discovered by applying advanced technologies, such as proton nuclear magnetic resonance spectrophotometry, electrophoretic mobility, acid hydrolysis, and other biological processes (Kurth et al. 2016).

13.2.1 CATECHOLATE TYPE

Siderophores with phenolate as a complex forming group belong to the category of catecholate and have 2, 3-dihydroxybenzoate (DHB) as their functional group. They are also referred to as pyrocatechols [C6H4(OH)2] (Wittmann et al. 2004). Generally, catecholate siderophores utilize two oxygen atoms that form a complex with the iron (Fe) ions by synthesizing a didentate ligand complex, and it can further result in the creation of a hexadentate octahedral coordination compound (Ali and Vidhale 2013). Catecholates generally occur in the environment in small quantities as a colourless compound. The orthoisomeric molecule is formed from three isomeric benzediols. Enterobactin, also referred to as enterochelin, is the most frequently studied catecholate, which constitutes a cyclic trimester complex functional group (2,3-dyhydroxyserine), generally synthesized by *Klebsiella pneumonia* and *Salmonella typhimurium* (Achard et al. 2013).

13.2.2 HYDROXAMATE TYPE

The hydroxamate class of siderophores is prevalent in the environment, and it constitutes C(=O) N-(OH) R, where R specifies the amino-alkanoic acids and its different functional forms that are primarily produced by microorganisms (Renshaw et al. 2002). This type of siderophore has a 1:1 ratio of ferric ions, which is similar to ferric-Ethylene-diamine-tetra-acetic acid coordination compounds (Mosa et al. 2016). Basically, hydroxamate siderophores are grouped under three broad categories, which include ferrioxamines, aerobactin, and ferrichrome based on the functional group (Winkelmann 2007). Ferrioxamines are composed of a straight chain structure having a chemical notation of C25H48N6O. The ferrichromes have a cyclic structure and mainly constitute amino acids, such as alanine, serine, or glycine with a peptide bond between 3 N-acyl-N-hydroxyl-L- glycine and ornithine (Ali et al. 2011). Another class of hydroxamates is the aerobactin with a chemical formula of C22H36N4O13 (Neilands1995). Its occurrence has been identified in various bacteria such as *E. coli*, *A. aerogenes*, *K. pneumoniae*, and *Pseudomonas* (Buyer et al. 1991).

13.2.3 CARBOXYLATE TYPE

An entirely new class of siderophores with a completely different complex forming group than hydromates, 2, 3-dihydroxybenzoate (DHB), has been reported. The chelation process in these kinds of siderophores occurs by the carboxyl or hydroxyl carboxylate functional unit (Schwyn and Neilands 1987). Isolation and purification of significant types of carboxyl-type siderophores was obtained from the bacterial strain DM4 *Rhizobium meliloti*. Rhizobactin is an important type of carbolic acid with an ethylene-diamine di-carboxyl and hydroxy carboxyl complex (Bergeron et al. 2014). Staphyloferrin A is also an example of a carboxyl-type siderophore produced by the strain DSM20459 (*Staphylococcus hyicus*). This type of siderophore constitutes two citrate residues and 1 D-ornithine that are linked by an amide linkage (Ali and Vidhale 2013). A similar example of carboxylate siderophore is Rhizoferrin, which is produced by the fungus belonging to the zygomycetes group (Al-Fakih 2014).

13.2.4 MIXED TYPE

This class of siderophores basically consists of two kinds of Fe-binding sites (AznarandDellagi 2015). Mixed complexes consist of the class of siderophores that are derivatives of lysine, ornithine, and histamine (Sah and Singh 2015). Ornitine acts as the precursor for pyoveridine, which is also called pseudobactin. It is generally synthesized by the bacteria *Pseudomonas* and its different spp. (Meneely and Lamb 2007). Mycobactin is another type of mixed siderophore that is formed from lysine in the bacterium *Mycobacterium tuberculosis* (Varma and Podila 2005). Histamine is reported to produce anguibactin in the aquatic pathogens *Vibrio anguillarum* (Naka et al. 2013). Table 13.1 shows siderophores' functions and their associated chemical moieties.

13.3 SIDEROPHORE BIOSYNTHESIS

Microbes require heavy metal ions to carry out basic metabolic activities; however, their demand remains quite low. When the availability of these metal ions reaches concentrations above a certain threshold level, they are known to adversely affect metabolism and physiological functions. However, to counteract heavy metal concentrations, these bacteria are known to develop resistance. The microbes generally uptake the metal ions of similar chemical and physical properties. The uptake usually occurs through two mechanisms: one is the fast method of transfer of metal ions through the concentration gradient, and the other method is passive, requiring ATP molecules for the substrate specific transfer. The microbes tolerate the toxic effect of the metal ions by immobilization, sequestration, compartmentation, or transformation of these ions to non-toxic forms. Metal ions are known to be bound by sulfhydryl, carboxyl, amine, and amide groups as well as by polymers, such as proteins and polysaccharides, and may also form complexes or may form less toxic compounds (Rajkumar et al. 2010). These phenomena, thus, reduce the availability of heavy metals and reduce their toxicity (Rajkumar et al. 2010). The microbes are also known to expel the toxic ions by metal efflux systems, the energy dependent systems, requiring ATP molecules for substrate specific transfer are reported for arsenic, cadmium and chromium resistant systems. The microbes also synthesize

TABLE 13.1

Siderophore Functions and Their Associated Chemical Moieties

S.No.	Siderophore	Source	Importance in plant system	References
1.	Pyrocatechols (Catecholate type)	*Enterobacterium*	Chelating agent for Fe^{3+} with agricultural applications	Wittmann et al. (2004); Gregory et al. (2012); Sah and Singh(2015)
		Klebsiella pneumoniae *Erwinia herbicola*	In response to oxidative stress conditions	Raymond et al. (2003)
		Salmonella typhimurium	Nitrogen fixation process	Ward et al. (1999)
	Agrobactin (Catecholate type)	*Agrobacterium tumefaciens*	Fe-complex binding ligand	Leong and Neilands (1982)
	Parabactin (catecholate type)	*Paracoccusdenitrificans*	Fe-chelator	Leong and Neilands (1982)
2.	Rhizobactin (diaminopropane-ac Rhizobactin) (diaminopropane-acytylated with citric acid)	Fungi, basically belong to zygomycetes family	Utilization for biotechnological process Metal-bindingability	Gregory et al. (2012) Sah and Singh (2015)
	Staphyloferrin Rhizoferrin	*Staphylococcus hyicus* *Rhizobium meliloti*	Naturally biodegradable Synthesizedrhizoferrin	Munzinger et al. (1999) Ali and Vidhale (2013)
3.	Ferrioxamines (Hydroxamate type)	Strains containing *Frankia* (Cpl2,HsIi2andHsIi4)	Siderophores decreases the Mg^{2+}, Cu^{2+}, and Zn^{2+} metal-dependent growth inhibition in Frankia bydecontaminatingmoderately affected polluted soil	Singh et al. (2010)
4.	Aerobactin (Hydroxamate type)	*Pseudomonas*	Increase in seed germination process followed by high chlorophyll content in foliage accompanied byrapid growth inroot length	Manwar et al. (2001)
5.	Aerobactin (Hydroxamate type)	*Escherichia coli, Klebsiella sp.* *Aerobacter aerogenes*	Sequester Fe in Fe-deficient conditions	Gregory et al. (2012) and Buyer et al. (1991) *(Continued)*

TABLE 13.1 (CONTINUED)
Siderophore Functions and Their Associated Chemical Moieties

S.No.	Siderophore	Source	Importance in plant system	References
6.	Ferrichromes (Hydroxamate type)	Pseudomonas strain GRP3	Vigna radiata inoculated with siderophore in response to Fe-deficient conditions showed improved chlorophyll content as compared to control, which was not inoculated	Sharma et al. (2003)
7.	Ferrichromes (Hydroxamate type)	Pseudomonas sp.	Provide Fe to plants for growth and development, sequester toxic metals in soil to increase microbial diversity, and activity further reduces contaminants from soil	Joshi et al. (2014) and Ahmad (2014)
8.	Pyoverdines (Mixed type)	Pseudomonas fluorescens C7	Improved plant growth in Arabidopsis thaliana plants because of uptake of Fe–pyoverdine complex and increase in Fe inside plant tissues	Vansuyt et al. (2007)
9.	Pseudobactin (Mixed type)	Pseudomonas sp. Pseudomonas aeruginosa	Protection of plants from various pathogenic fungi and other virulent organisms	Husen (2020) and Martin (2003)
10.	Pseudobactin (Mixed type)	Pseudomonas putida	Increases the yield of different crop plants with increased growth and developmental activities	Chaiharn et al. (2008)

the metallothioneins (MTs), which are an excellent example of low molecular weight metal-binding proteins; however, they find limited application in the metal-binding ability (Rajkumar et al. 2010). In addition to this, microbes are also known to lower the toxic potential of heavy metals by lowering the redox state of heavy metals (Jing et al. 2007).

Fe is one of the major nutrients required by microorganisms. Under natural conditions, Fe usually occurs in Fe3+ state and is linked with the hydroxyl group to form insoluble compounds, making the availability of Fe scarce. To acquire Fe efficiently, microbes have developed a mechanism for the solubilization of the Fe complexes by means of siderophores. Siderophores are metal chelators with low molecular mass and a high affinity for Fe complexes (Neilands1995Miethke and Marahiel 2007). Siderophores are also known to combine with metals such as aluminium, cadmium, lead, and Zn to form complexes.

Siderophores are classified into two groups based on the enzymes involved in their synthesis: the nonribosomal peptide synthetase (NRPS) dependent and the NRPS independent pathway.

13.3.1 NONRIBOSOMAL PEPTIDE SYNTHETASE MECHANISM

Synthesis of siderophores by the NRPS pathway occur through multiple enzyme systems. The NRPS consists of multiple enzymes that have been reported to have antibiotic and related product synthesis (Aggarwal et al. 1995; May et al. 2001; Keating and Walsh1999; Keating et al. 2001). These enzymes do not require an RNA template for the synthesis of peptides, instead the order of peptide synthesis is generally dependent on the NRPS domain (Kohli et al. 2001; May et al.2001; Walsh et al.1997). The NRPS is an assemblage of domains that are folded and bundled together, resulting in the selection of the monomer followed by activation, elongation, and finally termination.

The NRPS assembly line is organized in such a way that it catalyzes a number of reactions. The first set of reactions is initiated in the domains of peptidyl carrier proteins (PCPs), in which the enzyme phosphopantetheinyl transferase (PPTase) catalyzes the transfer of phosphopantathenate, the P-pant moiety, by establishing the phosphodiester bonds, resulting in the formation of 3',5' ADP (Lambalot et al. 1996). Thereafter, the selection, activation, and incorporation of the monomers to NRPS assembly result in the growth of the chain. The adenylation domains (A-domain) selects the monomers to be activated. The active site of the A domain selects and converts the amino acid to aminoacyl AMP. The monomer is activated thermodynamically and is transferred to the HS-pant-PCP domain, resulting in the formation of amino-thioesters.

The chain growth of the NRPS assembly occurs from the N-terminal to the C-terminal PCP domains. The peptide synthetase catalytic domain is called the C-domain. The typical NRPS assembly consists of the C-domain, A-domain, and the PCP region. The A-domain selects and activates the amino acids that are to be integrated. This is followed by the C-domain, the main function of which is the formation of peptide bonds. Ultimately, the amino-S-PCP acts as the attacking substrate. As a result of this step, a peptide is attached to the translocated chain, and it becomes ready for the addition of another monomer (Miethke and Marahiel 2007; VonDohren et al.1997; Welch et al.2000).

The process terminates with the liberation of a full length non ribosomal peptide (Keating et al. 2001). Even after full-length elongation of the monomer chain, the liberation of the chain requires enzymatic cleavage. In the NRPS assembly line, the carboxy-terminal region consists of a folding region called the thioesterase (TE). The liberation of the chain involves the transfer from the PCP domain to the active region of the thioester to form acyl-O-TE (Rusnak et al. 1989). The intermediate, so formed, can either undergo hydrolysis for the release of free acid or may be captured by a hydroxyl or amine group for the liberation of lactones.

The NRPS assembly line of the siderophore biosynthesis has many variations. In certain bacterial siderophores, the peptides of the N-terminal, at the initiation stage, are modified and are derived from aryl acids. These aryl acids may help in the first step of chain elongation by converting the aminoacid to work as N-acylated substrate. The incorporation of the salicylic or DBH groups have also been reported for their ability to initiate the chain elongation during the chelation of the Fe ions. The integration of these groups at the initiation of the assembly line generally occurs at the A-domain (Gehring et al. 1997; Keating and Walsh 1999; Rusnak et al. 1989). Further, the aryl-carrier protein domain should also be taken up in the assembly line by the salicyl-AMP or DBH-AMP domain. In another type of variation, the condensation, or the C-domain, may consist of the thiazoline or oxazoline rings, and, as a result, the C-domain is called the cyclization domain. The bis-heterocycle formation by the yersiniabactin (Ybt) synthetase and pyochelin synthetase has been reported (Keating et al. 2000; Quadri et al. 1999; Suo et al. 2001).

13.3.2 SIDEROPHORE BIOSYNTHESIS BY NRPS INDEPENDENT PATHWAY

The enzymes responsible for the NRPS independent pathway (NIS) belong to the family of synthetase enzymes of aerobactin referred to as Iuc A and Iuc C (Challis 2005; De Lorenzo et al. 1986). Based on the similarities in their sequences, the NIS synthetases are classified as type A, B, and C (Challis 2005). These synthetases have been reported to have varying specificity to carboxylic acids.

There are two major sets of enzymes that catalyze the synthesis of NIS. The first set of enzymes uses amino carboxylic acid as a substrate and, with the formation of amide bonds, it results in oligomerization. The second set of enzymes mainly focuses on petrobactin biosynthesis.

13.3.2.1 Oligomerization

The siderophores that bring about the oligomerization reaction, such as desferri-oxamine E, bisucaberin, and putrebactin are macrocylones that mainly consists of Fe-chelating hydroxamate groups (Kameyama et al.1987; Takahashi et al.1987; Ledyard and Butler 1997).

13.3.2.2 Desferrioxamine

When the biosynthesis of desferrioxamine E and deasferrioxamine B in *Streptomyces coelicolour* was taken into consideration, it was observed that a cluster of six genes was reported to be responsible for its biosynthesis. Out of these, the gene desD possesses a type C NIS synthetase that is responsible for synthesizing the desferrioxamine E from N-hydroxy-N-succinyl-cadaverine (HSC) and desferrioxamine B from its precursor HSC and N-acetylN-hydroxy-cadaverine (AHC) (Barona-Go et al. 2004). The desABC genes were reported to code for enzymes responsible for catalyzing the conversion of L-lysine, succinyl-CoA, and molecular oxygen for the formation of AHC and HSC. In an independent study, the purified products from the recombinant desD were found responsible for the formation of trishydroxamatedesferrioxamine from HSC with tris-hydroxymatedesferrioxamine as the intermediate of the process. The formation of the dimeric bishydroxamate during the formation of HSC to trishydroxamatedesferrioxamine was also reported (Kadi et al. 2007). The incorporation of desD, ATP, and Mg2+ to the reaction mixture that consisted of HSC and AHC resulted in the liberation of desferrioxamine B and desferrioxamine E, in whichthe ATP was converted to AMP in the process (Kadi et al. 2007). DesD was also reported to play a vital role in the dimerization and trimerization of N-succinylcadaverine.

13.3.2.3 Bisucaberin

Bisucaberin was isolated for the first time from *Alteromonas haloplanktis*, a deep-sea bacterium that causes cold water vibriosis in fish (Kameyama et al. 1987; Takahashi et al.1987; Winkelmann et al. 2002). When the desD gene was used as a probe for the identification of genome of *V. salmonicida*, the bibC gene was found to encode for enzymes with two terminals (the C- and N-terminal domains) possessing catalytic properties (Kadi et al. 2008). The C- and N-domains were found to be 44% and 32% similar to the desD and desC genes, respectively. This clearly indicates that the bibC gene is capable of synthesizing HSC. Further upstream of the bibC gene, other genes capable of biosynthesis were identified (Barona-Go et al.2004; Kadi et al. 2007). The His6-bibCC protein was synthesized from the C-terminal of *E. coli*. The incorporation of the His6-BibCC with synthetic HSC, ATP, and Mg 2+ resulted in the formation of bisucaberin with bis-hydroxamate acting as an intermediate in the formation of bisucaberin from HSC along with the utilization of ATP and the formation of AMP and pyrophosphate (Kadi et al. 2008).

13.3.2.4 Putrebactin

The macrocyclic, putrebactin, a dimer of N-hyroxy-Nsuccinyl-putrescine (HSP) is isolated from *Shewanella putrefaciens*. The identification of the sequence similarities of the various *Shewanella*species reported that the pubC gene encoded for enzymes with50% similarity to the desD gene (Kadi et al, 2008). The other genes identified upstream, having biosynthetic activity, were termed PubA, PubB, and PubC. All three were found to catalyze the synthesis of putrebactin from putrescine, succinyl CoA, and molecular oxygen. The inoculation of HSP, ATP, and Mg2+ mixture with recombinant putrescine resulted in the formation of putrebactin along with the utilization of ATP and its conversion to AMP and pyrophosphate.

The NRPS independent pathways discussed have some common considerations. The oligomerizing synthetase family of enzymes use ATP to catalyze the reactions and, thus, produce AMP and pyrophosphate. The desferrioxamine E, bisucaberin, and putrebactin, in the case of NIS, were observed to dissociate from the enzymes easily, whereas, in the case of the NRPS biosynthetic pathway, the intermediates remain attached to the enzymes by means of covalent bonds.

13.4   APPLICATIONS OF SIDEROPHORES

Bacteria and fungi produce chelating agents known as siderophores that are mainly concerned with Fe uptake when it is present in a limited concentration. An important distinctive feature of marine ecosystems is that they have a low Fe content. Although marine microorganisms differ in structure to their terrestrial counterparts, they perform a similar function to overcome Fe limitation through siderophore production. Therefore, these microbial siderophores can be utilized for the bioremediation of polluted areas due to metal and organic compound contamination.

13.4.1   SIDEROPHORE RESPONSE AGAINST ABIOTIC STRESS

Bacteria and fungi release siderophores into the rhizosphere, and these siderophores behave as heavy metal scavengers. The rhizosphere (shoot and root interface) is mainly involved in the phytoremediation of polluted soils. Due to the existence of siderophore-releasing bacteria in the rhizosphere, it is regarded as a metabolically active region (Rajkumar et al. 2010). These types of bacteria secrete various chelating agents, phosphate solubilizing complexes, and phytohormones that increase the bioavailability and mobility of heavy metals (i.e., cadmium, nickel, lead, Zn, etc.) (Schalk et al. 2011). The specificity of ligands, along with a stable complex between the metal and siderophore, is a prerequisite for increasing the phytoremediation using siderophores (Braud et al. 2006, 2007).

Several genetic engineering studies have been conducted by incorporating the host plant genome with the siderophore producing genes from bacteria and fungi. Murata et al. (2015) reported an increased concentration of Fe-phyto-siderophore complex along with its transporters in the roots of *P. hybrid -a* transgenic plants that were grown in soil that was Fe deficient. Plants also induce stress tolerance through the production of siderophores, including enterobactin facilitating *E. coli* colonization and commensalism (Searle et al. 2015). The production of multiple siderophores, along with the plant hormone indole-3-acetic acid, occurred in a Cd-resistant soil bacteria, i.e., *Enterobacter*. When applied to plants, the extracts of these bacteria resulted in 31% Cd accumulation (Chen et al. 2016).

13.4.2   MICROBIAL ECOLOGY

Microbial ecology involves a relationship between the microorganisms and their environment. It can be further divided as described in the following sections.

13.4.2.1 Growth Enhancement of Non-Cultivable Microbes in Artificial Media Supplemented with Siderophores

A group of microbes associated together in a habitat forms the microbial community, wherein the habitat represents a balance between the living and non-living organisms. Among such a large microbial population, only a small amount can be grown under artificial conditions (approximately 0.1%–1%), while most of the population cannot be cultivated under these conditions (Torsvik and Ovreas 2002). The increase in the unculturable microbial population may be attributed to the nonavailability of some of the basic growth requirements in their artificial growth environment, i.e., temperature, nutrient load, and pH (Vartoukian et al. 2010). Bacterial cultivation on the artificial growth medium is a prerequisite for evaluation of the bacterial physiology and further utilization in ecological and bioremediation studies (Stewart 2012). The growth of various anaerobic thermophilic bacteria belonging to Clostridiaceae family was reported when an extract from *Geobacillus toebii* was used on the artificial medium (Kim et al. 2008, 2011). Furthermore, Kaeberlein et al. (2002) demonstrated siderophore production by a microorganism that would serve as a growth factor and be responsible for the growth enhancement of various microorganisms that would otherwise be impossible to culture. Similarly, D' Onofrio et al. (2010) illustrated the production of five acyl- desferrioxamine siderophores in the *M. luteus* strain KLE1011, in which each one of these siderophores enhanced the growth of the *M. polysiphoniae* strain KLE1104.

In another study conducted by Kamino (2001) on the marine ecosystem, an addition of exogenous siderophores resulted in increased growth of non-cultivable marine bacterial species under laboratory conditions. Even though a large number of unculturable microbial strains were not able to grow under laboratory conditions, as they could not produce siderophores, their dependency on other organisms helped them survive in that environment (D'Onofrio et al.2010).

Therefore, through this approach, pure culture of many non-cultivable organisms can be obtained, which can be further utilized for research purposes. Figure 13.1 shows a pure culture of culturable as well as non-cultivable organisms obtained with or without (auxotrophs) exogenous siderophores in the growth medium.

13.4.2.2 Alteration in the Microbial Community

Different findings have clearly suggested the role of soil mineral supplementation for bringing about a change in its bacterial community (Carson et al. 2007). Fe is involved in the regulation of some essential cellular processes related to microorganisms, including energy production, and it can even alter the microbial community when added to the soil (Sullivan et al.2012; Eldridge et al. 2007; Jin et al. 2010,2014). The low availability of Fe and its high demand poses a threat to the microorganism's survival as well as its growth under aerobic conditions. So, in order to capture Fe molecules from the environment, numerous aerobic as well as facultative anaerobes produce siderophores (Chincholkar et al. 2007). This increased Fe concentration within the soil microorganism has resulted in an altered microbial community through the enhanced proliferation of the microbes (Sullivan et al. 2012).

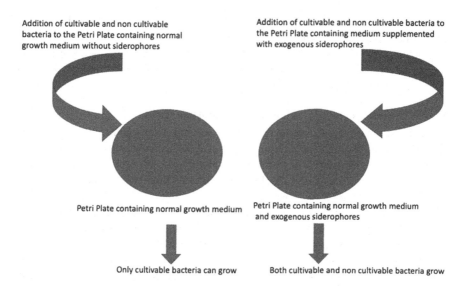

Addition of cultivable and non cultivable bacteria to the Petri Plate containing normal growth medium without siderophores

Addition of cultivable and non cultivable bacteria to the Petri Plate containing medium supplemented with exogenous siderophores

Petri Plate containing normal growth medium

Petri Plate containing normal growth medium and exogenous siderophores

Only cultivable bacteria can grow

Both cultivable and non cultivable bacteria grow

FIGURE 13.1 Growth enhancement of noncultivable microbes in artificial media supplemented with siderophores Obtaining a pure culture of culturable as well as noncultivable organisms via supplementing with or without (auxotrophs) exogenous siderophores in growth medium.

13.4.3 ROLE IN AGRICULTURE

As siderophores are eco-friendly, they can serve as a substitute to the harmful pesticides that are used in agriculture. Their use in the agricultural sector can be further divided as discussed in the following sections.

13.4.3.1 Siderophore Utilization as a Biocontrol Agent

Siderophores are crucial for suppressing the growth of various pathogens and can be utilized as an effective biocontrol agent, wherein they kill phytopathogens by reducing the Fe bioavailability of the pathogens by themselves binding with Fe (Beneduzi et al. 2012; Ahmed and Holmstrom 2014). Kloepper et al. (1980) demonstrated the role of siderophore producing *P. fluorescens* as a potential biocontrol against *E. carotovora*. Similarly, pseudomonas produce pyoverdine, which can be used to control potato wilt disease which is caused by *Fusarium oxysporum* (Schippers et al. 1987). The pyoverdine have proven to be effective in dealing with the pathogen related to the growth associated deficiencies of wheat, i.e., *G. graminis*. Additionally, it is also efficient against some of the disease-causing pathogens of maize and peanuts (Voisard et al. 1989; Pal et al. 2001).

Additionally, the siderophores of *B. subtilis* act as a biocontrol agent for *F. oxysporum*, which causes *Fusarium* wilt in peppers (Yu et al. 2011). Various neem siderophores undergo chelation with Fe that is present in the soil in order to hamper the growth of various fungal pathogens (Verma et al. 2011). In view of this, it was

illustrated that siderophores could further hamper the growth of some other pathogenic fungi including *P. ultimum, Phytophthora parasitica*, and *S. schlerotiorum*
(Park et al. 1988; Hamdan et al. 1991; McLoughlin et al. 1992). Consequently, these
studies clearly regard siderophores to be potent biocontrol agents that can be used
effectively against different pathogens.

13.4.3.2 Enhancement of Heavy Metal Bioremediation

Soil contamination due to extensive usage of pesticides and chemical fertilizers has
resulted in heavy metal accumulation(Zhang et al. 2010; Wuana and Okieimen 2011).
In addition to being used as an Fe chelator, siderophores are crucial for the elimination of toxic metals (i.e., Cu, lead, chromium, etc.) by binding with them (Nair et al.
2007; Rajkumar et al. 2010; O'Brien et al. 2014).

At their lower concentrations, many metals can facilitate bacterial growth but
may prove toxic to the bacteria as their concentration increases (Heldal et al. 1985).
Valko et al. (2005) described increased bacterial growth at a low Cu concentration
as it was utilized by the bacteria for enzyme functioning as well as in the electron
transport chain, whereas at a higher concentration, Cu caused DNA damage due to
oxidative stress generation. Further study demonstrated that arsenic detoxification
from its contaminated soil was carried out using siderophores that were produced
by *P. azotoformans* (Nair et al. 2007). Braud et al. (2009b) indicated the function of
another siderophore, namely pyochelin, that was produced from *P. aeruginosa* and
was used for the chelation of various toxic metals, including cadmium, Cu, and mercury. Siderophores can be further utilized for the mobilization of these toxic metals
(Ahmed and Holmstrom 2014). Wang et al. (2011) reported arsenic removal by the *A.
radiobacter* siderophore. Similarly, some siderophores were utilized in mine waste
detoxification, and various metals (e.g., Fe, cobalt, and nickel) were immobilized
from uranium contaminated mines utilizing *P. fluorescens* siderophores (Edberg
et al. 2010; Wang et al 2011).Burd et al. (2000) demonstrated that the siderophores of
the *Kluyveraa scorbata* mutant SUD165 decreased the heavy metal toxicity in soil
samples in Ontario. Therefore, siderophores can be considered to be effective agents
for the bioremediation of hazardous metals.

13.4.4 Role of Siderophore as a Biosensor

Biosensor refers to an instrument that consists mainly of a biorecognition component
(bioreceptor), a biotransducer, and an electronic component with an amplifier, pro-
cessor, and display for signalanalysis (Eggins 1996).

Using a biosensor involves interaction between a biological element and a bioreceptor that produces quantitative or semi-quantitative information that the transducer
recognizes (Thevenot et al. 1999; Gupta et al. 2008). Pyoverdine is a water-soluble
yellow-green siderophore released from *P. aeruginosa* and has been utilised as a
biosensor for Fe detection (Pesce and Kaplan 1990). Biosensors have been used to
determine the Fe concentration in water samples (Barrero et al. 1993). Similarly,
chemo-sensors, made up of N-methylanthranyldesferrioxamine, have a significant
role in detecting natural water sources (Palanche et al. 1999).

Biosensors based on *Pseudomonas fluorescens* siderophore are robust, user-friendly, and Fe specific (Gupta et al. 2008). In a further study, MA-DFB, a derivative of a fluorescent siderophore, was utilized as an Fe photoactive sensor for aquatic ecosystems (Orcutt et al. 2010).

Although the siderophores are effectively utilized for the detoxification of hazardous metals using the bioremediation process, their role in clinical studies is also significant, hence, they cannot be ignored. The process is outlined in the following sections

13.4.5 In Medicine

Siderophores play a key role against bacteria and can be used in the treatment of various human diseases, including the following.

13.4.5.1 Trojan Horse Strategy

Bacteria have an ability to show resistance toward the antimicrobial agents that makes it difficult to treat bacterial infections. In order to overcome drug resistance, a strategy using "Trojan horse" antibiotics can be employed (Mollmann et al. 2009). In this strategy, the siderophores help in the selective antibiotic transport to the antibiotic resistant bacteria (Huang et al. 2013).

Albomycin, an antibiotic that occurs naturally can bind to the siderophore (sideromycins) and inhibit the effects of both gram positive as well as gram negative bacteria (Pramanik and Braun 2006). The microorganism's uptake of this antibiotic can occur through the ferrichrome system, thereby, releasing its toxic part into the cell (Ali and Vidhale 2013). Braun et al. (2009) described incremental diffusion of the antibiotic within a bacterial cell through siderophore transport in the bacterial membrane. It was further reported that an increased antibiotic transport rate was observed using sideromycin in comparison to using antibiotics alone. Although naturally occurring, sideromycin was efficient and showed better antimicrobial activity, but at times it also showed low solubility, instability, insufficient absorption, and low penetration within the tissues (Rautio et al. 2008). So, in these cases, synthetic siderophore drugs can provide a promising solution against bacteria that are multi-drug resistant (Krewulak and Vogel 2008). The conjugate of these synthesized siderophores, when accompanied with beta-lactam antibiotics, proved effective against the microorganisms, as this complex is responsible for the growth retardation in bacteria, *i.e.*, gram negative (Brochu et al. 1992). Hydroxamate, along with catecholate siderophores, have been mainly used as antimicrobials in the case of membrane permeation of drugs. But the carboxylate-type, including staphyloferrin A, proved to be effective Fe-chelating agents in comparison to the hydroxamate and catecholate siderophores (Milner et al. 2013). Further, Milner et al. (2013) produced many staphyloferrin conjugates of Trojan horse antibiotics, thereby, testing them against microbes. It was revealed that one conjugate exhibited antimicrobial activity against *S. aureus*.

Nanotechnology is an emerging field that can be utilized for the production of the best therapeutic system, which can prove helpful in delivering the precise drug into the appropriate location. The nanoparticle therapeutic systems using Fe oxide can be used for efficient drug delivery (Gorska et al. 2014).

13.4.5.2 Iron Overload Remediation Through Siderophores

Siderophores also play a significant role in dealing with diseases involving Fe overload. Siderophore-based drugs can help in the treatment of these types of diseases. One such drug, desferal, can be used for the treatment of thalassemia and sickle cell anemia (Propper et al. 1977; Summers et al. 1979; Pietrangelo 2002).

13.4.5.3 Siderophores as Antimalarial Agents

Antimalarial activity is a characteristic feature that some of the siderophores possess. Tsafack et al. (1996) reported that the siderophore for *P. falciparum* dis played antimalarial activity. Similarly, *K. pneumoniae* and *S. pilosus* siderophores (desferrioxamine B) showed antimalarial activity for *P. falciparum* (Gysin et al. 1993; Nagoba and Vedpathak 2011). This siderophore causes intracellular Fe depletion after entering inside the parasite. It then forms a conjugate with methyl anthranilic acid, thereby, showing increased activity against *P. falciparum* (Loyevsky et al. 1993, 1999).

13.4.5.4 Transuranic Element Removal

There has been increased human exposure to the different transuranic elements (e.g., aluminium, Vn) (Nagoba and Vedpathak 2011). An overload of aluminium takes place in those patients who suffer from chronic renal failure. So, desferal could be utilized for the treatment of aluminum overload, as it can chelate aluminium and forms a water-soluble aluminoxamine complex, so it can be excreted in the urine or feces (Nagoba and Vedpathak 2011).

Desferal has also proven to be efficient in the elimination of another transuranic element, namely Vn, through its increased excretion in urine and feces (Ackrill et al. 1980; Arze et al. 1981; Pogglitsch et al. 1981; Hansen et al. 1982; Nagoba and Vedpathak 2011). Further study suggested that desferal decreased the concentration of Vn in different organs of rats, including the kidney, liver, and lungs up to 20%, 25%, and 26%, respectively (Nagoba and Vedpathak 2011).

13.5 CONCLUSION

Soil contaminated with heavy metals has been reported from most parts of the world. The removal of these toxic metals is not an easy task, hence, the use of natural means to combat this problem has become important. Siderophores result in bioremediation by converting these toxic metals to biodegradable forms. The action of enzymes produced by these organisms lower the toxicity of these compounds. Therefore, siderophores serve as a promising alternative to chemical means of soil remediation.

REFERENCES

Abou-Shanab, R. A. I., Angle, J. S., and Chaney, R. L.2006. Bacterial inoculants affecting nickel uptake by Alyssum murale from low, moderate and high Ni soils. *Soil Biology and Biochemistry*38:2882–2889.

Achard, M. E., Chen, K. W., Sweet, M. J., Watts, R. E., Schroder, K., Schembri, M. A., and McEwan, A. G.2013. An antioxidant role for catecholate siderophores in Salmonella. *Biochemical Journal*454: 543–549.

Ackrill, P. A. J. K., Ralston, A. J., Day, J. P., and Hodge, K. C.1980. Successful removal of aluminium from patient with dialysis encephalopathy. *The Lancet* 316:692–693.

Aggarwal, R., Caffrey, P., Leadlay, P. F., Smith, C. J., and Staunton, J.1995. The thioesterase of the erythromycin-producing polyketide synthase: mechanistic studies in vitro to investigate its mode of action and substrate specificity. *Journal of the Chemical Society, Chemical Communications*15:1519–1520.

Ahmed, E., and Holmström, S. J.2014. Siderophores in environmental research: roles and applications. *Microbial Biotechnology* 7:196–208.

Akafia, M. M., Harrington, J. M., Bargar, J. R., and Duckworth, O.W.2014. Metal oxyhydroxide dissolution as promoted by structurally diverse siderophores and oxalate. *Geochimica et Cosmochimica Acta* 141:258–269.

Al-Fakih, A. A.2014. Overview on the fungal metabolites involved in mycopathy. *Open Journal of Medical Microbiology* 2014.

Ali, S. S., and Vidhale, N.N.2013. Bacterial siderophore and their application: a review. *International Journal of Current Microbiology and Applied Sciences* 2:303–312.

Ali, T., Bylund, D., Essén, S. A., and Lundström, U. S.2011. Liquid extraction of low molecular mass organic acids and hydroxamate siderophores from boreal forest soil. *Soil Biology and Biochemistry* 43:2417–2422.

Allred, B. E., Rupert, P. B., Gauny, S. S., An, D. D., Ralston, C. Y., Sturzbecher-Hoehne, M., ... and Abergel, R. J.2015. Siderocalin-mediated recognition, sensitization, and cellular uptake of actinides. *Proceedings of the National Academy of Sciences* 112: 10342–10347.

Amin, S. A., Küpper, F. C., Green, D. H., Harris, W. R., and Carrano, C. J.2007. Boron binding by a siderophore isolated from marine bacteria associated with the toxic dinoflagellate *Gymnodiniumcatenatum*. *Journal of the American Chemical Society* 129: 478–479.

Arze, R. S., Parkinson, I. S., Cartlidge, N. E. F., Britton, P., and Ward, M. K.1981. Reversal of aluminium dialysis encephalopathy after desferriox-amine treatment. *Reversalofalumi niumdialysisencephalopathyafterdesferriox-aminetreatment* 2.

Aznar, A., and Dellagi, A.2015. New insights into the role of siderophores as triggers of plant immunity: what can we learn from animals?. *Journal of Experimental Botany* 66:3001–3010.

Aznar, A., Chen, N. W., Thomine, S., and Dellagi, A.2015. Immunity to plant pathogens and iron homeostasis. *Plant Science* 240:90–97.

Banik, S., Das, K. C., Islam, M. S., and SalimullahM.2014. Recent advancements and challenges in microbial bioremediation of heavy metals contamination. *JSM Biotechnology & Biomedical Engineering* 2:1035.

Barona-Gómez, F., Wong, U., Giannakopulos, A. E., Derrick, P. J., and Challis, G. L.2004. Identification of a cluster of genes that directs desferrioxamine biosynthesis in *Streptomyces coelicolor* M145. *Journal of the American Chemical Society* 126:16282–16283.

Barrero, J. M., Morino-Bondi, M. C., Pérez-Conde, M. C., and Cámara, C.1993. A biosensor for ferric ion. *Talanta* 40:1619–1623.

Beneduzi, A., Ambrosini, A., and Passaglia, L. M.2012. Plant growth-promoting rhizobacteria (PGPR): their potential as antagonists and biocontrol agents. *Genetics and Molecular Biology* 35 (Supplement):1044–1051.

Bergeron, R. J., Wiegand, J., McManis, J. S., and Bharti, N.2014. Desferrithiocin: a search for clinically effective iron chelators. *Journal of Medicinal Chemistry* 57:9259–9291.

Braud, A., Jézéquel, K., Vieille, E., Tritter, A., and Lebeau, T.2006. Changes in extractability of Cr and Pb in a polycontaminated soil after bioaugmentation with microbial producers of biosurfactants, organic acids and siderophores. *Water, Air, and Soil Pollution: Focus* 6:261–279.

Braud, A., Jézéquel, K., and Lebeau, T.2007. Impact of substrates and cell immobilization on siderophore activity by Pseudomonads in a Fe and/or Cr, Hg, Pb containing-medium. *Journal of Hazardous Materials* 144:229–239.

Braud, A., Jézéquel, K., Bazot, S., and Lebeau, T.2009. Enhanced phytoextraction of an agricultural Cr-and Pb-contaminated soil by bioaugmentation with siderophore-producing bacteria. *Chemosphere* 74:280–286.

Braun, V., Pramanik, A., Gwinner, T., Köberle, M., and Bohn, E.2009. Sideromycins: tools and antibiotics. *Biometals* 22: 3–13.

Brochu, A., Brochu, N., Nicas, T. I., Parr, T. R., Minnick, A. A., Dolence, E. K., and Malouin, F.1992. Modes of action and inhibitory activities of new siderophore-beta-lactam conjugates that use specific iron uptake pathways for entry into bacteria. *Antimicrobial Agents and Chemotherapy* 36: 2166–2175.

Burd, G. I., Dixon, D. G., and Glick, B. R.2000. Plant growth-promoting bacteria that decrease heavy metal toxicity in plants. *Canadian Journal of Microbiology* 46:237–245.

Butler, A., and Theisen, R. M.2010. Iron (III)–siderophore coordination chemistry: Reactivity of marine siderophores. *Coordination Chemistry Reviews* 254:288–296.

Buyer, J. S., De Lorenzo, V., and Neilands, J. B.1991. Production of the siderophore aerobactin by a halophilic pseudomonad. *Applied and Environmental Microbiology* 57:2246–2250.

Carson, J. K., Rooney, D., Gleeson, D. B., and Clipson, N.2007. Altering the mineral composition of soil causes a shift in microbial community structure. *FEMS Microbiology Ecology* 61:414–423.

Chaiharn, M., Chunhaleuchanon, S., Kozo, A., and Lumyong, S.2008. Screening of rhizobacteria for their plant growth promoting activities. *Current Applied Science and Technology* 8:18–23.

Challis, G. L.2005. A widely distributed bacterial pathway for siderophore biosynthesis independent of nonribosomal peptide synthetases. *Chembiochem* 6:601–611.

Chaturvedi, K. S., and Henderson, J. P.2014. Pathogenic adaptations to host-derived antibacterial copper. *Frontiers in Cellular and Infection Microbiology* 4:3.

Chaturvedi, K. S., Hung, C. S., Giblin, D. E., Urushidani, S., Austin, A. M., Dinauer, M. C., and Henderson, J. P.2014. *ACS Chemical Biology* 9:551–561.

Chen, Y., Chao, Y., Li, Y., Lin, Q., Bai, J., Tang, L. and Qiu, R.2016. Survival strategies of the plant-associated bacterium Enterobacter sp. strain EG16 under cadmium stress. *Applied and Environmental Microbiology* 82:1734–1744.

Chincholkar, S. B., Chaudhari, B. L., and Rane, M. R.2007. Microbial siderophore: a state of art. In *Microbial Siderophores*, 233–242. Springer, Berlin, Heidelberg.

de Lorenzo, V. I. C. T. O. R., Bindereif, A. L. B. R. E. C. H. T., Paw, B. H., and Neilands, J. B.1986. Aerobactin biosynthesis and transport genes of plasmid ColV-K30 in Escherichia coli K-12. *Journal of Bacteriology* 165:570–578.

DentonB.2007. Advances in phytoremediation of heavy metals using plant growth promoting bacteria and fungi. *MMG 445 Basic Biotechnology* 3:1–5.

D'Onofrio, A., Crawford, J. M., Stewart, E.J., Witt, K., Gavrish, E., Epstein, S., ... and Lewis, K.2010. Siderophores from neighboring organisms promote the growth of uncultured bacteria. *Chemistry and Biology* 17:254–264.

Duckworth, O. W., Akafia, M. M., Andrews, M. Y., and Bargar, J. R.2014. Siderophore-promoted dissolution of chromium from hydroxide minerals. *Environmental Science: Processes and Impacts* 16:1348–1359.

Edberg, F., Kalinowski, B. E., Holmström, S. J., and Holm, K. 2010. Mobilization of metals from uranium mine waste: the role of pyoverdines produced by Pseudomonas fluorescens. *Geobiology* 8:278–292.

Eggins, B. R. 1996. *Biosensors: An Introduction*. Wiley, Chichester, UK, pp 16–19.

Eldridge, M. L., Cadotte, M. W., Rozmus, A. E., and Wilhelm, S. W. 2007. The response of bacterial groups to changes in available iron in the Eastern subtropical Pacific Ocean. *Journal of Experimental Marine Biology and Ecology* 348:11–22.

Gadd, G. M., and Griffiths, A. J.1977. Microorganisms and heavy metal toxicity. *Microbial Ecology* 4: 303–317.

Gamalero, E., Lingua, G., Berta, G., and Glick, B. R.2009. Beneficial role of plant growth promoting bacteria and arbuscular mycorrhizal fungi on plant responses to heavy metal stress. *Canadian Journal of Microbiology* 55:501–514.

Gehring, A. M., Bradley, K. A., and Walsh, C. T.1997. Enterobactin biosynthesis in *Escherichia coli*: isochorismate lyase (EntB) is a bifunctional enzyme that is phosphopantetheinylated by EntD and then acylated by EntE using ATP and 2, 3-dihydroxybenzoate. *Biochemistry* 36:8495–8503.

Glick, B. R.2012. Plant growth-promoting bacteria: mechanisms and applications. *Scientifica* 2012: 1–15. Article ID 963401.

Górska, A., Sloderbach, A., and Marszałł, M. P.2014. Siderophore–drug complexes: potential medicinal applications of the 'Trojan horse'strategy. *Trends in Pharmacological Sciences* 35:442–449.

Gregory, J. A., Li, F., Tomosada, L. M., Cox, C. J., Topol, A. B., Vinetz, J. M., and Mayfield, S.2012. Algae-produced Pfs25 elicits antibodies that inhibit malaria transmission. *PloS One* 7:37179.

Gupta, G., Parihar, S. S., Ahirwar, N. K., Snehi, S. K., and Singh, V.2015. Plant growth promoting rhizobacteria (PGPR): current and future prospects for development of sustainable agriculture. *Journal of Microbial and Biochemical Technology* 7: 96–102.

Gupta, V., Saharan, K., Kumar, L., Gupta, R., Sahai, V., and Mittal, A.2008. Spectrophotometric ferric ion biosensor from *Pseudomonas fluorescens* culture. *Biotechnology and Bioengineering* 100:284–296.

Gysin, J., Crenn, Y., da Silva, L. P., and Breton, C.1993. U.S. Patent No. 5,192,807. Washington, DC: U.S. Patent and Trademark Office.

Hamdan, H., Weller, D. M., and Thomashow, L. S.1991. Relative importance of fluorescent siderophores and other factors in biological control of *Gaeumannomycesgraminis* var. tritici by *Pseudomonas fluorescens* 2–79 and M4-80R. *Applied and Environmental Microbiology* 57:3270–3277.

Hansen, T. V., Aaseth, J., and Alexander, J.1982. The effect of chelating agents on vanadium distribution in the rat body and on uptake by human erythrocytes. *Archives of Toxicology* 50:195–202.

Harrington, J. M., Parker, D. L., Bargar, J. R., Jarzecki, A. A., Tebo, B. M., Sposito, G., and Duckworth, O. W.2012. Structural dependence of Mn complexation by siderophores: donor group dependence on complex stability and reactivity. *Geochimica et Cosmochimica Acta* 88:106–119.

Heldal, M. I. K. A. L., Norland, S. V. E. I. N., and Tumyr, O.1985. X-ray microanalytic method for measurement of dry matter and elemental content of individual bacteria. *Applied and Environmental Microbiology* 50:1251–1257.

Hider, R. C., and Kong, X.2010. Chemistry and biology of siderophores. *Natural Product Reports* 27:637–657.

Hood, M. I., and Skaar, E. P.2012. Nutritional immunity: transition metals at the pathogen-host interface. *Nature Reviews Microbiology* 10:525–537.

Huang, Y., Jiang, Y., Wang, H., Wang, J., Shin, M. C., Byun, Y., ... and Yang, V. C.2013. Curb challenges of the "Trojan Horse" approach: smart strategies in achieving effective yet safe cell-penetrating peptide-based drug delivery. *Advanced Drug Delivery Reviews* 65:1299–1315.

Husen, E.2020. Screening of soil bacteria for plant growth promotion activities in vitro.

Jin, C. W., Li, G. X., Yu, X. H., and Zheng, S. J.2010. Plant Fe status affects the composition of siderophore-secreting microbes in the rhizosphere. *Annals of Botany* 105:835–841.

Jin, C. W., Ye, Y. Q., and Zheng, S. J.2014. An underground tale: contribution of microbial activity to plant iron acquisition via ecological processes. *Annals of Botany* 113:7–18.

Jing, Y. D., He, Z. L., and Yang, X. E.2007. Effects of pH, organic acids, and competitive cations on mercury desorption in soils. *Chemosphere* 69:1662–1669.

Johnstone, T.C., andNolana, E.M.2015. Beyond iron: non-classical biological functions of bacterial siderophores. *Dalton Transactions* 44:6320–6339.

Joshi, H., Dave, R., and Venugopalan, V. P.2014. Pumping iron to keep fit: modulation of siderophore secretion helps efficient aromatic utilization in *Pseudomonas putida* KT2440. *Microbiology* 160:1393–1400.

Kadi, N., Oves-Costales, D., Barona-Gomez, F., and Challis, G. L.2007. A new family of ATP-dependent oligomerization-macrocyclization biocatalysts. *Nature Chemical Biology* 3:652–656.

Kadi, N., Song, L., and Challis, G. L.2008. Bisucaberin biosynthesis: an adenylating domain of the BibC multi-enzyme catalyzes cyclodimerization of N-hydroxy-N-succinylcadaverine. *Chemical Communications* 41:5119–5121.

Kaeberlein, T., Lewis, K., and Epstein, S. S.2002. Isolating" uncultivable" microorganisms in pure culture in a simulated natural environment. *Science* 296:1127–1129.

Kameyama, T., Takahashi, A., Kurasawa, S., Ishizuka, M., Okami, Y., Takeuchi, T., and Umezawa, H.1987. Bisucaberin, a new siderophore, sensitizing tumor cells to macrophage-mediated cytolysis I. taxonomy of the producing organism, isolation and biological properties. *The Journal of Antibiotics* 40:1664–1670.

Kamino, K.2001. Bacterial response to siderophore and quorum-sensing chemical signals in the seawater microbial community. *BMC Microbiology* 1:1–11.

Keating, T. A., and Walsh, C. T.1999. Initiation, elongation, and termination strategies in polyketide and polypeptide antibiotic biosynthesis. *Current Opinion in Chemical Biology* 3:598–606.

Keating, T. A., Suo, Z., Ehmann, D. E., and Walsh, C. T.2000. Selectivity of the yersiniabactin synthetase adenylation domain in the two-step process of amino acid activation and transfer to a holo-carrier protein domain. *Biochemistry* 39:2297–2306.

Keating, T. A., Ehmann, D. E., Kohli, R. M., Marshall, C. G., Trauger, J. W., and Walsh, C. T. 2001. Chain termination steps in nonribosomal peptide synthetase assembly lines: Directed acyl-S-enzyme breakdown in antibiotic and siderophore biosynthesis. *ChemBioChem* 2:99–107.

Khan, A. R., Ghorai, A. K., and Singh, S. R.2000. Improvement of crop and soil sustainability through green manuring in a rainfed lowland rice ecosystem. *Agrochimica* 44:21–29.

Kim, J. J., Masui, R., Kuramitsu, S., Seo, J. H., Kim, K., and Sung, M. H.2008. Characterization of growth-supporting factors produced by *Geobacillustoebii* for the commensal thermophile Symbiobacteriumtoebii. *Journal of Microbiology and Biotechnology* 18:490–496.

Kim, K., Kim, J. J., Masui, R., Kuramitsu, S., and Sung, M. H.2011. A commensal symbiotic interrelationship for the growth of *Symbiobacteriumtoebii* with its partner bacterium, *Geobacillustoebii*. *BMC Research Notes* 4:1–9.

Kloepper, J. W., Leong, J., Teintze, M., and Schroth, M. N.1980. Enhanced plant growth by siderophores produced by plant growth-promoting rhizobacteria. *Nature* 286:885–886.

Kohli, R. M., Trauger, J. W., Schwarzer, D., Marahiel, M. A., and Walsh, C. T.2001. Generality of peptide cyclization catalyzed by isolated thioesterase domains of nonribosomal peptide synthetases. *Biochemistry* 40:7099–7108.

Kraemer, S. M.2004. Iron oxide dissolution and solubility in the presence of siderophores. *Aquatic Sciences* 66:3–18.

Kraemer, S. M., Crowley, D., and Kretzschmar, R.2006. Siderophores in plant iron acquisition: Geochemical aspects. *Advances in Agronomy* 91:1–46.

Krewulak, K. D., and Vogel, H. J.2008. Structural biology of bacterial iron uptake. *Biochimica et Biophysica Acta (BBA)-Biomembranes* 1778:1781–1804.

Kurth, C., Kage, H., and Nett, M.2016. Siderophores as molecular tools in medical and environmental applications. *Organic and Biomolecular Chemistry*14:8212–8227.

Lambalot, R. H., Gehring, A. M., Flugel, R. S., Zuber, P., LaCelle, M., Marahiel, M. A., ... and Walsh, C. T.1996. A new enzyme superfamily: The phosphopantetheinyl transferases. *Chemistry and Biology* 3:923–936.

Ledyard, K. M., and Butler, A.1997. Structure of putrebactin, a new dihydroxamate siderophore produced by *Shewanellaputrefaciens*. *JBIC Journal of Biological Inorganic Chemistry* 2:93–97.

Leong, S. A., and Neilands, J. B.1982. Siderophore production by phytopathogenic microbial species. *Archives of Biochemistry and Biophysics* 218:351–359.

Leyval, C., Turnau, K., and Haselwandter, K.1997. Effect of heavy metal pollution on mycorrhizal colonization and function: physiological, ecological and applied aspects. *Mycorrhiza* 7:139–153.

Loyevsky, M., Lytton, S. D., Mester, B., Libman, J., Shanzer, A., and Cabantchik, Z. I.1993. The antimalarial action of desferal involves a direct access route to erythrocytic (*Plasmodium falciparum*) parasites. *The Journal of Clinical Investigation* 91: 218–224.

Loyevsky, M., John, C., Dickens, B., Hu, V., Miller, J. H., and Gordeuk, V. R.1999. Chelation of iron within the erythrocytic *Plasmodium falciparum* parasite by iron chelators. *Molecular and Biochemical Parasitology* 101:43–59.

Madhaiyan, M., Poonguzhali, S., and Sa, T.2007. Metal tolerating methylotrophic bacteria reduces nickel and cadmium toxicity and promotes plant growth of tomato (*Lycopersicon esculentum* L.). *Chemosphere* 69:220–228.

Manwar, A. V., Khandelwal, S. R., Chaudhari, B. L., Kothari, R. M., andChincholkar, S. B.2001. Generic technology for assured biocontrol of groundnut infections leading to its yield improvement. *Chemical Weekly-Bombay* 46:157–158.

May, J. J., Wendrich, T. M., and Marahiel, M. A.2001. The dhb operon of bacillus subtilisEncodes the biosynthetic template for the catecholic siderophore 2, 3-dihydroxybenzoate-glycine-threonine trimeric ester bacillibactin. *Journal of Biological Chemistry* 276:7209–7217.

McGrath, S. P., Chaudri, A. M., andGiller, K. E.1995. Long-term effects of metals in sewage sludge on soils, microorganisms and plants. *Journal of Industrial Microbiology*14:94–104.

McLoughlin, T. J., Quinn, J. P., Bettermann, A., and Bookland, R.1992. Pseudomonas cepacia suppression of sunflower wilt fungus and role of antifungal compounds in controlling the disease. *Applied and Environmental Microbiology* 58:1760–1763.

Meneely, K. M., and Lamb, A. L.2007. Biochemical characterization of a flavin adenine dinculeotide-dependent monooxygenase, ornithine hydroxylase from *Pseudomonas aeruginosa*, suggests a novel reaction mechanism. *Biochemistry* 46:11930–11937.

Miethke, M., and Marahiel, M. A.2007. Siderophore-based iron acquisition and pathogen control. *Microbiology and Molecular Biology Reviews* 71:413–451.

Milner, S. J., Seve, A., Snelling, A. M., Thomas, G. H., Kerr, K. G., Routledge, A., and Duhme-Klair, A. K.2013. Staphyloferrin A as siderophore-component in fluoroquinolone-based Trojan horse antibiotics. *Organic and Biomolecular Chemistry* 11:3461–3468.

Möllmann, U., Heinisch, L., Bauernfeind, A., Köhler, T., and Ankel-Fuchs, D.2009. Siderophores as drug delivery agents: application of the "Trojan Horse" strategy. *Biometals* 22:615–624.

Mosa, K.A., Saadoun, I., Kumar, K., Helmy, M., Dhankher, O. P.2016. Potential biotechnological strategies for the cleanup of heavy metals and metalloids. *Frontiers in Plant Science* 7:303. doi. https://doi.org/10.3389/fpls.2016.00303

Münzinger, M., Taraz, K., Budzikiewicz, H., Drechsel, H., Heymann, P., Winkelmann, G., and Meyer, J. M.1999. S, S-rhizoferrin (enantio-rhizoferrin)–a siderophore of Ralstonia (Pseudomonas) pickettii DSM 6297–the optical antipode of R, R-rhizoferrin isolated from fungi. *Biometals*12:189–193.

Murata, Y., Itoh, Y., Iwashita, T., and Namba, K.2015. Transgenic petunia with the iron (III)-phytosiderophore transporter gene acquires tolerance to iron deficiency in alkaline environments. *PLoS One* 10:e0120227.

Nagoba, B., and Vedpathak, D.2011. Medical applications of siderophores. *European Journal of General Medicine* 8:229–235.

Nair, A., Juwarkar, A. A., and Singh, S. K.2007. Production and characterization of siderophores and its application in arsenic removal from contaminated soil. *Water, Air, and Soil Pollution* 180:199–212.

Naka, H., Liu, M., Actis, L. A., and Crosa, J. H.2013. Plasmid-and chromosome-encoded siderophore anguibactin systems found in marine vibrios: biosynthesis, transport and evolution. *Biometals* 26:537–547.

Neilands, J. B.1995. Siderophores: structure and function of microbial iron transport compounds. *Journal of Biological Chemistry* 270:26723–26726.

O'Brien, S., Hodgson, D. J., and Buckling, A.2014. Social evolution of toxic metal bioremediation in Pseudomonas aeruginosa. *Proceedings of the Royal Society B: Biological Sciences* 281:20140858.

Orcutt, K. M., Jones, W. S., McDonald, A., Schrock, D., and Wallace, K. J.2010. A lanthanide-based chemosensor for bioavailable Fe^{3+} using a fluorescent siderophore: an assay displacement approach. *Sensors* 10:1326–1337.

Pal, K. K., Tilak, K. V. B. R., Saxcna, A. K., Dey, R., and Singh, C. S.2001. Suppression of maize root diseases caused by *Macrophominaphaseolina, Fusarium moniliforme* and *Fusarium graminearum* by plant growth promoting rhizobacteria. *Microbiological Research* 156:209–223.

Palanché, T., Marmolle, F., Abdallah, M. A., Shanzer, A., and Albrecht-Gary, A. M.1999. Fluorescent siderophore-based chemosensors: iron (III) quantitative determinations. *JBIC Journal of Biological Inorganic Chemistry* 4:188–198.

Park, C. S., Paulitz, T. C., and Baker, R.1988. Attributes associated with increased biocontrol activity of fluorescent Pseudomonas. *Korean Journal of Plant Pathology (Korea R.)*.

Pesce, A. J., and Kaplan, L. A.1990. *MPtodosQubnicuClinica*. Buenos Aires.

PietrangeloA.2002. Mechanism of iron toxicity. In: HershkoC (ed) *Iron Chelation Therapy*, Kluwer Academic / Plenum Publishers, New York Vol. 509, 1 Ed pp 19–43.

Pogglitsch, H., Petek, W., Wawschinek, O., and Holzer, W.1981. Treatment of early stages of dialysis encephalopathy by aluminium depletion. *Treatment of Early Stages of Dialysis Encephalopathy by Aluminium Depletion* 2:1344–1345.

Pramanik, A. and BraunV.2006. Albomycin uptake via a ferric hydroxamate transport system of Streptococcus pneumoniae R6. *Journal of Bacteriology* 188:3878–3886.

Prasad, M. N. V., Freitas, H., Fraenzle, S., Wuenschmann, S., and Markert, B.2010. Knowledge explosion in phytotechnologies for environmental solutions. *Environmental Pollution* 158:18–23.

Prentice, A. M., Ghattas, H., and Cox, S. E.2007. Host-pathogen interactions: can micronutrients tip the balance?. *The Journal of Nutrition* 137:1334–1337.

Propper, R. D., Cooper, B., Rufo, R. R., Nienhuis, A. W., Anderson, W. F., Bunn, H. F., ... and Nathan, D. G.1977. Continuous subcutaneous administration of deferoxamine in patients with iron overload. *New England Journal of Medicine* 297:418–423.

Quadri, L. E., Keating, T. A., Patel, H. M., and Walsh, C. T.1999. Assembly of the *Pseudomonasaeruginosa*nonribosomal peptide siderophore pyochelin: In vitro reconstitution of aryl-4, 2-bisthiazoline synthetase activity from PchD, PchE, and PchF. *Biochemistry* 38:14941–14954.

Rajkumar, M., Ae, N., Prasad, M. N. V., and Freitas, H.2010. Potential of siderophore-producing bacteria for improving heavy metal phytoextraction. *Trends in Biotechnology* 28:142–149.

Rautio, J., Kumpulainen, H., Heimbach, T., Oliyai, R., Oh, D., Järvinen, T., and Savolainen, J.2008. Prodrugs: design and clinical applications. *Nature Reviews Drug Discovery* 7:255–270.

Renshaw, J. C., Robson, G. D., Trinci, A. P., Wiebe, M. G., Livens, F. R., Collison, D., and Taylor, R. J.2002. Fungal siderophores: structures, functions and applications. *Mycological Research* 106:1123–1142.

Rusnak, F., Faraci, W. S., and Walsh, C. T.1989. Subcloning, expression, and purification of the enterobactin biosynthetic enzyme 2, 3-dihydroxybenzoate-AMP ligase: demonstration of enzyme-bound (2, 3-dihydroxybenzoyl) adenylate product. *Biochemistry* 28:6827–6835.

Sah, S., and Singh, R.2015. Siderophore: Structural and functional characterisation–A comprehensive review. *Agriculture (Pol'nohospodárstvo)* 61:97–114.

Sandy, M., and Butler, A.2009. Microbial iron acquisition: marine and terrestrial siderophores. *Chemical Reviews* 109:4580–4595.

Schalk, I. J., Hannauer, M., and Braud, A.2011. New roles for bacterial siderophores in metal transport and tolerance. *Environmental Microbiology* 13:2844–2854.

Schippers, B., Bakker, A. W., and Bakker, P. A.1987. Interactions of deleterious and beneficial rhizosphere microorganisms and the effect of cropping practices. *Annual Review of Phytopathology* 25:339–358.

Schwyn, B., and Neilands, J. B.1987. Universal chemical assay for the detection and determination of siderophores. *Analytical Biochemistry* 160:47–56.

Searle, L. J., Méric, G., Porcelli, I., Sheppard, S. K., and Lucchini, S.2015. Variation in siderophore biosynthetic gene distribution and production across environmental and faecal populations of Escherichia coli. *PloS One* 10:e0117906.

Sheng, X., He, L., Wang, Q., Ye, H., and Jiang, C.2008. Effects of inoculation of biosurfactant-producing Bacillus sp. J119 on plant growth and cadmium uptake in a cadmium-amended soil. *Journal of Hazardous Materials* 155:17–22.

Singh, A., Singh, S. S., Pandey, P. C., and Mishra, A. K.2010. Attenuation of metal toxicity by frankial siderophores. *Toxicological and Environmental Chemistry* 92:1339–1346.

Stewart, E. J.2012. Growing unculturable bacteria. *Journal of Bacteriology* 194:4151–4160.

Sullivan, T. S., Ramkissoon, S., Garrison, V. H., Ramsubhag, A., and Thies, J. E.2012. Siderophore production of African dust microorganisms over Trinidad and Tobago. *Aerobiologia* 28:391–401.

Summers, M. R., Jacobs, A., Tudway, D., Perera, P., and Ricketts, C.1979. Studies in desferrioxamine and ferrioxamine metabolism in normal and iron-loaded subjects. *British Journal of Haematology* 42:547–555.

Suo, Z., Tseng, C. C., and Walsh, C. T.2001. Purification, priming, and catalytic acylation of carrier protein domains in the polyketide synthase and nonribosomal peptidyl synthetase modules of the HMWP1 subunit of yersiniabactin synthetase. *Proceedings of the National Academy of Sciences* 98: 99–104.

Takahashi, A., Nakamura, H., Kameyama, T., Kurasawa, S., Naganawa, H., Okami, Y., ... and Iitaka, Y.1987. Bisucaberin, a new siderophore, sensitizing tumor cells to macrophage-mediated cytolysis II. Physico-chemical properties and structure determination. *The Journal of Antibiotics* 40:1671–1676.

Thevenot, D. R., Toth, K., Durst, R. A., and Wilson, G. S.1999. Electrochemical biosensors: recommended definitions and classification. *Pure and Applied Chemistry* 71:2333–2348.

Torsvik, V., and Øvreås, L.2002. Microbial diversity and function in soil: from genes to ecosystems. *Current Opinion in Microbiology* 5:240–245.

Tsafack, A., Libman, J., Shanzer, A., and Cabantchik, Z. I.1996. Chemical Determinants of antimalarial activity of reversed siderophores. *Antimicrobial Agents and Chemotherapy* 40:2160–2166.

Valko, M. M. H. C. M., Morris, H., and Cronin, M. T. D.2005. Metals, toxicity and oxidative stress. *Current Medicinal Chemistry* 12:1161–1208.

Vansuyt, G., Robin, A., Briat, J. F., Curie, C., and Lemanceau, P.2007. Iron acquisition from Fe-pyoverdine by *Arabidopsis thaliana*. *Molecular Plant-Microbe Interactions* 20:441–447.

Varma, A., and Podila, G. K.2005. Siderophore their biotechnological application. *Biotechnological Applications of Microbes*177–199.

Vartoukian, S. R., Palmer, R. M., and Wade, W. G.2010. Strategies for culture of 'unculturable' bacteria. *FEMS Microbiology Letters* 309:1–7.

Verma, V. C., Singh, S. K., and Prakash, S.2011. Bio-control and plant growth promotion potential of siderophore producing endophytic Streptomyces from *Azadirachta indica* A. Juss. *Journal of Basic Microbiology* 51:550–556.

Vivas, A., Azcón, R., Biró, B., Barea, J. M., and Ruiz-Lozano, J. M.2003. Influence of bacterial strains isolated from lead-polluted soil and their interactions with arbuscular mycorrhizae on the growth of *Trifolium pratense* L. under lead toxicity. *Canadian Journal of Microbiology* 49:577–588.

Voisard, C., Keel, C., Haas, D., and Dèfago, G.1989. Cyanide production by Pseudomonas fluorescens helps suppress black root rot of tobacco under gnotobiotic conditions. *The EMBO Journal* 8:351–358.

von Döhren, H., Keller, U., Vater, J., and Zocher, R.1997. Multifunctional peptide synthetases. *Chemical Reviews* 97:2675–2706.

Walsh, C. T., Gehring, A. M., Weinreb, P. H., Quadri, L. E., and Flugel, R. S.1997. Post-translational modification of polyketide and nonribosomal peptide synthases. *Current Opinion in Chemical Biology* 1:309–315.

Wang, Q., Xiong, D., Zhao, P., Yu, X., Tu, B., and Wang, G.2011. Effect of applying an arsenic-resistant and plant growth–promoting rhizobacterium to enhance soil arsenic phytoremediation by Populus deltoides LH05-17. *Journal of Applied Microbiology* 111:1065–1074.

WardTR, ReasL, SergeP, ParelJE, PhilippG, PeterB, and ChrisO. 1999. An iron-based molecular redox switch as a model for iron release from enterobactin via the salicylate binding mode. *Inorganic Chemistry* 38:5007–5017.

Welch, T. J., Chai, S., and Crosa, J. H.2000. The overlapping angB and angG genes are encoded within the trans-acting factor region of the virulence plasmid in *Vibrio anguillarum*: essential role in siderophore biosynthesis. *Journal of Bacteriology* 182:6762–6773.

Williams, L. E., Pittman, J. K., and Hall, J. L.2000. Emerging mechanisms for heavy metal transport in plants. *Biochimica et Biophysica Acta: Biomembranes* 1465:104–126.

Winkelmann, G.2007. Ecology of siderophores with special reference to the fungi. *Biometals* 20:379.

Winkelmann, G., Schmid, D. G., Nicholson, G., Jung, G., and Colquhoun, D. J.2002. Bisucaberin–a dihydroxamate siderophore isolated from *Vibrio salmonicida*, an important pathogen of farmed Atlantic salmon (Salmo salar). *Biometals* 15:153–160.

Wittmann, S., Heinisch, L., Scherlitz-Hofmann, I., Stoiber, T., Ankel-Fuchs, D., and Möllmann, U.2004. Catecholates and mixed catecholate hydroxamates as artificial siderophores for mycobacteria. *Biometals* 17:53–64.

Wuana, R. A., and Okieimen, F. E.2011. Heavy metals in contaminated soils: a review of sources, chemistry, risks and best available strategies for remediation. *ISRN Ecology.*

Wyatt, M. A., Johnston, C. W., and Magarvey, N. A.2014. Gold nanoparticle formation via microbial metallophore chemistries. *Journal of Nanoparticle Research* 16:1–7.

Yu, X., Ai, C., Xin, L., and Zhou, G.2011. The siderophore-producing bacterium, Bacillus subtilis CAS15, has a biocontrol effect on Fusarium wilt and promotes the growth of pepper. *European Journal of Soil Biology* 47:138–145.

Zhang, M. K., Liu, Z. Y., and Wang, H.2010. Use of single extraction methods to predict bioavailability of heavy metals in polluted soils to rice. *Communications in Soil Science and Plant Analysis* 41:820–831.

Index

Taylor & Francis eBooks

www.taylorfrancis.com

A single destination for eBooks from Taylor & Francis
with increased functionality and an improved user
experience to meet the needs of our customers.

90,000+ eBooks of award-winning academic content in
Humanities, Social Science, Science, Technology, Engineering,
and Medical written by a global network of editors and authors.

TAYLOR & FRANCIS EBOOKS OFFERS:

A streamlined
experience for
our library
customers

A single point
of discovery
for all of our
eBook content

Improved
search and
discovery of
content at both
book and
chapter level

REQUEST A FREE TRIAL
support@taylorfrancis.com

Routledge
Taylor & Francis Group

CRC Press
Taylor & Francis Group

Printed in the United States
by Baker & Taylor Publisher Services